Ethical Practice of Statistics and Data Science

By

Rochelle E. Tractenberg, PhD, MPH, PhD, PStat®, FASA, FAAAS

Ethical Practice of Statistics and Data Science

By Rochelle E. Tractenberg, PhD, MPH, PhD, PStat®, FASA, FAAAS

This book first published 2022

Ethics International Press Ltd, UK

British Library Cataloguing in Publication Data

A catalogue record for this book is available from the British Library

Print Book ISBN: 978-1-80441-076-9

eBook ISBN: 978-1-80441-077-6

Dedication

This book is dedicated to people who encouraged me to write (Linda Vieira, Gary LaVigna, David Liebke, Clifford Singer, Leon Thal) and those who encouraged me to reason (Donald Tractenberg, Howard Asher, David Liebke, Kevin FitzGerald) in spite of those who would have preferred for me to play (David, Seven, Emma, Atticus, Ripley).

Acknowledgements

This book was first drafted during my sabbatical from Georgetown University (2019), and I am very grateful for that time. The work, and my thinking, benefitted greatly from extensive discussions about what constitutes ethical practice in statistics and data science with Ron Wasserstein and Donna LaLonde of the American Statistical Association. I am grateful to have had the opportunity since 2002 to grow my skills and interests at Georgetown, and for the sense of community that I have developed with the ASA.

I owe my ethical reasoning mentor, Father Kevin FitzGerald SJ, PhD, PhD a huge debt of gratitude for fostering my engagement with the ethical reasoning construct, and for encouraging me to subvert the dominant paradigm. It was no coincidence that we ended up seated side by side at our first task force meeting for "responsible conduct of research"! I hope this work does your mentorship and collaboration justice.

My seven years on the Committee on Professional Ethics (COPE) of the American Statistical Association, which I had the privilege of vice-chairing (2014-2016) and chairing (2017-2019), and particularly the intensive engagement with Committee members chairing (2014-2015; 2017) and co-chairing (2020) the Working Groups on Revising the ASA Ethical Guidelines for Statistical Practice, helped me to internalize the Guidelines and their importance in the ethical reasoning paradigm and in practice. I hope the book serves instructors, learners, and practitioners in the development of their professional identity and successful, fulfilling, and ethical practice of statistics and data science no matter what their job title.

Table of Contents

List of Cases ix
Introduction xix

Section 1: Ethical Reasoning KSAs, the ASA Guidelines, ACM Code of Ethics

Chapter 1.1 Ethical Reasoning (ER) Knowledge, Skills, and Abilities (KSAs). .1
Chapter 1.2 The ASA Ethical Guidelines for Statistical Practice 11
Chapter 1.3 ACM Code of Ethics 24
Chapter 1.4 The Data Science Ethics Checklist (DSEC) and Data Ethics Framework (DEFW) 39
Chapter 1.5 The "Universe of Statistics and Data Science": Tasks 52
Chapter 1.6 Exploring Guidance from ACM, ASA, DSEC, & DEFW 60
Chapter 1.7. Augmenting your "Prerequisite Knowledge": Stakeholder Analysis 92

Section 2. Stewardship of the Profession: Prequisite Knowledge

Chapter 2.1 Introduction to Section 2 101
Chapter 2.2 Planning/Designing 167
Chapter 2.3 Data Collection/Munging/Wrangling 175
Chapter 2.4 Analysis (perform or program to perform) 181
Chapter 2.5 Interpretation 188
Chapter 2.6 Documenting your Work 195
Chapter 2.7 Reporting your Results/Communication 201
Chapter 2.8 Engaging in Team Science/Teamwork 207
Chapter 2.9 Summary of Section 2 213

Section 3. Using all Six Ethical Reasoning KSAs with the GLs/CE in Practice (case analyses with discussion)

Chapter 3.1 Introduction to Section 3 217
Chapter 3.2. Planning/Designing 231
Chapter 3.3 Data Collection/Munging/Wrangling 274
Chapter 3.4 Analysis (Perform, or Program to Perform) 325
Chapter 3.5 Interpretation 399
Chapter 3.6 Documenting your work 451
Chapter 3.7 Reporting 511
Chapter 3.8 Engaging in team science/working with others 576
Chapter 3.9 Summary of Section 3 643
Chapter 3.10. Book Summary 647
References 653

List of Cases

Abbreviations:

DEFW is the Data Ethics Framework

DSEC is the Data Science Ethics Checklist

ASA is the American Statistical Association

ACM is the Association of Computing Machinery

Page	Case Title	Relevant Principles
	PLANNING/DESIGN	
232	**Case 1.** You recognize during the planning stage that there is/you have/the team has an incomplete understanding of the problem to be addressed	**DEFW:** 1 (inconsistent) Not addressed in **DSEC** **ASA** A, B, C, D **ACM:** 1, 2, 3
238	**Case 2.** You are asked to create one computational step in a multi-step process, and *no one will tell you what* will happen with your results	**DEFW:** 1 (inconsistent) **DSEC:** C, D, E (no way to answer) **ASA:** A, B, C, E, F, H **ACM:** 1, 2, 3
244	**Case 3.** You seek to incorporate sensitivity checks along the planning/development process but meet with resistance	**DEFW:** 4, 5, 6 (inconsistent) **DSEC:** C, D (impossible to answer) **ASA:** A, B, D, E, H **ACM:** 1, 2
250	**Case 4.** You recognize a better way to achieve a computational result than the proprietary way you were told to follow. Your way takes longer, so there is resistance to trying your method; but you can show it uses less data and results are less biased	**DEFW:** 1, 3, 4 **DSEC:** C, D **ASA:** A, B, C, D **ACM:** 1, 2

Page	Case Title	Relevant Principles
256	**Case 5.** You are asked to use a specific analysis or system design that is methodologically inappropriate given the research question or objective	**DEFW:** 1, 2 (not aligned) **DSEC:** C, D, E **ASA:** A, B, C, D, E, F, H **ACM:** 1, 2
263	**Case 6.** You are asked to design a study or system that will collect either implausible/unreasonably low amounts of data (small sample size) or unnecessarily high amounts of data	**DEFW:** 1, 3 **DSEC:** B, E **ASA:** A, B, C, D, H **ACM:** 1, 2
	COLLECT/MUNGE/WRANGLE DATA	
275	**Case 7.** A plan is created to collect data that cannot possibly be housed securely	**DEFW:** 1, 2 **DSEC:** B, E **ASA:** A, B, C, D, H **ACM:** 1, 2
284	**Case 8.** Data collection is carried out by scraping the Internet; you notice that at least some of the time, the results of confidentiality and privacy breeches get swept up in the scraping	**DEFW:** 2, 4 **DSEC:** a **ASA:** C, D, H **ACM:** 1, 2
292	**Case 9.** Your supervisor directs you to assume that if *any* of the data in your collection was obtained with any level of consent (whether none *or* opt-out), then treat *all* of the data as if it was obtained "with consent"	**DEFW:** 2, 4 **DSEC:** A **ASA:** C, D, H **ACM:** 1, 2
299	**Case 10.** Standard Operating Procedures (SOP) manuals direct you to ignore data provenance	**DEFW:** 2, 4 **DSEC:** A **ASA:** C, D, E, H **ACM:** 1, 2

Page	Case Title	Relevant Principles
307	**Case 11.** You discover that there has been no consent obtained for any of the data you are asked to collect/wrangle/munge	**DEFW:** 2, 4 **DSEC:** A ASA: A, C, D, H ACM: 1, 2
314	**Case 12.** You have collected/wrangled data from multiple sources and provenance information about the data is inconsistent – different people at work describe it differently and there's no real evidence about the provenance of *any* of the data	**DEFW:** 2, 4 **DSEC:** A, B, C, D ASA: C, D, H **ACM:** 1, 2
	ANALYSIS	
327	**Case 13.** You are told to implement an analysis plan that you suspect was written by someone else (who does not know it is being used) and for another problem/project	**DEFW:** 1, 2, 3, 4, 5, 6, 7 **DSEC:** A ASA: A, B, C, D, E, F, G, H ACM: 1, 2, 3
336	**Case 14.** Your supervisor ignores your requests for reviews of your work and tells you that no one else can review it either	**DEFW:**5, 6 **DSEC:** C, D ASA: A, C, F, G ACM: 2, 3
343	**Case 15.** You are asked to carry out an analysis you are confident that you do *not know how to do or interpret* (or troubleshoot)	**DEFW:** 2, 5, 6 Not addressed in **DSEC** ASA: A, B, C, D, E ACM: 2

Page	Case Title	Relevant Principles
351	**Case 16.** You are given code to execute and while the code runs, you discover a mistake in the program	**DEFW:** 5, 6 **DSEC:** C, D, E **ASA:** A, B, C, E, F, G, H **ACM:** 2
358	**Case 17.** You notice that at least some of the assumptions required for interpretable results, using the code you were asked to implement, are not supportable. The code does run and yield results, but the assumptions underpinning those results are not valid	**DEFW:** 5, 6 **DSEC:** C, D, E **ASA:** A, B, C, E, F, G, H **ACM:** 1, 2
366	**Case 18.** You are asked to evaluate a new system or analyze a data set, and told the results that your evaluation or analysis should generate	**DEFW:** 1, 2 **DSEC:** C, D, E **ASA:** A, B, C, E, F, H **ACM:** 1, 2
374	**Case 19.** Your analysis of your new system suggests that there is an unexpectedly high error rate, but only for a small subgroup of users. *Overall,* your system's results are exactly as expected; *for the subgroup,* the results are the opposite of the overall result	**DEFW:** 4, 5, 6 **DSEC:** A, C, D **ASA:** A, B, C, D **ACM:** 1
381	**Case 20.** You institute an interim check of results and discover that there is bias in the results. The interim check is literally the middle of a multi-part process that you are working on with several colleagues, so there's no way to immediately pinpoint the source of the bias	**DEFW:** 5, 6 **DSEC:** A, C, D **ASA:** A, B, C, D **ACM:** 1
387	**Case 21.** You are told that your results with new data must match original results (i.e., you must replicate other results), and your analyses/code are right, but they do not replicate earlier results	**DEFW:** 5, 6 **DSEC:** C, D, E **ASA:** A, B, C, E, F, H **ACM:** 2

xiii

Page	Case Title	Relevant Principles
	INTERPRETATION	
401	**Case 22.** You discover that prior (expected) results cannot be reproduced. Sensitivity analyses strongly suggest that earlier results were spurious; reading the team's report of that analysis confirms this: the results were improperly interpreted to favour the team's objective	**DEFW**: 5, 6 Not addressed in **DSEC** **ASA**: A, B, C, D, E, F, G, H & Appendix Not addressed in **ACM**
410	**Case 23.** At the end of a long project, you realize you made an error early on. The results cannot be interpreted in a valid way. Everything has to be redone	**DEFW**: 5, 6 **DSEC**:C, D, E **ASA**: A, B, C, E, F, G, H Not addressed in **ACM**
418	**Case 24.** At the end of a long project, you realize your supervisor made an error early on. The results cannot be interpreted in a valid way. Everything has to be redone	**DEFW**: 5, 6 Not addressed in **DSEC** **ASA**: A, B, C, E, F, G, H Not addressed in **ACM**
426	**Case 25.** You complete a very large set of analyses; one result happens to be "significant". A senior team member highlights this result, interpreting it without considering the context	**DEFW**: 4, 5, 6 **DSEC**: C, D, E **ASA**: A, B, C, D Not addressed in **ACM**
436	**Case 26.** Your supervisor singles out one "meaningful" result to demonstrate that whatever you've been doing "is working", even after you carry out multiple simulations that show their single, "favourite," result is totally spurious	**DEFW**: 4, 5, 6 **DSEC**: C, D, E **ASA**: A, B, C, E, F, G, H Not addressed in **ACM**

Page	Case Title	Relevant Principles
	DOCUMENTING YOUR WORK	
452	**Case 27.** It takes as long to fully and transparently document your work as it does to do the work itself. Since this is just *your* job, not documenting it will only affect you (for the foreseeable future) –and is faster	Not addressed in **DEFW** Not addressed in **DSEC** **ASA:** A, B, D, E, F, G, H & Appendix **ACM:** 1, 2
458	**Case 28.** You failed to fully document your work a few months ago and now your supervisor is requesting your comprehensive documentation so that another person can replicate your work. You really only have time for minimal documentation	Not addressed in **DEFW** Not addressed in **DSEC** **ASA:** A, B, D, E, F, G, H & Appendix **ACM:** 1, 2
464	**Case 29.** You receive documentation of an ongoing program/analysis that lacks all information about data provenance	**DEFW:** 2, 4 **DSEC:** A, B, C, D **ASA:** A, B, C, D, E, F, G, H & Appendix **ACM:** 1, 2
472	**Case 30.** Prior documentation of an organization-wide method is complete and correct. The method development did not include sensitivity analyses. You do a few and identify two important errors in the method	**DEFW:** 5, 6 Not addressed in **DSEC** **ASA:** A, B, C, G, H & Appendix **ACM:** 1

Page	Case Title	Relevant Principles
479	**Case 31.** You are given documentation that is not complete: it lacks details about exactly what methods and in what order were used	**DEFW:** 5, 6 **DSEC:** C, D, E **ASA:** A, B, C, E, F, G & Appendix **ACM:** 2
486	**Case 32.** You provide complete and correct documentation, and this gets "edited" by a supervisor so that it is now no longer complete or correct	**DEFW:** 6 **DSEC:** C, D, E **ASA:** A, B, C, E, F, G, H **ACM:** 1, 2
499	**Case 33.** The documentation you receive specifies an analysis method that is not appropriate for the specific question that must be addressed	**DEFW:** 5, 6 **DSEC:** C, D, E **ASA:** A, B, C, D, E, F **ACM:** 1, 2
	REPORTING	
512	**Case 34.** You discover that incorrect results (yours and/or your team's) are going to be featured in a high-profile publication	Not addressed in **DEFW** Not addressed in **DSEC** **ASA:** A, B, C, D, E, F, G, H & Appendix **ACM:** 1, 2
522	**Case 35.** You follow SOP and the GLs/CE, and report your methods and results fully, but the final report has incorrect methods and results that were "edited" to suit a senior member of the team without your knowledge or agreement	**DEFW:** 6 **DSEC:** C, D, E **ASA:** A, B, C, D, E, F, G, H & Appendix **ACM:** 1, 2

Page	Case Title	Relevant Principles
533	**Case 36.** Stakeholders (donors, funders, employers) are given a misleading summary of your methods and results	**DEFW:** 1, 6 **DSEC:** C, D, E **ASA:** A, B, C, D, E, F, G, H & Appendix **ACM:** 1, 2
545	**Case 37.** Your sensitivity analyses that pinpoint the next logical step in your work are omitted from a final report to funders because "we could get a grant to support the team for another 5 years to figure that out!"	Not addressed in **DEFW** Not addressed in **DSEC** **ASA:** A, B, C, E, H **ACM:** 2
553	**Case 38.** If you report your method fully and transparently, then you will lose the opportunity to patent it	Not addressed in **DEFW** Not addressed in **DSEC** **ASA:** A, B, C, F **ACM:** 1
560	**Case 39.** If you report your method fully and transparently, then a reviewer might notice that you are not the original developer of this method – although the same method was published over 30 years ago and in *another* field	Not addressed in **DEFW** Not addressed in **DSEC** **ASA:** A, B, C, E, F, H **ACM:**1
568	**Case 40.** You prepare a report identifying difficulties you encountered in your evaluation of a system your organization wants to deploy or an analysis that was done. The organization does not have a mechanism for submitting or sharing this report (or peer review of any type) either internally or with stakeholders	Not addressed in **DEFW** Not addressed in **DSEC** **ASA:** A, B, C, E, F, G, H & Appendix **ACM:** 2, 3

Page	Case Title		Relevant Principles
	TEAM WORK/TEAM SCIENCE		
577	**Case 41.** You notice a pattern of bullying by a senior team member		Not addressed in **DEFW** Not addressed in **DSEC** **ASA**: A, G, H & Appendix **ACM**: 1, 2
587	**Case 42.** You are asked to do some coding/analysis by someone who is prevented from acknowledging that you helped. Your contribution cannot be recognized		Not addressed in **DEFW** Not addressed in **DSEC** **ASA**: A, E, F, G, H **ACM**: 1, 2
596	**Case 43.** Your supervisor tells you that you "only need to read/review your own work" and you are not allowed to see the final/full document or work product		**DEFW**: 2, 5, 6 **DSEC**: C, D **ASA**: A, B, C, D, E, F, G, H & Appendix **ACM**: 1, 2, 3
604	**Case 44.** You complete the analysis plan/system design, oversee its operation, and draft the report. You suddenly receive a "new draft" of the report that excludes all of the work you did, nor does any of the documentation of the system or work from your original report appear. You can tell without carefully reading it that the "new draft" has obvious errors in the design/analysis, results, and other reported elements, but you are asked to "approve" the new draft – and agree to be/remain a co-author on the report – within the next two days. You also have another project deadline in two days		Not addressed in **DEFW** Not addressed in **DSEC** **ASA**: A, B, C, E, F, G, H & Appendix **ACM**: 1, 2, 3

xviii

Page	Case Title	Relevant Principles
613	**Case 45.** Someone on your team suggests a technical method to overcome a lack of consent from data contributors and collect their data even if they do not consent	**DEFW:** 2 **DSEC:** A **ASA:** A, B, C, D, E, F, G, H & Appendix **ACM:** 1, 2, 3
622	**Case 46.** You recognize the potential for "dual use" of your team's code, data, and/or results	Not addressed in **DEFW** Not addressed in **DSEC** **ASA:** A, B, C, D **ACM:** 1, 2, 3
630	**Case 47.** You inadvertently discover that a proprietary "new method" that you were told to prepare for publication/patent application was actually published decades ago and was apparently unnoticed until you found it	Not addressed in **DEFW** Not addressed in **DSEC** **ASA:** A, B, E, G, H & Appendix **ACM:** 1

Introduction

"Upon entry into practice, all professionals assume at least a tacit responsibility for the quality and integrity of their own work and that of colleagues. They also take on a responsibility to the larger public for the standards of practice associated with the profession". (Golde & Walker 2006: 10) This book is intended to describe these responsibilities, and prepare people to fully assume these responsibilities whenever they use or practice statistics and data science. Although one focus is on those about to embark on a career, the book is also intended for practitioners and mentors or supervisors of practitioners. This book is intended to support *the ethical practice* of statistics, data science, and "statistics and data science", with an emphasis on how to earn the designation of, and recognize, "the ethical practitioner".

While the field of Statistics has long-established guidelines for ethical practice (Hogan & Steffey, 2014), data science is too new of a domain for there to be "ethical practice norms" that are specific to it. Statisticians and data scientists may be required to complete traditional training in "responsible conduct of research", but not all will actually be going into *research*. Moreover, the traditional paradigm for training in "ethics" or "responsible research" is limited: a) it tends to focus on the avoidance of falsification, fraud, and plagiarism; b) it tends to focus on human and animal subjects and their rights in ways that are quite specific for clinical trials and experimental studies in the traditional sense, rather than in the data mining/machine learning applications many statisticians and data scientists might be engaged in the more modern sense. By contrast, a great deal of current discussion (July 2022) around ethics and data science is focused on privacy and confidentiality, and security of data. This book is not focused on just data ethics, or ethical issues relating to data; nor is it about memorizing rules and precedents. Instead, the book emphasizes understanding community-based practice standards and how to use these in reasoning to identify and respond ethically to real world, complex situations with often subtle, competing, prioritization requirements. Keeping in mind that the practice standards are likely to change over time, the book utilizes ethical reasoning, which is a set of knowledge, skills, and abilities, that can be improved over time no matter what material (guidelines, codes, policies) are being reasoned about or with.

The subjects of the book are a) how to utilize professional practice standards in the workplace, and b) how to reason ethically when how best to use the practice standards isn't obvious. Section 1 outlines background material including complete presentations of the Ethical Guidelines (GLs) for Statistical Practice of the American Statistical Association (**ASA**) and the Code of Ethics (CE) of the Association of Computing Machinery (**ACM**). Section 1 also introduces checklists, series of questions designed to support "data science ethics" and "data ethics". Section 1 ends with the contextualization of these checklists within a wider professional setting, demonstrating how project-based decisions and decision making do not prepare individuals to practice statistics and data science ethically.

One of the core contents of the book is the importance of ethical reasoning (ER). Derived from the Framework for ethical decision making published by the Markkula Center for Applied Ethics at Santa Clara University (https://www.scu.edu/ethics/ethics-resources/ethical-decision-making/a-framework-for-ethical-decision-making/) six aspects of knowledge, skill, and ability (KSAs) have been identified as core to ER. The author, a cognitive scientist, and ASA-Accredited Professional Statistician has published extensively on how the KSAs that are required for effective ER can be developed. Rather than being a book about ER itself, or about applied ethics or even growing and developing these KSAs on their own, this book is organized following the KSAs to optimize the chances that a reader will learn "the standards of practice associated with the profession", which is the focus of Section 2. Currently, "professional ethics" are not taught in programs in statistics, data science, or statistics and data science; if they are taught, it is not consistent across programs. The way federally mandated training in research ethics is currently taught in the United States, the focus tends to be on:

a. obtaining consent in experimental settings, and principles of autonomy, justice, beneficence, and nonmaleficence that are standard for clinical and human experimental research, derived from Beauchamp & Childress' (1989) core principles of bioethics; but are not helpful for statistics and data science practice.

b. case analyses that feature interesting vignettes, and how to avoid falsification, fabrication, and plagiarism. This focus means that case analyses all involve some ethical dilemma – so training is focused on resolving ethical dilemmas with no training or practice, or even discussion, about how to simply *practice* ethically. Moreover, falsification, fabrication, and plagiarism are core aspects of unethical *research*, but not all statisticians and data scientists are engaged in research. Furthermore, as will become apparent in the chapters on the practice standards, unethical practice arises from far more than these three activities.

Other favoured topics in courses or workshops on ethics in academics/research are highly academic-research focused, and are not informative about statistical malpractice and other challenges that are common (and becoming more so) in data science, in government, in business, and in collaborative enterprises that cross these boundaries. For these reasons, the emphasis in Section 2 is on how everyday practice can be guided by understanding, and utilizing, the ethical standards associated with the professions of statistics, computing, and statistics and data science, as well as "data science ethics" and "data ethics" checklists. Developing familiarity with the roles these practice standards can play to promote ethical practice – rather than only in cases where misconduct or wrongdoing is described – is critical so that readers will be able to meet Golde & Walker's description, and assume "responsibility for the quality and integrity of their own work and that of colleagues". This responsibility is an essential one for a quantitative practitioner, whether they work primarily with others who are similarly trained/qualified or as parts of multi- and inter-disciplinary teams.

In Section 3, a series of vignettes representing authentic challenges that arise (have arisen) in the workplace where statistics and data science are utilized will give practice with the ER KSAs as well as with the Ethical Guidelines for Statistical Practice of the American Statistical Association (**ASA)** and the Code of Ethics of the Association of Computing Machinery (**ACM)**. The purpose of Section 3 is to give practice in identifying when behaviors– whether the readers' or the behaviors of others - may not actually *be ethical,* and to teach and give practice in communicating the fact that there is a decision to be made, what decision is made, and that/how that particular decision optimizes particular stakeholders over others. The entirety of ethical reasoning knowledge, skills, and abilities must be brought to bear in the making and justification of decisions in response to ethical challenges, which must be identified so that appropriate responses can be articulated and evaluated. Section 3 enlists and models all of the ER KSAs, leveraging the work of accumulating the prerequisite knowledge (KSAs 1-2) that is needed to reason ethically as a practitioner of statistics and data science. Exercises in Section 3 include evaluating the case analyses provided and role playing to better understand exactly how they would broach the topic of having identified (and determined on a course of action to resolve) an ethical issue, how they or others might respond to such a conversation, and how they might respond in cases where there is resistance to resolving the ethical issue.

The book could be used by a self-directed reader in or outside of formal education; for a stand-alone course (aiming for students closest to entering practice); or, as an adjunct text for a consulting course. The book could also be used alongside an internship and/or capstone course (for advanced undergraduates or graduate students). More extensive introduction to, and background material for, ethical reasoning is the topic of another book, called "Ethical Reasoning for a Data Centered World" (Tractenberg, 2022).

The 2018 Guidelines for Data Science curriculum contents (National Academy of Sciences, 2018; see also DeVeaux et al. 2017) include, but have no specifics relating to how to teach, "ethics". They also do not describe what must be demonstrated by any student to demonstrate that they have learned what was intended by the curricular suggestions to include "ethics". To offset the lack of specifics, this book can be used to structure a course or a self-directed program of study, and also lead to evidence that an individual has at least embarked on a program to develop sufficient ethical reasoning abilities.

Outlining a learnable and improvable skill set, and then explicitly demonstrating how the prerequisite knowledge, which is extensive, can support ethical professional practice (Sections 1 and 2) or the identification and resolution of ethical challenges in the workplace (Section 3), the book also includes opportunities for students to discuss their options, thoughts, and answers – and to critique the analyses of the cases that are given. These features allow the individual (self-directed learner) to track whether or not they are improving, and gives an instructor options for actually grading student performance.

This book is laid out in three sections: 1. Ethical reasoning KSAs, the ASA Guidelines and ACM Code of Ethics, and data/data science ethics checklists; 2. Using practice standards and checklists as the prerequisite knowledge for ethical practice; 3. Using ethical reasoning with the Guidelines in practice (47 case analyses with discussion). The author was the Chair (2017-2019) and Vice-Chair (2014-2016) of the ASA Committee on Professional Ethics, and has chaired (2014-2016; 2018) or co-chaired (2021) Working Groups on revising the ASA Ethical Guidelines. The perspectives and opinions represented in these case analyses are solely those of the author and do not reflect the opinions/perspectives of the Committee on Professional Ethics or the ASA. The book reflects the 2018 ACM CE and the 2022 ASA Ethical Guidelines.

Options for using the materials in class or ongoing conversations include discussing the analysis of each vignette (like journal club) or having students critique them (what else would you have done/suggested and why?) with reflection comprising an assigned reflective essay, for example:

- Why is "do nothing" a decision in this case? Is it justifiable? How so/why not?

- Why should you do or say something (in a case like this) when no one else seems to think there is a problem?
- Why is "no one else seems to have a problem (with this)" NOT a good answer?
- What justification is there for a claim that "ethical practice is just common sense applied at work"? Is that true (in this case)?
- Why do people say that there is "no right answer" to problems/questions of ethical practice? (Does this case support that claim or not?)
- Explain whether/how the application of the ASA Guidelines or the ACM Code in this case encourages ethical conduct in research (or practice, as appropriate)
- Do the ASA Guidelines or the ACM Code of Ethics promote *professionalism* in this case? How/how not?

Additionally, those new to practice are encouraged to consider role-playing to ensure that they have some practice thinking about how to initiate conversations about the existence of ethical challenges, and what to do to resolve any that are identified. Those with more experience, or in leadership roles, can utilize these materials to structure "open door" or "office hour" policies. Encouraging a standard, formal examination of what is known, what seems to be a problem, and what might be useful solutions to such problems, will facilitate conversations about not only what might be "going wrong", but also how to address them, rectify them, and ensure that they're not repeated in future.

Section 1: Ethical Reasoning KSAs, the ASA Guidelines, ACM Code of Ethics

Chapter 1.1. Ethical Reasoning (ER) Knowledge, Skills, and Abilities (KSAs)

Chapter 1.2 The ASA Ethical Guidelines for Statistical Practice

Chapter 1.3 The ACM Code of Ethics

Chapter 1.4 The Data Science Ethics Checklist (DSEC) and Data Ethics Framework (DEFW)

Chapter 1.5 The "Universe of Statistics and Data Science": Tasks

Chapter 1.6 Exploring Guidance from ACM, ASA, DSEC, & DEFW

Chapter 1.7 Augmenting your "Prerequisite Knowledge": Stakeholder Analysis

Chapter 1.1
Ethical Reasoning (ER) Knowledge, Skills, and Abilities (KSAs)[1]

A steward is defined as one to whom "we can entrust the vigor, quality, and integrity of the field" (Golde & Walker 2006: p. 5).

"Being stewardly involves both ethical practice and professional integrity; if you are stewardly and practice data science ethically, these serve to make your professional integrity *observable*. Together, stewardship and ethical practice promote the integrity of the field." (Tractenberg, RE. (2019-b, April 23). Becoming a steward of data science. https://doi.org/10.31235/osf.io/j7h8t)

"Awareness of the potential for unethical behavior by data scientists has led to the profession "data science" becoming a target requiring ethical guidelines of some kind (e.g., Simonite 2018; Loukides et al. 2018). Public trust in this discipline was harmed by each of these events, and legislation in the United States and European Union was created specifically to limit the potential for data science to engage in harmful behaviors." (Tractenberg, RE. (2019-b, April 23). Becoming a steward of data science. https://doi.org/10.31235/osf.io/j7h8t).

The development of a discipline or profession may be initiated, or cemented, by a statement of what the community agreement represents "norms" for behavior. That is, a typical code of professional conduct or ethics may evolve to notify consumers or colleagues of what they should expect from anyone in the profession; or to orient current and future practitioners to what behavior is expected from those who do qualify (reviewed in Tractenberg, et al. 2015). Data science is a new field that attracts practitioners from diverse backgrounds and with different types and levels of formal (and informal) training, which means that the development of a *single* disciplinary or professional code of conduct or

---

[1] The material in this chapter is adapted from Tractenberg RE. (2019-a, April 30). Strengthening the practice and profession of statistics and data science using ethical guidelines. Published in the *Open Archive of the Social Sciences* (SocArXiv), https://doi.org/10.31235/osf.io/58umw and from Tractenberg RE. (2020, February 19). Concordance of professional ethical practice standards for the domain of Data Science: A white paper. Published in the *Open Archive of the Social Sciences* (SocArXiv), 10.31235/osf.io/p7rj2

ethics is unlikely. However, there are existing – relevant – practice standards that can be utilized and may in fact be descriptive/supportive of "data science" (Hogan et al. 2017). Data science as a discipline has emerged at the intersection of computing and statistics – two disciplines with long standing guidance for ethical practice that feature professional integrity and responsibility. The Association of Computing Machinery (ACM) revised their professional ethical practice standards in 2018, and the American Statistical Association (ASA) revised theirs most recently in 2022. Both sets of guidance represent the perspectives of experienced professionals in their respective domains, but both organizations explicitly state that the guidelines apply to – should be utilized by – *all* who employ the domain in their work, irrespective of job title or training/professional preparation. Given that both statistics and computing are essential foundations for data science, their ethical guidance should therefore be a starting point for the community as it contemplates what "ethical data science" looks like. The ASA Ethical Guidelines (https://www.amstat.org/docs/default-source/amstat-documents/ethicalguidelines.pdf) are already widely accepted practice standards for statistics, with the 2018 Guidelines translated into Chinese and endorsed by the International Chinese Statistical Association (the 2022 translation is a work in progress as of October 2022). Meanwhile, the ACM Code of Ethics has been translated into Spanish and Chinese, and a booklet on their utilization has also been created (https://www.acm.org/binaries/content/assets/about/acm-code-of-ethics-booklet.pdf).

The Ethical Guidelines (GLs) for Statistical Practice of the American Statistical Association (**ASA**) and the Code of Ethics (CE) of the Association of Computing Machinery (**ACM**) are therefore two community-based standards of practice, each of which has important guidance that is relevant for statisticians who use computationally intensive methods, computing professionals who integrate statistics into their work, and those who consider themselves "data scientists". Other tools include checklists that comprise series of questions designed to support "data science ethics" and "data ethics" – but which themselves are not community derived, and which do not seek to orient either the practitioner or the consumer/collaborator to the standards of ethical practice that can help identify trustworthy – stewardly - practitioners.

A *steward of the discipline or profession* is defined as one to whom "we can entrust the vigor, quality, and integrity of the field" (Golde & Walker 2006: p. 5). These individuals must be recognized – by consumers, collaborators, and others within the profession – as exhibiting and embodying the core features of the professional. The ASA and ACM standards of practice do not entail exhaustive lists of behaviors that do or do not describe ethical practice. Instead, they offer guidance that must be utilized in daily/usual practice, and that can also be leveraged in the (hopefully) *unusual* circumstances where behavior or practice is – or could be/become – unethical. Striving for stewardship supports ethical practice (Rios et al. 2019) by the individual while simultaneously strengthening the discipline/profession.

Part of stewardly practice is the ability to reason ethically in the workplace. Ethical reasoning (ER) is an approach that can be utilized to guide professional decision-making. Currently, "professional ethics" are not taught in programs in statistics, data science, or statistics and data science. This is due, in part, to the fact that "data science" is not yet a "profession" in the sense that there is no code of conduct that the international community agrees on. However, no matter what your degree or job title, the ethical practice of "statistics", "data science", and "statistics and data science" can be guided by understanding, and utilizing, the standards of practice associated with the professions of statistics and computing. Developing familiarity with these practice standards can (and should) promote ethical practice, which is a responsibility for a quantitative practitioner, whether they work primarily with others who are similarly trained/qualified or as parts of multi- and inter-disciplinary teams. Both the ASA and ACM articulate clearly that anyone – regardless of job title or whether they are members of these organizations - who utilizes their tools, techniques, and methods has responsibility to do so ethically. However, simply reading or acknowledging the practice standards is insufficient for ethical practice. As you can imagine, stewardship of the discipline or profession requires embodiment, not memorization, of the practice standards. This book is intended to enable professional stewardship in statistics, data science, and "statistics and data science". Practitioners in any/all of these domains are "quantitative practitioners", whether or not "statistics", "computing", or "data science" are in their degrees, job titles, or principal responsibilities.

As will be outlined in Chapter 1.4 (and Sections 2 and 3), the Data Science Ethics Checklist (DSEC; http://deon.drivendata.org/) and Data Ethics Framework (DEFW; https://www.gov.uk/government/publications/data-ethics-workbook/data-ethics-workbook) are tools first developed in 2018 to support *decision making* in data science projects. Both include lists of questions relating to key decisions relating to data. The DSEC focuses the practitioner's attention on five facets of a project that involve data: A. Data Collection; B. Data Storage; C. Analysis; D. Modeling; and E. Deployment. The DSEC asks the practitioner 20 questions across these five features. The original (2018) DEFW includes 46 questions across seven principles relating to data specifically (i.e., not relating to statistical analysis or computational statistics):

Principle 1. Start with clear user need and public benefit
Principle 2. Be aware of relevant legislation and codes of practice
Principle 3. Use data that is proportionate to the user need
Principle 4. Understand the limitations of the data
Principle 5. Ensure robust practices and work within your skillset
Principle 6. Make your work transparent and be accountable
Principle 7. Embed data use responsibly

Neither DSEC nor DEFW was intended/designed for professional qualification, but instead as tools to support the consideration of ethics into the facets (DSEC) and principles (DEFW) relating to data that are or have been collected. Neither discusses whether the practitioner has a special responsibility or obligation to answer any/all questions in the affirmative; by contrast, the practice standards do state (ASA) or imply (ACM) that the practitioner has responsibilities - and to multiple stakeholders – to follow the relevant guidelines. That is, the practice standards suggest that "the ethical practitioner" can be recognized as that professional who accepts these responsibilities, and acts to follow the practice standards. However, because these two tools capture or relate directly to common decisions across projects and other work involving data, they are included in the discussions of practice standards as examples of the reach of these practice standards. Importantly, the DEFW and DSEC are *not* designed by "data ethicists" or "data science ethicists" – but rather, by practitioners involved with data and data science. As will be shown in later chapters, the overlap between the ASA and ACM practice standards with these checklists is considerable – all behaviors included in the checklists are also discussed in the practice standards. Thus, as argued above, useful guidance for "data science" can be derived from the ASA and ACM guidelines – even though they were not developed specifically for this relatively new "profession".

As noted, stewardly practice involves/requires the ability to reason ethically in the workplace. Ethical reasoning (ER) is an approach to identifying "responsible" responses to an identified ethical challenge, and choosing among the possible responses in a defensible – and reproducible - manner. While the practice standards are *learnable*, the six elements of Knowledge, Skills, and Abilities (KSAs) of ER are *learnable* and ***improvable*** (Tractenberg & FitzGerald, 2012). That means that there are ER-specific KSAs to learn, and the level at which you perform each KSA can change as you become more sophisticated and capable. The ER KSAs are actually steps you can walk through:

1. Identification and assessment of one's prerequisite knowledge;
2. identification of relevant decision-making frameworks;
3. identification/recognition of an ethical issue;

4. identification and evaluation of alternative actions;
5. making & justifying a decision (about the ethical issue); and
6. reflection on the decision.

The ASA and ACM practice standards are recognized in their respective contexts as being subject to change over time and as technology develops. By contrast, the ER KSAs *are fixed* –but you (the user of the KSAs) actually grow and change over time in your ability to use them (Tractenberg & FitzGerald, 2012). The KSAs can be deployed in a planning stage, i.e., before any unethical behaviors can be executed, or at a later stage to evaluate what may have gone wrong or what may be going on in an existing situation. Moreover, the KSAs can be applied to *any practice standard or policy*. This book will help a reader learn the KSAs and also how to utilize ethical reasoning with these two major ethical practice standards that are so essential to ethical practice of statistics and data science. Because of their importance to statistics and data science, the ethical reasoning KSAs are presented utilizing these practice standards. However, additional materials are also included (the DSEC and DEFW), to achieve two objectives. Firstly, to familiarize the reader with the ways in which diverse documents that purport to represent "ethical" practice agree on what should be considered "ethical". Secondly, by working through cases where some of these reference materials do – while others of the materials do not – support ethical reasoning or formulating a justifiable response to an ethical challenge, the reader will see how to utilize the reference that is most supportive of consistent, ethical practice. For example, the ASA Purpose states, "If an unexpected ethical challenge arises, the ethical practitioner seeks guidance, not exceptions, in the Guidelines. To justify unethical behaviors, or to exploit gaps in the Guidelines, is unprofessional, and inconsistent with these Guidelines." If a reader finds that one reference does not offer specific guidance, that constitutes a "gap" and the ASA GLs state plainly that exploiting that gap is unethical. As readers will see, the ASA GLs are very extensive, covering much of the same type of activities as the ACM CE – as well as a great deal that is not contemplated in the ACM CE. Similarly, DSEC and the 2018 DEFW that are utilized throughout the book represent only a very small part of professional work with data. The unethical practitioner will take advantage of the differences, choosing the reference document with the least guidance to exploit the gap between it and the ASA GLs, if one exists.

The KSA are basic steps in identifying ethical challenges and then reasoning through them to decide what to do. Although they are numbered, and generally reflect how reasoning would happen (see https://www.scu.edu/ethics/ethics-resources/ethical-decision-making/a-framework-for-ethical-decision-making/), they do not have to be followed in the specific order given above (and below). Moreover, Section 2 is dedicated to just KSAs 1-2, understanding the prerequisite knowledge that the practice standards from ASA and ACM, and ancillary materials (the DSEC and DEFW) represent for general practice. Understanding and using these guidance resources should be sufficient for ethical and stewardly work at least 90% of the time. For the other 10% (hopefully less!!) of time, the full set of ER KSAs will be essential, which is why the entirety of Section 3 is dedicated to applying all 6 KSAs in 47 different case analyses. Understanding that ethical reasoning is a set of KSAs like any other, which requires instruction and practice, can both support ethical practice by the individual and also empower practitioners to identify problems in the workplace, and respond in ethical – justifiable – ways.

Elaborating on the six KSAs in ethical reasoning (ER KSAs):

KSA 1. Identify and 'quantify' your prerequisite knowledge.

To practice ethically, there are professional practice standards – norms of professional behavior – that you should become familiar with. In this book we will focus on the Ethical Guidelines for Statistical Practice from the American Statistical Association (ASA) and the Code of Ethics of the Association for Computing Machinery (ACM). There are also laws and local rules, including policies of the workplace –or governing those with whom you work or contract; all of these must be known by the ethical practitioner in order to follow guidelines and particularly, to ensure that all work is consistent with both professional ethical standards and the law/local rules. One key exception is when there are cultural norms, or standards that other practitioners may follow but that are inconsistent with your professional ethical standards and/or with local laws. As you will see in later chapters, other practice standards cannot be prioritized over the professional ethical standards for your discipline.

Other prerequisite knowledge includes an understanding of the stakeholders in any work you complete: *you,* (the decider/practitioner); your boss/ supervisor; your colleagues; the individuals (human or animal) that gave or generated the data (i.e., subjects); society at large, and sometimes, cultures or groups; your profession; the public trust in your institution/organization/ profession; and in some cases, the environment. In addition to recognizing these stakeholders, you must know whether or not benefits or harms accrue to each of these stakeholders as a result of your work.

Importantly, this KSA (prerequisite knowledge) will include slightly different information in the event that you encounter an ethical challenge than it does for you to simply practice as an ethical professional. This will become more apparent in Section 2 (focused on ensuring ethical practice) and Section 3 (focused on identifying ethical challenges, and making decisions about what to do in response).

KSA 2. Identify decision-making frameworks.

Two philosophical perspectives on decision making are featured in this book: a "virtue ethics" perspective (or framework) and a "utilitarian" perspective. In the virtue perspective, decisions are made so that the decider follows the "virtuous" or "ideal" practitioner. By contrast, the utilitarian perspective encourages decisions that minimize harms and maximize benefits. We will explore harms and benefits more fully in subsequent chapters, but identifying to whom harms and benefits accrue is part of the collection/identification of prerequisite knowledge (KSA 1).

The ASA Ethical Guidelines for Statistical Practice, or "GLs", represent a "virtue ethics" perspective, because as you will see in Chapter 1.2, ethical practice is described as what "the ethical statistical practitioner" does. That is, the GLs describe "what the ethical practitioner would do" or how the ethical practitioner does his/her work. The ACM Code of Ethics, or "CE" (Chapter 1.3), happens to be more like a utilitarian framework; it directs "computing professionals" to focus attention on the positive and negative effects of a decision – and also features stakeholder perspectives explicitly. In making decisions about how to practice ethically, the ACM CE emphasizes decisions that avoid or minimize harms; however, because the ACM CE also specifies what "a computing professional *should*" do (in order to comply with the CE), it can also be interpreted from a virtue perspective. Importantly, behaving in a manner that is consistent with either of these practice standards could be described as following a "virtue" perspective – even if "the ethical statistical practitioner" does not describe you **per se**, after reading this book, "the ethical practitioner" **should** describe you.

As we will see throughout Section 2, the GLs and CE are comprehensive and complex. "Learning" them requires effort and practice, which all the contents and activities in Section 2 are designed to support. Additional prerequisite knowledge will include an understanding of the construct of "stewardship of the profession", and how to execute a stakeholder analysis. Attention is focused on KSAs 1-2 throughout Sections 1 and 2 because this prerequisite knowledge will support decisions that lead to ethical professional practice – making up 90% (hopefully more!) of your working life.

However, sometimes, ethical challenges will arise; these can often seem to be raised or created *by others*, but we all must still face them **and respond**. These responses also require decisions, and the remaining KSAs in the ethical reasoning process – the focus of Section 3 – lead us through this process.

KSA 3. Identify or recognize the ethical issue.

This is almost universally a response (decision) required by the practitioner in response to a specific order, request, or situation. The problem that needs your response can be identified using your prerequisite knowledge – for example, if someone behaves, or asks you to behave, in a manner that is inconsistent with the GLs or CE (or the law), or in a manner that may cause, or has the potential to cause, harms to any stakeholder, this creates an ethical challenge. By recognizing exactly what the challenge is, it can help you to execute the next KSA. It is difficult to do any of the remaining KSAs if you simply "have a bad feeling" about a specific order, request, or situation; putting a description on what exactly the problem is will help.

KSA 4. Identify and evaluate alternative actions (on the ethical issue).

There are **always two** decisions that can be made in any circumstance: a) do *nothing*; or b) do *something*. It is essential to recognize that doing nothing *is a response – your decision to do nothing when faced with an ethical challenge* must be considered a possible alternative. However, a decision to "do something" is not specific enough to formulate an actual response. Through Section 3 we will explore alternative actions, considering what the virtue and utilitarian perspectives – as well as the prerequisite knowledge – tell us about which of the alternatives is "best" (i.e., evaluate them). Thus, in addition to helping to identify the ethical challenge that needs your response (KSA 3), your prerequisite knowledge and the decision-making frameworks (KSAs 1-2) are also essential in KSA 4 –particularly in the evaluation of alternatives. The evaluation will feature practice standards (like the GLs and CE), as well as the impacts of the alternatives on stakeholders.

KSA 5. Make and justify a decision.

Articulating the decision –what to do in the face of the ethical challenge that was identified – will utilize the alternatives and their evaluations from KSA 4. The justification of the decision (of which of the KSA 4 alternative actions you choose to take) will involve the evaluation of the alternatives as well as the stakeholder analysis. The decision-making frameworks (KSA 2) will also provide information about which action (KSA 4) to choose.

KSA 6. Reflect on the decision.

Reflection on the decision you have made can include consideration of "what to do next"; how to prevent the same sort of situation in the future; and how to communicate the "story" of what happened, describing the event that you identified using KSA 3, and the rest of the process. Conversely, consider three features of your decision can help to structure a reflection: *universality, justice,* and *publicity*. These features will help you to contextualize your decision (and its justification) with respect to the perspectives of others. Your reflection can feature a discussion of whether or not your decision would work in every similar situation (a test of the *universality* of your decision). If you consider the universality and determine that your decision would *not* work as a universal rule, does it suggest that your decision, or your justification, needs to change? Or are there features of the **case** that make *it*, rather than the decision, unique (not universal)? This is called the "test of universality". The "test of justice" involves consideration of whether everyone is equally affected by the decision. This may be particularly relevant for decisions driven by ACM CE Principles, since the CE includes a Principle (3) specifically for leaders, so consideration of the justice of a decision may include reflections on its applicability to those earlier, vs later, in their careers; conversely, the effects of a decision on others in the profession vs the public (or other stakeholders). A final test to reflect on is the "test of publicity" – would it be acceptable to you if your decision and its justification were made public, published, or reported on the news? Stadler (1986) [2] discusses these three tests for any decision (universality, justice, publicity); these three tests were derived from the writings of ethicist/ philosopher Immanuel Kant.

Finally, you may find it helpful to reflect on the case and its decision in terms of whether it has ever happened – or could ever happen - to you. Do you hold a position that protects you from, or exposes you to, cases like the one you've just reasoned about? There may be differences between (for example) academic and government, or academic and business, contexts and your decision or justification should be contemplated differently for different contexts. In the context of government work, decisions may need to be fully transparent and publicized or may be top secret and not shareable. In the context of a business, decisions may need to be shared only with supervisors or more junior team

---

[2] Stadler, H. A. (1986). Making hard choices: Clarifying controversial ethical issues. Counseling & Human Development, 19, 1-10. See also
www.walsallsocialcareworkforce.co.uk/ckfinder/userfiles/files/refStadler.docx

members. Note that these reflections relate, directly or indirectly, to the universality, fairness, and publicity of decisions and their justifications.

Communicating your *intention* to be ethical in the professional setting can serve as the grass roots culture change that is needed because of the pervasiveness of quantitation in the world. As noted, ER KSAs 1-2 are featured in Sections 1 and 2, while ER KSAs 3-6 are discussed more fully – and practiced - in Section 3 (see also Tractenberg 2022 -Ethical Reasoning for a Data Centered World). In this book, we will first explore the ethical practice standards with which statisticians and data scientists need to be familiar (Chapters 1.1 and 1.2). Following that, we will explore the impacts of our decisions and actions on others (stakeholders).

Chapter 1.2
The ASA Ethical Guidelines for
Statistical Practice

The quantitative world – wherever data are collected or utilized - requires careful consideration of the impacts of our work on the decisions – and sometimes the wellbeing – of others. Guidelines (GLs) and Codes of Ethics (CE) are established for "good" or ethical professional practice (see, e.g., Tractenberg et al. 2015). Just as there is more to "learning statistics" than applying "the right" formulae or running software, there is also more to "being an ethical statistical practitioner/data scientist" than learning the GLs/CE from a given professional association. There are multiple methods for accomplishing many statistics and data science tasks, and the "best" method may depend on what resources are available and the priorities that the data themselves cause/create (e.g., hard to collect; hard to verify; prone to measurement error; confounded with other variables in the data set, etc.). Similarly, there are often multiple ways to identify ethical problems or questions that can arise in practice, before or after they have arisen. There are definitely multiple ways to address ethical problems, including "ignore it". Factual knowledge (i.e., memorizing the GLs/CE) is unlikely to help you develop the ability to formulate, and then choose from, multiple possible responses to a breach of ethics - particularly if *all* of the options are justifiable. Once you are familiar with the contents of the GLs/CE, including their specific elements, their use in your everyday decision-making will typically involve some prioritization of competing obligations or solutions. This chapter focuses on the Ethical Guidelines for Statistical Practice of the American Statistical Association (ASA), while the next chapter focuses on the Code of Ethics from the Association for Computing Machinery (ACM).

The ASA Ethical Guidelines for Statistical Practice were first approved by the ASA Board of Directors and published in 1995. They were not revised until 2016, updated in 2018, and fully revised in 2022. A regular review-revision cycle every five years was established in 2016 by the ASA Board. The Guidelines include eight principles, each of which has 4-12 elements, plus an Appendix.

The 2022 ASA Guidelines (GLs) appear below in their entirety. You can download the current .pdf from the ASA website (https://www.amstat.org/ASA/Your-Career/Ethical-Guidelines-for-Statistical-Practice.aspx).

Ethical Guidelines for Statistical Practice

Prepared by the Committee on Professional Ethics of the American Statistical Association

Approved by ASA Board of Directors February 1, 2022.

Purpose of the Guidelines

The American Statistical Association's Ethical Guidelines for Statistical Practice are intended to help statistical practitioners make decisions ethically. In these Guidelines, "statistical practice" includes activities such as: designing the collection of, summarizing, processing, analyzing, interpreting, or presenting, data; as well as model or algorithm development and deployment. Throughout these Guidelines, the term "statistical practitioner" includes all those who engage in statistical practice, regardless of job title, profession, level, or field of degree. The Guidelines are intended for individuals, but these principles are also relevant to organizations that engage in statistical practice.

The Ethical Guidelines aim to promote accountability by informing those who rely on any aspects of statistical practice of the standards that they should expect. Society benefits from informed judgments supported by ethical statistical practice. All statistical practitioners are expected to follow these Guidelines and to encourage others to do the same.

In some situations, Guideline principles may require balancing of competing interests. If an unexpected ethical challenge arises, the ethical practitioner seeks guidance, not exceptions, in the Guidelines. To justify unethical behaviors, or to exploit gaps in the Guidelines, is unprofessional, and inconsistent with these Guidelines.

Principle A: Professional integrity and accountability

Professional integrity and accountability require taking responsibility for one's work. Ethical statistical practice supports valid and prudent decision making with appropriate methodology. The ethical statistical practitioner represents their capabilities and activities honestly, and treats others with respect.

The ethical statistical practitioner:

1. Takes responsibility for evaluating potential tasks, assessing whether they have (or can attain) sufficient competence to execute each task, and that the work and timeline are feasible. Does not solicit or deliver work

for which they are not qualified, or that they would not be willing to have peer reviewed.

2. Uses methodology and data that are valid, relevant, and appropriate, without favoritism or prejudice, and in a manner intended to produce valid, interpretable, and reproducible results.

3. Does not knowingly conduct statistical practices that exploit vulnerable populations or create or perpetuate unfair outcomes.

4. Opposes efforts to predetermine or influence the results of statistical practices, and resists pressure to selectively interpret data.

5. Accepts full responsibility for their own work; does not take credit for the work of others; and gives credit to those who contribute. Respects and acknowledges the intellectual property of others.

6. Strives to follow, and encourages all collaborators to follow, an established protocol for authorship. Advocates for recognition commensurate with each person's contribution to the work. Recognizes that inclusion as an author does imply, while acknowledgement may imply, endorsement of the work.

7. Discloses conflicts of interest, financial and otherwise, and manages or resolves them according to established policies, regulations, and laws.

8. Promotes the dignity and fair treatment of all people. Neither engages in nor condones discrimination based on personal characteristics. Respects personal boundaries in interactions and avoids harassment including sexual harassment, bullying, and other abuses of power or authority.

9. Takes appropriate action when aware of deviations from these Guidelines by others.

10. Acquires and maintains competence through upgrading of skills as needed to maintain a high standard of practice.

11. Follows applicable policies, regulations, and laws relating to their professional work, unless there is a compelling ethical justification to do otherwise.

12. Upholds, respects, and promotes these Guidelines. Those who teach, train, or mentor in statistical practice have a special obligation to promote behavior that is consistent with these Guidelines.

Principle B: Integrity of data and methods

The ethical statistical practitioner seeks to understand and mitigate known or suspected limitations, defects, or biases in the data or methods and communicates potential impacts on the interpretation, conclusions, recommendations, decisions, or other results of statistical practices.

The ethical statistical practitioner:

1. Communicates data sources and fitness for use, including data generation and collection processes and known biases. Discloses and manages any conflicts of interest relating to the data sources. Communicates data processing and transformation procedures, including missing data handling.
2. Is transparent about assumptions made in the execution and interpretation of statistical practices including methods used, limitations, possible sources of error, and algorithmic biases. Conveys results or applications of statistical practices in ways that are honest and meaningful.
3. Communicates the stated purpose and the intended use of statistical practices. Is transparent regarding a priori versus post hoc objectives and planned versus unplanned statistical practices. Discloses when multiple comparisons are conducted, and any relevant adjustments.
4. Meets obligations to share the data used in the statistical practices, for example, for peer review and replication, as allowable. Respects expectations of data contributors when using or sharing data. Exercises due caution to protect proprietary and confidential data, including all data that might inappropriately harm data subjects.
5. Strives to promptly correct substantive errors discovered after publication or implementation. As appropriate, disseminates the correction publicly and/or to others relying on the results.
6. For models and algorithms designed to inform or implement decisions repeatedly, develops and/or implements plans to validate assumptions and assess performance over time, as needed. Considers criteria and mitigation plans for model or algorithm failure and retirement.
7. Explores and describes the effect of variation in human characteristics and groups on statistical practice when feasible and relevant.

Principle C: Responsibilities to stakeholders

Those who fund, contribute to, use, or are affected by statistical practices are considered stakeholders. The ethical statistical practitioner respects the interests of stakeholders while practicing in compliance with these Guidelines.

The ethical statistical practitioner:

1. Seeks to establish what stakeholders hope to obtain from any specific project. Strives to obtain sufficient subject-matter knowledge to conduct meaningful and relevant statistical practice.

2. Regardless of personal or institutional interests or external pressures, does not use statistical practices to mislead any stakeholder.

3. Uses practices appropriate to exploratory and confirmatory phases of a project, differentiating findings from each so the stakeholders can understand and apply the results.

4. Informs stakeholders of the potential limitations on use and re-use of statistical practices in different contexts and offers guidance and alternatives, where appropriate, about scope, cost, and precision considerations that affect the utility of the statistical practice.

5. Explains any expected adverse consequences from failing to follow through on an agreed-upon sampling or analytic plan.

6. Strives to make new methodological knowledge widely available to provide benefits to society at large. Presents relevant findings, when possible, to advance public knowledge.

7. Understands and conforms to confidentiality requirements for data collection, release, and dissemination and any restrictions on its use established by the data provider (to the extent legally required). Protects the use and disclosure of data accordingly. Safeguards privileged information of the employer, client, or funder.

8. Prioritizes both scientific integrity and the principles outlined in these Guidelines when interests are in conflict.

Principle D: Responsibilities to research subjects, data subjects, or those directly affected by statistical practices

The ethical statistical practitioner does not misuse or condone the misuse of data. They protect and respect the rights and interests of human and animal subjects. These responsibilities extend to those who will be directly affected by statistical practices.

The ethical statistical practitioner:

1. Keeps informed about and adheres to applicable rules, approvals, and guidelines for the protection and welfare of human and animal subjects. Knows when work requires ethical review and oversight.[ASA 1]

2. Makes informed recommendations for sample size and statistical practice methodology in order to avoid the use of excessive or inadequate numbers of subjects and excessive risk to subjects

---

[ASA 1] Examples of ethical review and oversight include an Institutional Review Board (IRB), an Institutional Animal Care and Use Committee (IACUC), or a compliance assessment.

3. For animal studies, seeks to leverage statistical practice to reduce the number of animals used, refine experiments to increase the humane treatment of animals, and replace animal use where possible.

4. Protects people's privacy and the confidentiality of data concerning them, whether obtained from the individuals directly, other persons, or existing records. Knows and adheres to applicable rules, consents, and guidelines to protect private information.

5. Uses data only as permitted by data subjects' consent when applicable or considering their interests and welfare when consent is not required. This includes primary and secondary uses, use of repurposed data, sharing data, and linking data with additional data sets.

6. Considers the impact of statistical practice on society, groups, and individuals. Recognizes that statistical practice could adversely affect groups or the public perception of groups, including marginalized groups. Considers approaches to minimize negative impacts in applications or in framing results in reporting.

7. Refrains from collecting or using more data than is necessary. Uses confidential information only when permitted and only to the extent necessary. Seeks to minimize the risk of re-identification when sharing de-identified data or results where there is an expectation of confidentiality. Explains any impact of de-identification on accuracy of results.

8. To maximize contributions of data subjects, considers how best to use available data sources for exploration, training, testing, validation, or replication as needed for the application. The ethical statistical practitioner appropriately discloses how the data is used for these purposes and any limitations.

9. Knows the legal limitations on privacy and confidentiality assurances and does not over-promise or assume legal privacy and confidentiality protections where they may not apply.

10. Understands the provenance of the data, including origins, revisions, and any restrictions on usage, and fitness for use prior to conducting statistical practices.

11. Does not conduct statistical practice that could reasonably be interpreted by subjects as sanctioning a violation of their rights. Seeks to use statistical practices to promote the just and impartial treatment of all individuals.

Principle E: Responsibilities to members of multi-disciplinary teams

Statistical practice is often conducted in teams made up of professionals with different professional standards. The statistical practitioner must know how to work ethically in this environment.

The ethical statistical practitioner:

1. Recognizes and respects that other professions may have different ethical standards and obligations. Dissonance in ethics may still arise even if all members feel that they are working towards the same goal. It is essential to have a respectful exchange of views.
2. Prioritizes these Guidelines for the conduct of statistical practice in cases where ethical guidelines conflict.
3. Ensures that all communications regarding statistical practices are consistent with these Guidelines. Promotes transparency in all statistical practices.
4. Avoids compromising validity for expediency. Regardless of pressure on or within the team, does not use inappropriate statistical practices.

Principle F: Responsibilities to fellow statistical practitioners and the profession

Statistical practices occur in a wide range of contexts. Irrespective of job title and training, those who practice statistics have a responsibility to treat statistical practitioners, and the profession, with respect. Responsibilities to other practitioners and the profession include honest communication and engagement that can strengthen the work of others and the profession.

The ethical statistical practitioner:

1. Recognizes that statistical practitioners may have different expertise and experiences, which may lead to divergent judgments about statistical practices and results. Constructive discourse with mutual respect focuses on scientific principles and methodology and not personal attributes.
2. Helps strengthen, and does not undermine, the work of others through appropriate peer review or consultation. Provides feedback or advice that is impartial, constructive, and objective.
3. Takes full responsibility for their contributions as instructors, mentors, and supervisors of statistical practice by ensuring their best teaching

and advising -- regardless of an academic or non-academic setting -- to ensure that developing practitioners are guided effectively as they learn and grow in their careers.

4. Promotes reproducibility and replication, whether results are "significant" or not, by sharing data, methods, and documentation to the extent possible.

5. Serves as an ambassador for statistical practice by promoting thoughtful choices about data acquisition, analytic procedures, and data structures among non-practitioners and students. Instills appreciation for the concepts and methods of statistical practice.

Principle G: Responsibilities of leaders, supervisors, and mentors in statistical practice

Statistical practitioners leading, supervising, and/or mentoring people in statistical practice have specific obligations to follow and promote these Ethical Guidelines. Their support for – and insistence on – ethical statistical practice is essential for the integrity of the practice and profession of statistics as well as the practitioners themselves.

Those leading, supervising, or mentoring statistical practitioners are expected to:

1. Ensure appropriate statistical practice that is consistent with these Guidelines. Protect the statistical practitioners who comply with these Guidelines, and advocate for a culture that supports ethical statistical practice.

2. Promote a respectful, safe, and productive work environment. Encourage constructive engagement to improve statistical practice.

3. Identify and/or create opportunities for team members/mentees to develop professionally and maintain their proficiency.

4. Advocate for appropriate, timely, inclusion and participation of statistical practitioners as contributors/collaborators. Promote appropriate recognition of the contributions of statistical practitioners, including authorship if applicable.

5. Establish a culture that values validation of assumptions, and assessment of model/algorithm performance over time and across relevant subgroups, as needed. Communicate with relevant stakeholders regarding model or algorithm maintenance, failure, or actual or proposed modifications.

Principle H: Responsibilities regarding potential misconduct

The ethical statistical practitioner understands that questions may arise concerning potential misconduct related to statistical, scientific, or professional practice. At times, a practitioner may accuse someone of misconduct, or be accused by others. At other times, a practitioner may be involved in the investigation of others' behavior. Allegations of misconduct may arise within different institutions with different standards and potentially different outcomes. The elements that follow relate specifically to allegations of statistical, scientific, and professional misconduct.

The ethical statistical practitioner:

1. Knows the definitions of, and procedures relating to, misconduct in their institutional setting. Seeks to clarify facts and intent before alleging misconduct by others. Recognizes that differences of opinion and honest error do not constitute unethical behavior.

2. Avoids condoning or appearing to condone statistical, scientific, or professional misconduct. Encourages other practitioners to avoid misconduct or the appearance of misconduct.

3. Does not make allegations that are poorly founded, or intended to intimidate. Recognizes such allegations as potential ethics violations.

4. Lodges complaints of misconduct discreetly and to the relevant institutional body. Does not act on allegations of misconduct without appropriate institutional referral, including those allegations originating from social media accounts or email listservs.

5. Insists upon a transparent and fair process to adjudicate claims of misconduct. Maintains confidentiality when participating in an investigation. Discloses the investigation results honestly to appropriate parties and stakeholders once they are available.

6. Refuses to publicly question or discredit the reputation of a person based on a specific accusation of misconduct while due process continues to unfold.

7. Following an investigation of misconduct, supports the efforts of all parties involved to resume their careers in as normal a manner as possible, consistent with the outcome of the investigation.

8. Avoids, and acts to discourage, retaliation against or damage to the employability of those who responsibly call attention to possible misconduct.

Appendix: Responsibilities of organizations/institutions

Whenever organizations and institutions design the collection of, summarize, process, analyze, interpret, or present, data; or develop and/or deploy models or algorithms, they have responsibilities to use statistical practice in ways that are consistent with these Guidelines, as well as promote ethical statistical practice.

Organizations and institutions engage in, and promote, ethical statistical practice by:

1. Expecting and encouraging all employees and vendors who conduct statistical practice to adhere to these Guidelines. Promoting a workplace where the ethical practitioner may apply the Guidelines without being intimidated or coerced. Protecting statistical practitioners who comply with these Guidelines.
2. Engaging competent personnel to conduct statistical practice, and promote a productive work environment.
3. Promoting the professional development and maintenance of proficiency for employed statistical practitioners.
4. Supporting statistical practice that is objective and transparent. Not allowing organizational objectives or expectations to encourage unethical statistical practice by its employees.
5. Recognizing that the inclusion of statistical practitioners as authors, or acknowledgement of their contributions to projects or publications, requires their explicit permission because it may imply endorsement of the work.
6. Avoiding statistical practices that exploit vulnerable populations or create or perpetuate discrimination or unjust outcomes. Considering both scientific validity and impact on societal and human well-being that results from the organization's statistical practice.
7. Using professional qualifications and contributions as the basis for decisions regarding statistical practitioners' hiring, firing, promotion, work assignments, publications and presentations, candidacy for offices and awards, funding or approval of research, and other professional matters.

Those in leadership, supervisory, or managerial positions who oversee statistical practitioners promote ethical statistical practice by following Principle G and:

1. Recognizing that it is contrary to these Guidelines to report or follow only those results that conform to expectations without explicitly acknowledging competing findings and the basis for choices regarding which results to report, use, and/or cite.
2. Recognizing that the results of valid statistical studies cannot be guaranteed to conform to the expectations or desires of those commissioning the study or employing/supervising the statistical practitioner(s).
3. Objectively, accurately, and efficiently communicating a team's or practitioners' statistical work throughout the organization.
4. In cases where ethical issues are raised, representing them fairly within the organization's leadership team.
5. Managing resources and organizational strategy to direct teams of statistical practitioners along the most productive lines in light of the ethical standards contained in these Guidelines.

These 72 elements are general – capturing eight dimensions (Principles) that represent distinct features of decisions that all quantitative practitioners face at some point in a career (if not the typical workweek). The eight principles are summarized below.

A. **Professional integrity and accountability** points out the need for respectful and professional conduct as well as competence, judgment, diligence, self-respect, and worthiness of the respect of other people.

B. **Integrity of data and methods** addresses the need to report sufficient information to give readers, including other practitioners, a clear understanding of the intent of the work, the provenance of the data, how and by whom any analysis was performed, and any limitations on the validity of results. This principle not only ensures that reporting will include sufficient information for competent evaluation, but also that the ethical practitioner uses appropriate methods, and ensures that the use of data is appropriate – in terms of sample size as well as consent.

C. **Responsibilities to stakeholders** discusses the practitioner's responsibility for assuring that statistical work is appropriate and also suitable to the needs and resources of those who are paying for it -and affected by it – and that these stakeholders all understand the capabilities and limitations of statistics in addressing their problem.

D. Responsibilities to research subjects, data subjects, or those directly affected by statistical practices describes requirements for respectful professional conduct generally and also protecting the interests of human and animal subjects of research-not only during data collection but also in the analysis, interpretation, and publication of the resulting findings.

E. Responsibilities to members of multidisciplinary teams addresses respectful professional conduct and the mutual responsibilities of professionals participating in multidisciplinary research teams.

F. Responsibilities to fellow statistical practitioners and the Profession notes the interdependence of practitioners doing similar work, whether in the same or different organizations. Basically, they must contribute to the dignity and strength of the Profession of statistics and data science overall, by not degrading the Profession's integrity, and also by sharing nonproprietary data and methods, participating in peer review, and respecting differing professional opinions.

G. Responsibilities of leaders, supervisors, and mentors in statistical practice encourages employers and clients to recognize the highly interdependent nature of statistical ethics and statistical validity. Employers and clients must not pressure practitioners to produce a particular "result," regardless of its statistical validity. They must avoid the potential social harm that can result from the dissemination of false or misleading statistical work.

H. Responsibilities regarding potential misconduct addresses the sometimes-painful process of investigating (or initiating the investigation of) potential ethical or legal violations, and treating those involved with both justice and respect.

The **Appendix** builds on Principle G, but expands the responsibilities of organizations and institutions to act, and support their employees and contractors to act, in a manner consistent with the ASA GLs whenever the organization or institution, or its employees or contractors, engage in statistical practices, i.e., design the collection of, summarize, process, analyze, interpret, or present, data; or develop and/or deploy models or algorithms.

The Guidelines are scheduled to be updated every 5 years starting in 2021 after the 2016 revisions. The 2018 revisions to the 2016 Guidelines were solely focused on ensuring that statistical practice standards explicitly denounced intimidation and harassment, and spelled out every practitioner's obligation to prevent scientific and professional misconduct (which includes, but is not limited to, intimidation and harassment). The revisions which began in 2021 took the entire year to complete, and were not approved until February 2022. There are therefore referred to as "the 2022 revisions".

The ASA has created a method for eliciting input from the wider (ASA) community in order to maintain a set of professional practice standards that are reflective of the full spectrum of statistical practitioners. Thus, the ASA Guidelines (GLs) are dynamic; more to the point, the describe how all quantitative practitioners should consider their qualifications and integrity, as well as the impacts of their decisions on others – in an ongoing manner. Just as the GLs are dynamic, the practitioner is, too: Principle G is specifically focused on statistics practitioners in leadership, supervisory, or mentorship roles. While an individual may start out their career in none of those roles, as the career progresses, any or all of these roles may become part of the practitioner's job or work. Early in the career, a practitioner can utilize Principle G to gauge, or select, mentors or groups with supervisors or leaders who follow Principle G (and the rest of the GLs). By the time the practitioner is filling those roles themselves, they would be familiar enough with them to be the kind of leader, supervisor, or mentor that new practitioners would want to work with or for, or to learn from.

The purposes of the GLs are stated to be: "intended to help statistics practitioners make decisions ethically. Additionally, the Ethical Guidelines aim to promote accountability by informing those who rely on statistical analysis of the standards that they should expect." Decision-making is a large part of all statistical practice. The GLs seek to make this decision making *ethical*, but rather than describe all the various (and uncountable) ways that decisions can be made ethically, the ASA GLs describe, in general terms, what it would look like when a statistician makes a decision that is ethical. For example, Principle A2 states that "(the ethical statistical practitioner) Uses methodology and data that are valid, relevant, and appropriate, without favoritism or prejudice, and in a manner intended to produce valid, interpretable, and reproducible results." Thus, every **ethical** decision about methodology and data will focus on assuring that the methods and data are valid, relevant, and appropriate (relevance and propriety are up to the practitioner to establish). The ethical *use* of that relevant and appropriate data is described as well: "a manner intended to produce valid, interpretable, and reproducible results." Principle A also specifies two specific features of ethical statistical practice relating to following the GLs: A9: "(the ethical statistical practitioner) Takes appropriate action when aware of deviations from these Guidelines by others." And A12, "(the ethical statistical practitioner) Upholds, respects, and promotes these Guidelines. Those who teach, train, or mentor in statistical practice have a special obligation to promote behavior that is consistent with these Guidelines."

Chapter 1.3
ACM Code of Ethics

Along a similar timeline followed by the ASA Ethical Guidelines for Statistical Practice, the Association for Computing Machinery (ACM) also formulated (1992) and revised (2018) a code of ethics (https://www.acm.org/binaries/content/assets/about/acm-code-of-ethics-and-professional-conduct.pdf). Four main areas are elaborated on with specific behaviors that "a computing professional should" do. The full ACM Code of Ethics (CE) is presented below, and is Copyrighted (c) 2018 by the Association for Computing Machinery.

Adopted by ACM Council 6/22/18.

Preamble

Computing professionals' actions change the world. To act responsibly, they should reflect upon the wider impacts of their work, consistently supporting the public good. The ACM Code of Ethics and Professional Conduct ("the Code") expresses the conscience of the profession.

The Code is designed to inspire and guide the ethical conduct of all computing professionals, including current and aspiring practitioners, instructors, students, influencers, and anyone who uses computing technology in an impactful way. Additionally, the Code serves as a basis for remediation when violations occur. The Code includes principles formulated as statements of responsibility, based on the understanding that the public good is always the primary consideration. Each principle is supplemented by guidelines, which provide explanations to assist computing professionals in understanding and applying the principle.

Section 1 outlines fundamental ethical principles that form the basis for the remainder of the Code. Section 2 addresses additional, more specific considerations of professional responsibility. Section 3 guides individuals who have a leadership role, whether in the workplace or in a volunteer professional capacity. Commitment to ethical conduct is required of every ACM member, and principles involving compliance with the Code are given in Section 4.

The Code as a whole is concerned with how fundamental ethical principles apply to a computing professional's conduct. The Code is not an algorithm for solving ethical problems; rather it serves as a basis for ethical decision-making. When thinking through a particular issue, a computing professional may find that multiple principles should be taken into account, and that different principles will have different relevance to the issue. Questions related to these kinds of issues can best be answered by thoughtful consideration of the fundamental ethical principles, understanding that the public good is the paramount consideration. The entire computing profession benefits when the ethical decision-making process is accountable to and transparent to all stakeholders. Open discussions about ethical issues promote this accountability and transparency.

1. General Ethical Principles.

A computing professional should:

1.1 Contribute to society and to human well-being, acknowledging that all people are stakeholders in computing.

This principle, which concerns the quality of life of all people, affirms an obligation of computing professionals, both individually and collectively, to use their skills for the benefit of society, its members, and the environment surrounding them. This obligation includes promoting fundamental human rights and protecting each individual's right to autonomy. An essential aim of computing professionals is to minimize negative consequences of computing, including threats to health, safety, personal security, and privacy. When the interests of multiple groups conflict, the needs of those less advantaged should be given increased attention and priority.

Computing professionals should consider whether the results of their efforts will respect diversity, will be used in socially responsible ways, will meet social needs, and will be broadly accessible. They are encouraged to actively contribute to society by engaging in pro bono or volunteer work that benefits the public good.

In addition to a safe social environment, human well-being requires a safe natural environment. Therefore, computing professionals should promote environmental sustainability both locally and globally.

1.2 Avoid harm.

In this document, "harm" means negative consequences, especially when those consequences are significant and unjust. Examples of harm include unjustified physical or mental injury, unjustified destruction or disclosure of information, and unjustified damage to property, reputation, and the environment. This list is not exhaustive.

Well-intended actions, including those that accomplish assigned duties, may lead to harm. When that harm is unintended, those responsible are obliged to undo or mitigate the harm as much as possible. Avoiding harm begins with careful consideration of potential impacts on all those affected by decisions. When harm is an intentional part of the system, those responsible are obligated to ensure that the harm is ethically justified. In either case, ensure that all harm is minimized.

To minimize the possibility of indirectly or unintentionally harming others, computing professionals should follow generally accepted best practices unless there is a compelling ethical reason to do otherwise. Additionally, the consequences of data aggregation and emergent properties of systems should be carefully analyzed. Those involved with pervasive or infrastructure systems should also consider Principle 3.7.

A computing professional has an additional obligation to report any signs of system risks that might result in harm. If leaders do not act to curtail or mitigate such risks, it may be necessary to "blow the whistle" to reduce potential harm. However, capricious or misguided reporting of risks can itself be harmful. Before reporting risks, a computing professional should carefully assess relevant aspects of the situation.

1.3 Be honest and trustworthy.

Honesty is an essential component of trustworthiness. A computing professional should be transparent and provide full disclosure of all pertinent system capabilities, limitations, and potential problems to the appropriate parties. Making deliberately false or misleading claims, fabricating or falsifying data, offering or accepting bribes, and other dishonest conduct are violations of the Code.

Computing professionals should be honest about their qualifications, and about any limitations in their competence to complete a task. Computing professionals should be forthright about any circumstances that might lead to either real or perceived conflicts of interest or otherwise tend to undermine the independence of their judgment. Furthermore, commitments should be honored.

Computing professionals should not misrepresent an organization's policies or procedures, and should not speak on behalf of an organization unless authorized to do so.

1.4 Be fair and take action not to discriminate.

The values of equality, tolerance, respect for others, and justice govern this principle. Fairness requires that even careful decision processes provide some avenue for redress of grievances.

Computing professionals should foster fair participation of all people, including those of underrepresented groups. Prejudicial discrimination on the basis of age, color, disability, ethnicity, family status, gender identity, labor union membership, military status, nationality, race, religion or belief, sex, sexual orientation, or any other inappropriate factor is an explicit violation of the Code. Harassment, including sexual harassment, bullying, and other abuses of power and authority, is a form of discrimination that, amongst other harms, limits fair access to the virtual and physical spaces where such harassment takes place.

The use of information and technology may cause new, or enhance existing, inequities. Technologies and practices should be as inclusive and accessible as possible and computing professionals should take action to avoid creating systems or technologies that disenfranchise or oppress people. Failure to design for inclusiveness and accessibility may constitute unfair discrimination.

1.5 Respect the work required to produce new ideas, inventions, creative works, and computing artifacts.

Developing new ideas, inventions, creative works, and computing artifacts creates value for society, and those who expend this effort should expect to gain value from their work. Computing professionals should therefore credit the creators of ideas, inventions, work, and artifacts, and respect copyrights, patents, trade secrets, license agreements, and other methods of protecting authors' works.

Both custom and the law recognize that some exceptions to a creator's control of a work are necessary for the public good. Computing professionals should not unduly oppose reasonable uses of their intellectual works. Efforts to help others by contributing time and energy to projects that help society illustrate a positive aspect of this principle. Such efforts include free and open-source software and work put into the public domain. Computing professionals should not claim private ownership of work that they or others have shared as public resources.

1.6 Respect privacy.

The responsibility of respecting privacy applies to computing professionals in a particularly profound way. Technology enables the collection, monitoring, and exchange of personal information quickly, inexpensively, and often without the knowledge of the people affected. Therefore, a computing professional should become conversant in the various definitions and forms of privacy and should understand the rights and responsibilities associated with the collection and use of personal information.

Computing professionals should only use personal information for legitimate ends and without violating the rights of individuals and groups. This requires taking precautions to prevent re- identification of anonymized data or unauthorized data collection, ensuring the accuracy of data, understanding the provenance of the data, and protecting it from unauthorized access and accidental disclosure. Computing professionals should establish transparent policies and procedures that allow individuals to understand what data is being collected and how it is being used, to give informed consent for automatic data collection, and to review, obtain, correct inaccuracies in, and delete their personal data.

Only the minimum amount of personal information necessary should be collected in a system. The retention and disposal periods for that information should be clearly defined, enforced, and communicated to data subjects. Personal information gathered for a specific purpose should not be used for other purposes without the person's consent. Merged data collections can compromise privacy features present in the original collections. Therefore, computing professionals should take special care for privacy when merging data collections.

1.7 Honour confidentiality.

Computing professionals are often entrusted with confidential information such as trade secrets, client data, nonpublic business strategies, financial information, research data, pre-publication scholarly articles, and patent applications. Computing professionals should protect confidentiality except in cases where it is evidence of the violation of law, of organizational regulations, or of the Code. In these cases, the nature or contents of that information should not be disclosed except to appropriate authorities. A computing professional should consider thoughtfully whether such disclosures are consistent with the Code.

2. Professional Responsibilities.

A computing professional should:

2.1 Strive to achieve high quality in both the processes and products of professional work.

Computing professionals should insist on and support high quality work from themselves and from colleagues. The dignity of employers, employees, colleagues, clients, users, and anyone else affected either directly or indirectly by the work should be respected throughout the process. Computing professionals should respect the right of those involved to transparent communication about the project. Professionals should be cognizant of any serious negative consequences affecting any stakeholder that may result from poor quality work and should resist inducements to neglect this responsibility.

2.2 Maintain high standards of professional competence, conduct, and ethical practice.

High quality computing depends on individuals and teams who take personal and group responsibility for acquiring and maintaining professional competence. Professional competence starts with technical knowledge and with awareness of the social context in which their work may be deployed. Professional competence also requires skill in communication, in reflective analysis, and in recognizing and navigating ethical challenges. Upgrading skills should be an ongoing process and might include independent study, attending conferences or seminars, and other informal or formal education. Professional organizations and employers should encourage and facilitate these activities.

2.3 Know and respect existing rules pertaining to professional work.

"Rules" here include local, regional, national, and international laws and regulations, as well as any policies and procedures of the organizations to which the professional belongs. Computing professionals must abide by these rules unless there is a compelling ethical justification to do otherwise. Rules that are judged unethical should be challenged. A rule may be unethical when it has an inadequate moral basis or causes recognizable harm. A computing professional should consider challenging the rule through existing channels before violating the rule. A computing professional who decides to violate a rule because it is unethical, or for any other reason, must consider potential consequences and accept responsibility for that action.

2.4 Accept and provide appropriate professional review.

High quality professional work in computing depends on professional review at all stages. Whenever appropriate, computing professionals should seek and utilize peer and stakeholder review. Computing professionals should also provide constructive, critical reviews of others' work.

2.5 Give comprehensive and thorough evaluations of computer systems and their impacts, including analysis of possible risks.

Computing professionals are in a position of trust, and therefore have a special responsibility to provide objective, credible evaluations and testimony to employers, employees, clients, users, and the public. Computing professionals should strive to be perceptive, thorough, and objective when evaluating, recommending, and presenting system descriptions and alternatives. Extraordinary care should be taken to identify and mitigate potential risks in machine learning systems. A system for which future risks cannot be reliably predicted requires frequent reassessment of risk as the system evolves in use, or it should not be deployed. Any issues that might result in major risk must be reported to appropriate parties.

2.6 Perform work only in areas of competence.

A computing professional is responsible for evaluating potential work assignments. This includes evaluating the work's feasibility and advisability, and making a judgment about whether the work assignment is within the professional's areas of competence. If at any time before or during the work assignment the professional identifies a lack of a necessary expertise, they must disclose this to the employer or client. The client or employer may decide to pursue the assignment with the professional after additional time to acquire the necessary competencies, to pursue the assignment with someone else who has the required expertise, or to forgo the assignment. A computing professional's ethical judgment should be the final guide in deciding whether to work on the assignment.

2.7 Foster public awareness and understanding of computing, related technologies, and their consequences.

As appropriate to the context and one's abilities, computing professionals should share technical knowledge with the public, foster awareness of computing, and encourage understanding of computing. These communications with the public should be clear, respectful, and welcoming. Important issues include the impacts of computer systems, their limitations, their vulnerabilities, and the opportunities that they present. Additionally, a computing professional should respectfully address inaccurate or misleading information related to computing.

2.8 Access computing and communication resources only when authorized or when compelled by the public good.

Individuals and organizations have the right to restrict access to their systems and data so long as the restrictions are consistent with other principles in the Code. Consequently, computing professionals should not access another's computer system, software, or data without a reasonable belief that such an action would be authorized or a compelling belief that it is consistent with the public good. A system being publicly accessible is not sufficient grounds on its own to imply authorization. Under exceptional circumstances a computing professional may use unauthorized access to disrupt or inhibit the functioning of malicious systems; extraordinary precautions must be taken in these instances to avoid harm to others.

2.9 Design and implement systems that are robustly and usably secure.

Breaches of computer security cause harm. Robust security should be a primary consideration when designing and implementing systems. Computing professionals should perform due diligence to ensure the system functions as intended, and take appropriate action to secure resources against accidental and intentional misuse, modification, and denial of service. As threats can arise and change after a system is deployed, computing professionals should integrate mitigation techniques and policies, such as monitoring, patching, and vulnerability reporting. Computing professionals should also take steps to ensure parties affected by data breaches are notified in a timely and clear manner, providing appropriate guidance and remediation.

To ensure the system achieves its intended purpose, security features should be designed to be as intuitive and easy to use as possible. Computing professionals should discourage security precautions that are too confusing, are situationally inappropriate, or otherwise inhibit legitimate use.

In cases where misuse or harm are predictable or unavoidable, the best option may be to not implement the system.

3. Professional Leadership Principles.

Leadership may either be a formal designation or arise informally from influence over others. In this section, "leader" means any member of an organization or group who has influence, educational responsibilities, or managerial responsibilities. While these principles apply to all computing professionals, leaders bear a heightened responsibility to uphold and promote them, both within and through their organizations.

A computing professional, especially one acting as a leader, should:

3.1 Ensure that the public good is the central concern during all professional computing work.

People—including users, customers, colleagues, and others affected directly or indirectly— should always be the central concern in computing. The public good should always be an explicit consideration when evaluating tasks associated with research, requirements analysis, design, implementation, testing, validation, deployment, maintenance, retirement, and disposal. Computing professionals should keep this focus no matter which methodologies or techniques they use in their practice.

3.2 Articulate, encourage acceptance of, and evaluate fulfillment of social responsibilities by members of the organization or group.

Technical organizations and groups affect broader society, and their leaders should accept the associated responsibilities. Organizations—through procedures and attitudes oriented toward quality, transparency, and the welfare of society—reduce harm to the public and raise awareness of the influence of technology in our lives. Therefore, leaders should encourage full participation of computing professionals in meeting relevant social responsibilities and discourage tendencies to do otherwise.

3.3 Manage personnel and resources to enhance the quality of working life.

Leaders should ensure that they enhance, not degrade, the quality of working life. Leaders should consider the personal and professional development, accessibility requirements, physical safety, psychological well-being, and human dignity of all workers. Appropriate human-computer ergonomic standards should be used in the workplace.

3.4 Articulate, apply, and support policies and processes that reflect the principles of the Code.

Leaders should pursue clearly defined organizational policies that are consistent with the Code and effectively communicate them to relevant stakeholders. In addition, leaders should encourage and reward compliance with those policies, and take appropriate action when policies are violated. Designing or implementing processes that deliberately or negligently violate, or tend to enable the violation of, the Code's principles is ethically unacceptable.

3.5 Create opportunities for members of the organization or group to grow as professionals.

Educational opportunities are essential for all organization and group members. Leaders should ensure that opportunities are available to computing professionals to help them improve their knowledge and skills in professionalism, in the practice of ethics, and in their technical specialties. These opportunities should include experiences that familiarize computing professionals with the consequences and limitations of particular types of systems. Computing professionals should be fully aware of the dangers of oversimplified approaches, the improbability of anticipating every possible operating condition, the inevitability of software errors, the interactions of systems and their contexts, and other issues related to the complexity of their profession—and thus be confident in taking on responsibilities for the work that they do.

3.6 Use care when modifying or retiring systems.

Interface changes, the removal of features, and even software updates have an impact on the productivity of users and the quality of their work. Leaders should take care when changing or discontinuing support for system features on which people still depend. Leaders should thoroughly investigate viable alternatives to removing support for a legacy system. If these alternatives are unacceptably risky or impractical, the developer should assist stakeholders' graceful migration from the system to an alternative. Users should be notified of the risks of continued use of the unsupported system long before support ends. Computing professionals should assist system users in monitoring the operational viability of their computing systems, and help them understand that timely replacement of inappropriate or outdated features or entire systems may be needed.

3.7 Recognize and take special care of systems that become integrated into the infrastructure of society.

Even the simplest computer systems have the potential to impact all aspects of society when integrated with everyday activities such as commerce, travel, government, healthcare, and education. When organizations and groups develop systems that become an important part of the infrastructure of society, their leaders have an added responsibility to be good stewards of these systems. Part of that stewardship requires establishing policies for fair system access, including for those who may have been excluded. That stewardship also requires that computing professionals monitor the level of integration of their systems into the infrastructure of society. As the level of adoption changes, the ethical responsibilities of the organization or group are likely to change as well. Continual monitoring of how society is using a system will allow the organization or group to remain consistent with their ethical obligations outlined in the Code. When appropriate standards of care do not exist, computing professionals have a duty to ensure they are developed.

4. Compliance with the Code.

A computing professional should:

4.1 Uphold, promote, and respect the principles of the Code.

The future of computing depends on both technical and ethical excellence. Computing professionals should adhere to the principles of the Code and contribute to improving them. Computing professionals who recognize breaches of the Code should take actions to resolve the ethical issues they recognize, including, when reasonable, expressing their concern to the person or persons thought to be violating the Code.

4.2 Treat violations of the Code as inconsistent with membership in the ACM.

Each ACM member should encourage and support adherence by all computing professionals regardless of ACM membership. ACM members who recognize a breach of the Code should consider reporting the violation to the ACM, which may result in remedial action as specified in the ACM's Code of Ethics and Professional Conduct Enforcement Policy, which can be found here: https://ethics.acm.org/wp-content/uploads/2018/07/2018 -ACM-Code-of-Ethics-Enforcement-Procedure.pdf

The ACM CE is shorter, with roughly half the number of elements, as compared to the ASA GLs. These documents were created completely independently – we are only exploring or comparing them because data science, as a field, has naturally arisen from the increasing computational power that can be deployed in statistical work, and the potential for computing to find and munge/wrangle data from an incredibly wide range of sources. The ACM CE principles are summarized briefly below.

1. General Moral Principles

A computing professional should...

1.1 Contribute to society and to human well-being, acknowledging that all people are stakeholders in computing.

1.2 Avoid harm. *In this document, "harm" means negative consequences to any stakeholder, especially when those consequences are significant and unjust.*

1.3 Be honest and trustworthy.

1.4 Be fair and take action not to discriminate.

1.5 Respect the work required to produce new ideas, inventions, creative works, and computing artifacts.

1.6 Respect privacy.

1.7 Honor confidentiality.

It is important to note that the ACM CE prioritizes making contributions "to society and human well-being" – implying that anything that detracts from either society or human well-being is a violation of this code. They go further, stating that the computing professional should "avoid harm" – which is an important part of the stakeholder analysis we focus on in the next chapter. The DEFW has a similar focus, such that every harm or potential harm can be evaluated against the public good that whatever data science/data collection *caused the harm.* In fact, the reasons why the ACM Code of Ethics (like the ASA GLs) includes items like 1.3 – 1.7 is because when individuals fail to follow these key principles, harm will or can easily accrue to others. That is, most of the elements of the CE follow from the priority of ensuring that all computing professionals' behaviors do not detract from society and human well-being, and that their work avoids harm. Principles 1.3-1.7 simply articulate some of the ways that harm can result from our work. You may recognize that the focus on a *public good* is present in both the DEFW and the ACM CE; this emphasis is notably missing from the DSEC and ASA GLs.

2. Professional Responsibilities

A computing professional should...

2.1 Strive to achieve high quality in both the process and products of professional work.

2.2 Maintain high standards of professional competence, conduct, and ethical practice.

2.3 Know, respect, and apply existing rules pertaining to professional work.

2.4 Accept and provide appropriate professional review.

2.5 Give comprehensive and thorough evaluations of computer systems and their impacts, including analysis of possible risks.

2.6 Have the necessary expertise, or the ability to obtain that expertise, for completing a work assignment before accepting it. Once accepted, that commitment should be honored.

2.7 Improve public awareness and understanding of computing, related technologies, and their consequences.

2.8 Access computing and communication resources only when authorized to do so.

2.9 Design and implement systems that are robustly and usably secure.

The longest list of ACM CE principles is categorized as "professional responsibilities". This is where the ASA and ACM have their greatest overlap. As we will see in the next chapter, the DEFW also has a Principle focusing on the practitioner, while DSEC does not.

3. Professional Leadership Principles

A computing professional should...

3.1 Ensure that the public good is the central concern during all professional computing work.

3.2 Articulate, encourage acceptance of, and evaluate fulfillment of the social responsibilities of members of an organization or group.

3.3 Manage personnel and resources to enhance the quality of working life.

3.4 Articulate, apply, and support policies and processes that reflect the principles in the Code.

3.5 Create opportunities for members of the organization or group to learn and be accountable for the scope, functions, limitations, and impacts of systems.

3.6 Retire legacy systems with care.

3.7 Recognize when a computer system is becoming integrated into the infrastructure of society, and adopt an appropriate standard of care for that system and its users.

Importantly, the ACM CE identifies additional and specific responsibilities that may only be relevant for those in leadership roles. Note that *all* of the Principles up until section 3 are possibly more important for those in leadership roles to both exhibit and encourage in all those with whom they work – and whom they may supervise, train, or mentor. Moreover, you should also keep in mind that, even if you play a junior or supporting role on a team or in a workplace, the "Professional Leadership Principles" may still pertain to you: specifically, 3.1 ("Ensure that the public good is the central concern during all professional computing work") speaks to the responsibility of every computing professional to ensure that the public good is prioritized at every career stage. In this sense, it is an echo of Principle 1.1.

4. Compliance with the Code

A computing professional should...

4.1 Uphold, promote, and respect the principles of the Code.
4.2 Treat violations of the Code as inconsistent with membership in the ACM.

ACM CE Principle 4, outlining the computing professional's responsibilities to both uphold the code and recognize that those who violate the code are also acting in a manner that is "inconsistent with membership in the ACM" highlights every computing professional's obligation to follow the code if they wish to be seen and treated as a "computing professional". That is, failures to follow these practice standards identify an individual as *unprofessional.*

[3]You may have noticed that the ACM Code represents a "utilitarian ethics" perspective: it directs "computing professionals" to focus attention on the positive and negative effects of their decisions – and emphasizes decisions that avoid or minimize harms. In the preamble, it is stated that computing professionals act responsibly when "consistently supporting the public good". However, because the ACM Code also specifies what "a computing professional *should*" do (in order to comply with the Code), it can also be interpreted from a virtue perspective, i.e., by describing the ideal professional (an ethical one). There are four main areas with specific points elaborated in terms of what "a computing professional should" do.

Like the ASA Guidelines, the ACM Code seeks to support ethical decision making rather than to establish punishment, or describe explicitly what is right or wrong about aspects of practice. In the CE preamble, it is noted that,

"The Code as a whole is concerned with how fundamental ethical principles apply to a computing professional's conduct. The Code is not

---

[3] This material is adapted from Tractenberg RE. (2020, February 19). Concordance of professional ethical practice standards for the domain of Data Science: A white paper. Published in the *Open Archive of the Social Sciences* (SocArXiv), 10.31235/osf.io/p7rj2

an algorithm for solving ethical problems; rather it serves as a basis for ethical decision-making. When thinking through a particular issue, a computing professional may find that multiple principles should be taken into account, and that different principles will have different relevance to the issue."

While the wording (and some of the manifestations of the specific feature of ethical practice) is different, both organizations specify similar features as foundational in ethical practice with the methods and materials of their respective disciplines.

Chapter 1.4
The Data Science Ethics Checklist (DSEC) and Data Ethics Framework (DEFW)

Many conversations seem to have erupted in 2018 about the need for ethics in data science.[4] Data breaches are increasing at a worrying rate – which has focused attention on the constructs of "data ethics" and the stewardship of data. In their "year in review" of artificial intelligence, The AI Now Institute created a fairly depressing timeline:

---

[4] Material in this chapter adapted from Tractenberg, RE. (2019-b, April 23). Becoming a steward of data science. https://doi.org/10.31235/osf.io/j7h8t

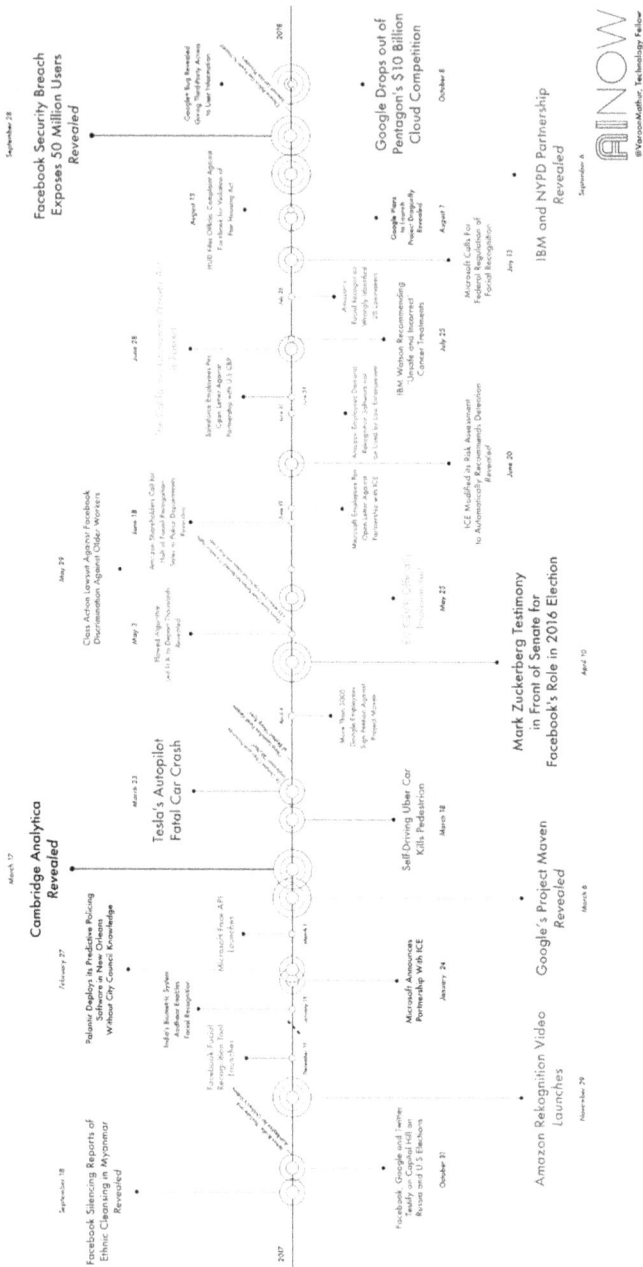

Figure 1.4.1. *Significant data-related events in 2018. Image attributed to Varoon Mathur; captured 25 March 2019 from https://medium.com/@AINowInstitute/ai-in-2018-a-year-in-review-8b161ead2b4e*

Some of these events represent illegal breaches of ostensible data security measures, i.e., whether or not collectors or holders of data implemented the strictest possible security around the data they captured, other data scientists hacked in to steal or otherwise obtain that data illegitimately – and illegally in some cases. In some of these cases, data scientists overstepped permissions and took data they did not obtain consent for. In other cases, computing systems failed to protect privacy, confidentiality, or human lives. These are, of course, only the situations worldwide that rose to the level of what is considered "newsworthy"; many other situations must have occurred that represented violations of ethics or laws but were not sufficiently newsworthy to be mentioned -and so were excluded from this figure. Still other violations have not yet been discovered or reported. This figure must be seen as a sample, and not the entirety, of "data breaches". This is an opportunity to briefly consider the tests of publicity, justice, and universality. Firstly, each scandal in the figure resulted from public recognition that was not desired by the perpetrators. If you are doing something you feel would create a problem for you, or your employer, if the public found out about it, that activity clearly fails the test of publicity. You (or others) might not want the activity to become known because some people are harmed while others benefit; if so, that activity likely fails the test of justice. Activities where some people are harmed while others benefit are also likely to fail a test of universality, because when the pattern of harms-to-some/benefits-to-others is perpetuated, it can create imbalances and unfairness that are challenging to eliminate.

In this context, it is important to consider **your role** in preventing unethical work, as well as in responding to it if you ever encounter it in the workplace. Section 2 focuses on helping every practitioner be as ethical as they can be in daily work, while Section 3 is focused on recognizing and responding to ethical challenges that can arise (and have arisen) in the workplace for statisticians and data scientists.

As we saw in the first chapter, ethical statistics and data science practice starts with prerequisite knowledge. A premise of this book is that the ethical practice standards outlined by the Association of Computing Machinery (ACM) in their Code of Ethics (CE), and the American Statistical Association (ASA) in their Ethical Guidelines (GLs) for Statistical Practice, are essential prerequisite knowledge for competent, professional practice in the domain of statistics and data science (and in the domains of statistics and of data science, in case you identify with or have a job title representing only one of these disciplines).

The Data Science Ethics Checklist (DSEC)

The Data Science Ethics Checklist (DSEC) includes 20 questions grouped into five areas that correspond to different facets of a statistical and data scientific project. These five areas are A. data collection; B. data storage; C. (data) analysis; D. (data) modeling; and E. deployment (of the results of modeling and analysis). The relevance of each question may vary depending on whether the project is specifically scientific, specifically business, or some mixture.

A. Data Collection

A.1 Informed consent: If there are human subjects, have they given informed consent, where subjects affirmatively opt-in and have a clear understanding of the data uses to which they consent?

A.2 Collection bias: Have we considered sources of bias that could be introduced during data collection and survey design and taken steps to mitigate those?

A.3 Limit PII exposure: Have we considered ways to minimize exposure of personally identifiable information (PII) for example through anonymization or not collecting information that isn't relevant for analysis?

B. Data Storage

B.1 Data security: Do we have a plan to protect and secure data (e.g., encryption at rest and in transit, access controls on internal users and third parties, access logs, and up-to-date software)?

B.2 Right to be forgotten: Do we have a mechanism through which an individual can request their personal information be removed?

B.3 Data retention plan: Is there a schedule or plan to delete the data after it is no longer needed?

C. Analysis

C.1 Missing perspectives: Have we sought to address blind spots in the analysis through engagement with relevant stakeholders (e.g., checking assumptions and discussing implications with affected communities and subject matter experts)?

C.2 Dataset bias: Have we examined the data for possible sources of bias and taken steps to mitigate or address these biases (e.g., stereotype perpetuation, confirmation bias, imbalanced classes, or omitted confounding variables)?

C.3 Honest representation: Are our visualizations, summary statistics, and reports designed to honestly represent the underlying data?

C.4 Privacy in analysis: Have we ensured that data with PII are not used or displayed unless necessary for the analysis?

C.5 Auditability: Is the process of generating the analysis well documented and reproducible if we discover issues in the future?

D. Modeling

D.1 Proxy discrimination: Have we ensured that the model does not rely on variables or proxies for variables that are unfairly discriminatory?

D.2 Fairness across groups: Have we tested model results for fairness with respect to different affected groups (e.g., tested for disparate error rates)?

D.3 Metric selection: Have we considered the effects of optimizing for our defined metrics and considered additional metrics?

D.4 Explainability: Can we explain in understandable terms a decision the model made in cases where a justification is needed?

D.5 Communicate bias: Have we communicated the shortcomings, limitations, and biases of the model to relevant stakeholders in ways that can be generally understood?

E. Deployment

E.1 Redress: Have we discussed with our organization a plan for response if users are harmed by the results (e.g., how does the data science team evaluate these cases and update analysis and models to prevent future harm)?

E.2 Roll back: Is there a way to turn off or roll back the model in production if necessary?

E.3 Concept drift: Do we test and monitor for concept drift to ensure the model remains fair over time?

E.4 Unintended use: Have we taken steps to identify and prevent unintended uses and abuse of the model and do we have a plan to monitor these once the model is deployed?

Note that, in addition to data collection (A) and security (B), the DSEC suggests that the ethical quantitative practitioner is **also** concerned with what happens to the data *after* it has been collected and secured (assuming these were also done ethically). Thus, while the DSEC emerged in 2018 at least partly in response to the worldwide rise in worries about data safety, privacy, and confidentiality, it does go further than just those three topics.

The Data Ethics Framework (DEFW)

According to its website, the DSEC was developed in 2018. In June 2018, the UK Government created and made publicly available a Data Science Ethics Framework (DEFW) and Workbook, which were updated in 2020. This framework and workbook have a specific target audience: "Public sector organisations should use the Data Ethics Framework to guide the appropriate use of data to inform policy and service design." Since that is only a single segment of the target audience for the ASA and ACM practice standards, and because the DSEC casts a wider net in terms of the range of projects to which it is applicable (i.e., not just those supporting policy or government projects), we briefly explore the Framework's *original* (2018) seven "principles" and their associated questions, which are outlined here:
http://api.ethicscodescollection.org/files/57adc8fa-072a-4d43-8296-00e1ad5d4923.pdf.[5]

Principle 1 - Start with clear user need and public benefit

Describe the user need.

Does everyone in the team understand the user need?
How does this benefit the public?
What would be the harm in not using data science - what needs might not be met?
Do you have supporting evidence for the approach being likely to meet a user need or provide public benefit?

Principle 2 - Be aware of relevant legislation and codes of practice

List the pieces of legislation, codes of practice and guidance that apply to your project.

Do all team members understand how relevant laws apply to the project?
If necessary, have you consulted with relevant experts?
Have you spoken to your information assurance team?
If using personal data, do you understand your obligations under data protection legislation?

---

5 The original 2018 DEFW was simpler than the 2020 update which added overarching principles of transparency, accountability, and fairness. The 2020 version (and presumably, the most current version) can be found here: https://www.gov.uk/government/publications/data-ethics-framework/data-ethics-framework

Do you have plans in place to handle any potential security breach?

Principle 3 - Use data that is proportionate to the user need

Describe how the data being used is proportionate to the user need.

Could you clearly explain why you need to use this data to members of the public?
Does this use of data interfere with the rights of individuals?
If yes, is there a less intrusive way of achieving the objective?
Is there a fair balance between the rights of individuals and the interests of the community?
Has the data you're using been specifically provided for your analysis?
By using data that the public has freely volunteered, would your project jeopardise people providing this again in the future?
How can you meet the project aim using the minimum personal data possible?
Is there a way to achieve the same aim with less identifiable data?
Can you use synthetic data?
If using personal data is unavoidable, have you answered the questions for determining proportionality?
If using personal data identifying individuals, what measures are in place to control access? How widely are you searching personal data?

Principle 4 - Understand the limitations of the data

Identify the potential limitations of the data source(s) and how they are being mitigated.

What data source(s) is being used?
Are all metadata and field names clearly understood?
What processes do you have in place to ensure and maintain data integrity?
Is there a plan in place to identify errors and biases?
What are the caveats?
How will the caveats be taken into account for any future policy or service which uses this work as an evidence base?

Principle 5 - Ensure robust practices and work within your skillset

Explain the relevant expertise and approaches that are being employed to maximise the efficacy of the project. Describe the disciplines involved and why.

Is there expertise that the project requires that you don't currently have?
Have you designed the approach with the policy team or a subject matter expert?

Has all subject matter context, from policy experts or otherwise, been taken into account when determining the appropriate loss function for the model?

If necessary, how can you (or external scrutiny) check that the algorithm is achieving the right output decision when new data is added?

How has reproducibility been ensured? Could another analyst repeat your procedure based on your documentation?

How confident are you that the algorithm is robust, and that any assumptions are met?

What is the quality of the model outputs, and how does this stack up against the project objectives?

If using data about people, is it possible that a data science technique is basing analysis on proxies for protected variables which could lead to a discriminatory policy decision?

Principle 6 - Make your work transparent and be accountable

Describe how you have considered making your work transparent and your team accountable.

Have you spoken to your organisation to find out if you can speak about your project openly?

Have you considered how both internal and external engagement could benefit your project?

How interpretable are the outputs of your work?

How are you explaining how approaches were designed in plain English to other practitioners, policy makers and if appropriate, the public?

Can you openly publish your methodology, metadata about your model, and/or the model itself e.g., on Github?

Can you get peers to review your Pull Requests?

Principle 7 - Embed data use responsibly

Describe the steps taken to ensure any insight is managed responsibly.

How many people will be affected by the new model, insight or service?

Who are the users of the insight, model, or new service?

Do users have the appropriate support and training to maintain the new technology?

Have future events been planned for?

Is your implementation plan correlated with the impact of a particular model?

How often will you report on these plans to Senior Responsible Officers?

Summary

DSEC and DEFW were created originally in 2018 (DEFW was updated in 2020; DSEC does not appear to have been updated as of October 2022). These reference materials are not the only such checklists or frameworks, they are simply two exemplars. Since 2019 (when this book was started), not only was the DEFW updated (in 2020), but also the United States also created a data ethics framework (https://resources.data.gov/assets/documents/fds-data-ethics-framework.pdf) which is intended to be revisited every 24 months to ensure that it is up to date.

Two important notes to point out: the UK government (DEFW) and US government created documents that are focused on *data*. The DSEC is about ethical data science – i.e., it is not just about data, but about working with data as a data scientist. In this book, the 2018 versions of DSEC and DEFW are utilized in analyses to demonstrate how ethical reasoning KSAs work with different types of reference materials. Any practitioner **in any field** should be able to do the KSAs of ethical reasoning in order to both *identify ethical challenges* that arise and to *make and justify decisions about how to respond* to those challenges. Note that the DSEC or DEFW may support your identifying and assessing your prerequisite knowledge (ER KSA #1): if you do not know how to determine if data were obtained with informed consent (DSEC item A. 1), then you recognize a gap in your prerequisite knowledge and that this gap must be addressed before any decisions can be made ethically. However, the items on the DSEC Checklist are all yes/no questions. If you answer no to any of them, does it constitute an ethical issue? "Any NO answer creates an ethical problem" is implied, but not stated on the DSEC. More to the point, *what to do* if you answer NO to any of the items on either DSEC or DEFW is unspecified by either of these sources.

The ACM and ASA practice standards do not state exactly *how* you should do any of the things on the DSEC/DEFW – that is left up to your judgment, ability level, and the specific context in which you practice. Importantly, this underscores the point that ethical quantitative **practice** (in statistics and data science) **involves decisions**, and it is the practitioner's responsibility to recognize when these decisions are needed, make them ethically, and justify explicitly how these decisions were made. The last two KSAs (make and justify a decision; and reflect on the decision) are specific about this, and this is part of what makes the ER KSAs so compatible with the ACM Code of Ethics (CE) and ASA Guidelines for Ethical Statistical Practice (GLs): all are concerned with making ethical decisions.

Not all of the questions in the DSEC and DEFW will be answerable – not on every project, and not in every case (e.g., not every project will have a timeframe that requires monitoring concept drift, DSEC E3; and not every company has a "Senior Responsible Officer"; DEFW Principle 7). Note again that DSEC is nominally for ethical "data science", but answering any of the questions requires that there is a specific project (about which questions can be answered). The UK DEFW, by contrast, targets "data ethics" and is not focused on the practice of data science *per se*. The DEFW relies on there being a specific project like the DSEC, but it also relies on the ability to balance potential harms that might accrue *to anyone* as a result of either the intended or the *unintended* consequences of collecting, analyzing, and/or drawing insights from the data. The DSEC does not discuss harms or benefits and, because of this, implies that harms and benefits are either all exchangeable and equally important, or that they are not/should not be considerations in the decisions made throughout data science projects. That particular perspective is in contrast to DEFW and the ACM CE specifically, as we have already seen.

It is worth considering how the project-specific DSEC and DEFW checklists promote **stewardship** *of data*. In the traditional use of the term, "stewardship" is defined as "the job of supervising or taking care of something, such as an organization or property." (Merriam-Webster). "Data stewardship" can therefore be interpreted, and seen in the DEFW in several principles, as ensuring that the data that are collected are both justified (benefits are seen to outweigh risks) and also well looked-after, secured, and destroyed once no longer needed. By contrast, when applied to statistics and data science, **stewardship of** *the profession* (or discipline) is defined by Golde & Walker (2006) and Rios et al. (2019) as comprising much more than *just the data*. While the ethical statistical practitioner/data scientist is as stewardly as possible of any data entrusted to their care – by definition of the profession – stewardship of the profession itself implies greater responsibilities. These are featured in the practice standards for statistics (ASA) and computing (ACM) (Chapters 1.2 and 1.3). The Venn diagram in the figure below describes the universe of ethical statistical and data science practice, and how the ASA and ACM ethical practice standards cover distinct and complementary aspects. The inclusion of DSEC and DEFW shows that, while **engagement with** the ASA and ACM practice standards *will* promote a fully ethical practitioner, use of DSEC or DEFW alone (or both!) *will not*.

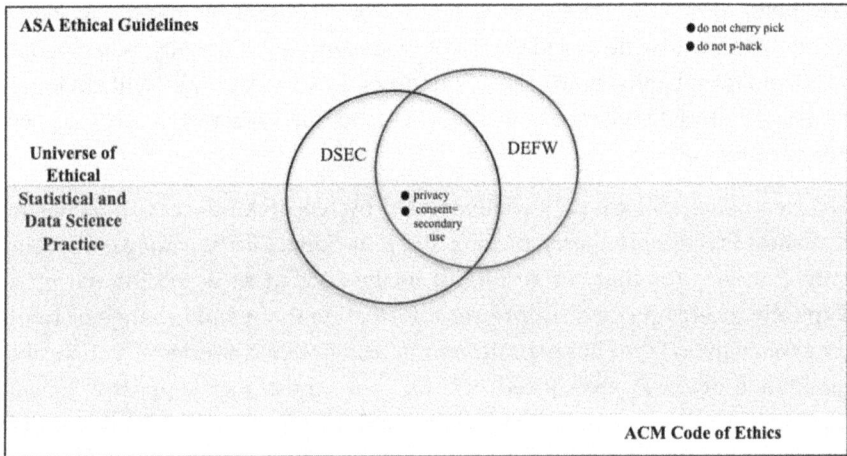

Figure 1.4.2. *Unions and disjunctions: ASA, ACM, DSEC, DEFW*

Four points in Figure 1.4.2 are labelled: "privacy", "consent=secondary use", "cherry-picking" and "p-hacking". For clarity, *privacy* is defined as the basic human right – i.e., a right specific to humans (not organizations) to a freedom from unwanted, or unintended sharing of information (see, for example, the United Nations Declaration of Human Rights). Confidentiality also pertains to information (that should not be shared without permission), but that information is not limited to humans (and so confidentiality can pertain to organizations or people). In some contexts, privacy is differentiated from confidentiality by describing privacy as pertaining to people or their activities, while confidentiality pertains to *information*. However, Title 13 of the US Code, which controls the activities of the United States Census Bureau, describes information only – and references both private information and its confidentiality. According to the Census, addresses and telephone numbers *that identify individuals or businesses* are "private information"; while the Census can (and does) collect such information every 10 years, they are legally required to "maintain the confidentiality of (your) data." (https://www.census.gov/history/www/reference/privacy_confidentiality/title_13_us_code.html). Thus, confidentiality and privacy are sometimes used synonymously.

The other issue that all four references address is *"consent"* -particularly with respect to use of data that was collected for one reason, but is used for a different ("secondary") reason. For example, if you apply for a car loan, you provide information for the purposes of demonstrating to a lender that you can afford the loan repayment. If that information is used by the lender for a different reason (or, sold or shared with another entity), that constitutes secondary use of your data. For private information (like your name, address, telephone number), its reuse without your consent is becoming better recognized worldwide as a violation of your rights. Additionally, when people steal data from banks, credit cards, and other data holders, they will obviously be using your data without your consent – and, for a secondary use it was not intended for.

The two issues that are only contemplated by the ASA Ethical Guidelines for Statistical Practice are cherry-picking and p-hacking. *Cherry-picking* is choosing only evidence (or data, or results of analyses) that fit a specific narrative. Typically, cherry-picking suppresses information that would change or refute (or explain away) the chosen or desired outcome. When a series of statistical or qualitative analyses are carried out, but only those that support a specific outcome are reported and all the others (and the fact that other analyses were done) is hidden or suppressed, results are being cherry-picked. In a similar vein, *p-hacking* refers to a variety of misuses of the p-values that are obtained from statistical analyses. Similar to cherry-picking, p-hacking involves running analyses until the desired p-value is obtained, and then either pretending that no other analyses had been done (like cherry-picking) or stopping analyses at that point, and pretending that had been the point all along (see Head et al. 2015 for a discussion of p-hacking in scientific research). These two types of statistical misconduct are not actually named specifically in the GLs because they are *only two* examples of inappropriate interpretation, and inappropriate methods use, respectively. These tend to be much bigger problems for designed data collection (like experiments or surveys), but are also difficulties in policy making.

All four are commonly considered aspects of ethical practice of statistics (so are all covered only by the ASA GLs), while the two at the center (privacy, consent/secondary use) are commonly emphasized in all conversations around data ethics or data science ethics, and so are contemplated by all four reference materials. These single points within the "universe" of ethical statistical and data science practice highlights that ethical practice is much, much more than just privacy and consent – although these two issues are important, and so are featured in all four references shown in this diagram. Moreover, these single points are identified to underscore the fact that there are many other issues to consider beyond just these four issues, when contemplating what constitutes ethical practice of statistics and data science.

Figure 1.4.2 was created to highlight the fact that there is a great deal of the *universe of ethical practice in statistics and data science* that is not addressed by the DSEC and DEFW checklists. Because these checklists do not contemplate other common challenges, it may erroneously send the message that, as long as the checklists are completed, the practitioner has discharged their obligations to practice statistics and data science ethically. Clearly, the ASA GLs and ACM CE disagree – and the extent to which the GLs and CE offer additional guidance can be seen in Figure 1.4.2.

This is not meant to criticize, only to point out that neither is sufficient for ethical practice. Thus, while the DSEC and/or DEFW may be useful characteristics for any given project, data scientists and those who take courses in the domain need to be aware of the complexities – and responsibilities – inherent in the practice of statistics and data science. The use of these checklists cannot be considered acceptable alternatives to training with the ethical practice standards outlined by the ASA and ACM. However, instruction in "ethical statistics" or "ethical data science" may benefit from the inclusion of these checklists by engaging students in conversations about how exactly to use the Guidelines and Code to answer each checklist question.

We can see from the figure that the checklists, DSEC and DEFW, do overlap some, but not fully. Moreover, this figure reflects the fact that those checklists capture a relatively small subset of the universe that is ethical practice of statistics and data science. Note that all four of these reference materials (Guidelines, Code, DSEC and DEFW) address privacy and secondary use of data/consent for data use. "Cherry-picking" and "p-hacking" are only addressed in the ASA Guidelines – although not specifically by name. These are important aspects of limiting *false discoveries* – spurious results that cannot be replicated and so, represent unreliable outcomes, whether they arise from experiments, surveys, or algorithms. Knowingly providing results that may represent false or spurious outcomes to stakeholders is inherently unethical, but a better understanding of one's ethical obligations to be transparent in reporting as well as utilizing appropriate methodology -arising from familiarity with the ASA Ethical Guidelines as well as the ACM Code of Ethics– can help to counter the effects of spurious results.

In summary, the figure shows how the *universe of ethical practice in statistics and data science* is more comprehensively covered by the ASA Ethical Guidelines for Statistical Practice (Chapter 1.2) and the ACM Code of Ethics (Chapter 1.3) than it is by the DSEC or DEFW. To better understand what these practice standards are, and how it can be claimed that they describe the universe of ethical practice – even though neither of them is specific to "data science" - we revisit this figure again in Chapter 1.6.

Chapter 1.5
The "Universe of Statistics and Data Science": Tasks

As we saw in the Venn diagram (Figure 1.4.2) and lists of checklists from the previous chapter, there are two specific sources of guidance, a data science ethics checklist (DSEC) and a data ethics framework (DEFW). As you will have noticed, one is for "data science" and one is for "data ethics". There is a bit of overlap, as indicated in the Venn diagram, which we explore below as we walk through the seven common tasks for statistics and data science[6]. These tasks are:

1. Planning/Designing
2. Data collection/munging/wrangling
3. Analysis (perform or program to perform) *(this means different things for computing professionals and statisticians and data scientists; more on that in Section 2.)*
4. Interpretation *(this is not a separable component of computing professionals' ethical practice, but is an essential feature of statistics and data science.)*
5. Documenting your work
6. Reporting your results/communication
7. Engaging in team science/teamwork

These tasks are discussed briefly below, and the relevant aspects of DSEC and DEFW are also considered.

Planning/Designing

All statistics and data science projects involve planning and design. Whether the individual is the planner/designer or receives a plan to execute, the planning and design process requires specific attention to ethical practice.

DSEC: As we saw in the previous chapter, the DSEC does not specifically start with the decisions that must be made (ethically, of course!) when engaging in plans or designs for analyses or systems that will collect and/or analyze data. However, many of the questions on the DSEC require that elements from all

---

[6] These seven tasks are introduced and featured in (Tractenberg, 2022)

five of the DSEC areas must be considered at the planning stage (and in fact, DSEC E1 is the articulation of a plan!).

DEFW: Principles 1, 2 and 3, as well as 7 must all be considered in the planning/design phase of any project, primarily because if the plan is not sufficiently concrete, then the tradeoff between risk or harm and the public good or user need cannot be estimated, and this tradeoff is essential to all decision-making and justification in the DEFW.

Data collection/munging/wrangling

Data is obviously the crucial component in data analysis and data science. The ways in which data are collected or synthesized in simulations, munged (transformed), and wrangled (manipulated, integrated), are all essential in statistics and data science.

DSEC: As we saw in the previous chapter, the DSEC has an entire section about "data collection" (section A), but storage and security (DSEC B) must also be considered in the collection and wrangling or munging of data. DSEC C2 relates to dataset bias, and this can arise in either collection or manipulation (munging/wrangling) of data. Because data collection requires consideration of security, E4 is also an important consideration for this task.

DEFW: Principles 2 and 3, and to some extent, 4 are important for data collection and any manipulation of that data (including security) going forward. Although the DEFW is about "data ethics", security of that data is implied, with only one question (plan for response to security breach) under Principle 2.

Analysis (perform or program to perform)

Analysis is an essential part of the statistician's job and, while the data scientist may consider the analysis to be the role of an algorithm, program, or system, the analysis is critical to the use, utility, and relevance of any system/program/algorithm. For computing professionals, "analysis" is less about statistics and more about an evaluation of the system itself. In either perspective, the analysis of a system or of data (or of a system that collects/analyses/uses data) is an *essential* function of the statistician and data scientist.

DSEC: Item A2, relating to bias in data collection, would be uncovered with analysis (and interpretation). Although section C is labelled "Analysis", DSEC C1, C2, C4 are directly related to analysis while DSEC C3 (representation of results) and DSEC C5 (documentation of analysis) are *indirectly* related. Four of five questions in D (modeling) are relevant in analysis, as is E3 (concept drift) – another question that would be answered using (planned) analyses.

DEFW: Like DSEC, the DEFW Principles apply widely to, or are answered via, analyses. Principle 4 (understand the limitations of the data) begins with planning, but includes analysis; Principle 5 (ensure robust practices) also requires analysis.

Interpretation

The interpretation of results from a system or analysis is an essential feature of statistical practice. Even when analyses are "provided" to a requestor, there must be preliminary evaluation of the results by the analyst to ensure that assumptions were not violated, and to detail the limitations and/or bias that might be present. Without even just preliminary interpretation, the expert's evaluation will unlikely be useful to a non-expert; and the expert's interpretation is a crucial part of communication to other experts in the field.

DSEC: The identification of bias in a data set (A2) derives from interpretation of analyses of that dataset. as does the identification of missing perspectives (C1) and whether any personally identifying information (PII) is needed for an analysis is also based on interpretation. The determination that variables in an analysis are actually proxies that may lead to discrimination (D1), unfairness across groups (D2), or concept drift across time (D3) are all interpretive and based on analyses. Finally, the selection or evaluation of any metric used in analyses also requires interpretation.

DEFW: Appreciation of the limitations of data (Principle 4), and the identification of robust practices (Principle 5) require interpretation of the analysis or functioning of a model. Principle 7, Embed data use responsibly, also requires interpretation.

Documenting your work

The documentation of work – outlining what and how the system or analysis works, and including methodological features and assumptions - is essential to all practice, even if only to ensure that mistakes that might occur are not repeated (and that the most efficient approach to any problem is identified).

DSEC: The documentation of a data science project should begin with the plan; the first point where it can be seen to be relevant in the DSEC is C5, auditability. Clearly, a system or program (or analysis) cannot be audited if it hasn't been fully documented. Similarly, the explainability of any decision made by a system (D4) requires careful documentation, as does the communication of limitations and biases of any model that a system implements (D5). DSEC section E (deployment) requires full documentation, otherwise, none of the aspects mentioned (redress; roll back; concept drift; unintended use) can be addressed.

DEFW: Understanding the limitations of the data (Principle 4), ensuring robust practices (Principle 5), and making work transparent (Principle 6) clearly require full documentation of the system and the justifications behind decisions in all aspects of its design and implementation. Principle 7 (embed data use responsibly) also requires documentation, so that potential tradeoffs between data use and harms can be considered.

Reporting your results/communication

Separate from documentation of all work, and the interpretations that accompany results, the communication of function, findings, or both; together with methods, assumptions, and caveats that describe what was done and why, are a fundamental and crucial part of statistics and data science practice. Without the expert's communication, non-experts are unlikely to transmit and report the results correctly and completely.

DSEC: Specific attention must be given to privacy (C4) and the auditability (C5) of the report of an analysis. Consideration of the potential for bias and discrimination that a model or system creates (D1) and its fairness (D2) and the explainability of a model's decisions (D4) are critical features of how these results are reported and communicated to stakeholders (D5). Whether or not a model can be rolled back (E2) is also a feature of how it is reported, and how its results would be communicated to stakeholders.

DEFW: Reporting and communication of the functioning and outputs of any system permit the creators and maintainers – as well as all stakeholders – to understand the limitations of the data (Principle 4), ensure that robust practices were used (Principle 5), that all work (system functioning and results) are transparent (Principle 6) and how/that the use of the data was necessary, and that users know how the system works (Principle 7).

Engaging in team science/teamwork

While there may be only one statistician or data scientist involved, statistics and data science are rarely done by individuals outside of teams. Whether the overall objective is to contribute new knowledge to the scientific literature (team science) or to complete projects in the workplace (teamwork), ethical practice by the statistician/data scientist is essential to cooperative – as well as individual – efforts. Similar to communication, consideration of the expertise of others on the team, particularly when it is not in statistics and data science, is an essential feature of ethical teamwork.

DSEC: Similar to communication, the statistician/data scientist must consider how to describe how the system works to team members who bring diverse stakeholder perspectives to the system's design so that missing perspectives can be identified (C1). The system and its results must be communicated clearly and honestly to members of the team (C3) –as well as other stakeholders, so that its effects can be fully appreciated and evaluated. Considerations of privacy (C4) are relevant even for communication within the team, and the auditability of the system (C5) and explainability of decisions (D4) must be accessible to team members who are, as well as those who are not, experts in the field. The team or organization will also need to consider a plan for responding if users or other stakeholders are harmed by a program, algorithm, or system (E1).

DEFW: Data work should always be transparent, and workers must consider themselves responsible and accountable (Principle 6).

As you can see, the DSEC and DEFW questions are useful in every one of these seven tasks. The alignment of these two checklists with the tasks – as well as gaps – are explored in the following two tables. Each "x" in these tables indicates at least some degree of overlap between DSEC and DEFW questions (columns) and the task involved in statistics and data science (rows).

DSEC areas: Seven Typical Data adjacent tasks[7]:	A. Data Collection	B. Data Storage	C. Analysis	D. Modeling	E. Deploy-ment
Planning/ Designing	x	x	x	x	x
Data collection/ munging/ wrangling	x	x	x		
Analysis (perform or program to perform)	x		x	x	x
Interpretation	x		x	x	x
Documenting your work			x		x
Reporting your results/ communication			x	x	x
Engaging in team science/ teamwork			x	x	x

Table 1.5.1. *DSEC areas (columns) x TASKS (rows) alignment*

You can see there is overlap between DSEC and the seven tasks, meaning that there are considerations from each of the five DSEC areas for carrying out each of the tasks (all five DSEC areas should be considered in the planning stage, for example). The table underscores the point that a checklist like DSEC is not something to be done at the start or end of a project, because there are DSEC considerations in every task along the statistics and data science pipeline. The next table explores the alignment between the DEFW and these tasks.

---

[7] These seven tasks are discussed in Tractenberg (2022).

DEFW Principles (2018):	P1 Clear need/ benefit	P2 Relevant laws/ codes	P3 Data proportionate to need	P4 Understand limitations of data	P5 robust practices	P6 Transparent and accountable	P7 Embed data use responsibly
Seven Typical Data adjacent tasks[4]:							
Planning/ Designing	X	X	X				X
Data collection/ munging/ wrangling		X	X	X			
Analysis (perform or program to perform)				X	X		
Interpretation				X	X		X
Documenting your work				X	X	X	X
Reporting your results/ communication				X	X	X	X
Engaging in team science/ teamwork					X	X	

Table 1.5.2. *DEFW 2018 Principles (P, columns) x TASKS (rows) alignment*

The empty cells in each of these two tables are important because they indicate gaps – aspects of daily tasks (rows) where these lists of questions do not pertain, i.e., do not provide guidance or support for ethical conduct/practice. Just like with DSEC and the previous table, the task x DEFW table also shows how important the considerations in each of these two tools are *throughout* the most common tasks in statistics and data science. These tables show clearly that neither the DSEC nor the DEFW can be legitimately utilized *at the end* of a study: both have important guidance to offer during planning or design, as well as during the collection and then munging or wrangling of obtained data. Note also that "engaging in team science/ teamwork" is listed as the "last" task in the statistics and data science pipeline, but in fact one typically engages in the other six tasks in the pipeline as part of a team. No matter which of the pipeline tasks represents the start or end of your specific role on a project, ethical considerations are relevant **throughout the tasks and the pipeline**.

Clearly, DSEC and DEFW are very different – their questions are different, and the areas of their organization are also quite distinct. As can be seen from the discussion above, both DSEC and DEFW questions can be useful in many of these tasks – throughout any project- which means that it is not possible to answer the questions *at the beginning* and then claim that the project had been done ethically. Importantly, while it is possible to complete the checklist (DSEC) or questions (DEFW) at the *end* of the project to describe that it was done ethically, for it to be true that the project was done ethically, the questions had to be used throughout the entirety of the project. So, the questions can be "finally" answered at the end, and serve as a complete record of the ethical decision making that went on for the whole project, but the point is that these questions would have had to be kept in mind from the outset.

Another difficulty with the questions on the checklists is that you cannot answer any of the questions without *one specific project that you can define* (so you can answer) in mind.

But what do you do the rest of the time?

This chapter was intended to demonstrate how DSEC and DEFW do overlap, but not completely, and in spite of their differences, there is a great deal more to ethical practice than just answering these particular questions – even if you used both of these resources. That is, this discussion helps us understand how our Venn diagram does reflect the scope of ethical practice – and that it goes beyond these sets of questions. The ethical practice standards from the ASA (Chapter 1.2) and ACM (Chapter 1.3) were developed to support the entirety of practice, not just project-by-project decisions; we will see more of this in Section 2. Moreover, while the 2018 DEFW questions speak to data stewardship, and the DSEC generally describes ethical practices that involve data, the ASA and ACM practice standards – explored together in the next two chapters) describe every aspect of practice – i.e., professional stewardship. That is, the ASA and ACM practice standards support stewardship of the practice and profession of statistics and data science.

Chapter 1.6
Exploring Guidance from ACM, ASA, DSEC, & DEFW[8]

As they are representative of two essential constituent disciplines for data science ASA and ACM offer guidance for ethical practice in statistics and data science; clearly the ASA offers guidance for ethical statistical practice. As noted, these practice standards are presented in full in previous chapters. The Table below explores the thematic alignment of the ASA Guidelines with the ACM Code; the ASA Principles (A-H) and each ACM element are shown and alignment is indicated with the specific ASA element ("A" means it is the A Principle statement; "A1" means it is the first element under Principle A, etc.) if any of the elements within that Principle is a thematic or exact match to the specific element of the ACM Code. If the alignment is abstract (e.g., if the core idea is "professional integrity" but this is demonstrated with computing machinery – so the principle is described in terms of computing specifically in the ACM Code but would be demonstrated with other aspects of statistical practice and described fundamentally differently by the ASA Guidelines, then the alignment with the particular ASA element is indicated by parenthesis "(element)". If there is neither exact nor thematic alignment, then the cell is left blank.

---

[8] This material is adapted from Tractenberg RE. (2020, February 19). Concordance of professional ethical practice standards for the domain of Data Science: A white paper. Published in the *Open Archive of the Social Sciences* (SocArXiv), 10.31235/osf.io/p7rj2, and adapted from Tractenberg, RE. (2019, May 1). Strengthening the practice and profession of statistics and data science using ethical guidelines. Retrieved from https://doi.org/10.31235/osf.io/93wuk.

ASA:	A Professional Integrity and Accountability	B Integrity of data and methods	C Stakeholders	D Research Subjects/ Data Subjects and those affected by statistical practices	E Inter-disciplinary Team Members	F Other Practitioners /Profession	G Leader/ supervisor/ mentor and **APPENDIX**	H Allegations of Potential Misconduct
				RESPONSIBILITIES TO/REGARDING:				
ACM:			ACM 1. GENERAL MORAL PRINCIPLES. *A computing professional should…*					
1.1 Contribute to society and to human well-being, acknowledging that all people are stakeholders in computing.	A2 and preamble	B6	C; C2, C6	D; D2, D6, D8, D11	E4	F4	H2, G5 Appendix 6	
1.2 Avoid harm. *In this document, "harm" means negative consequences to any stakeholder, especially when those consequences are significant and unjust.*	A; A1-A4, A7-A9	B1-B3, B5, B6	C; C1-C5, C7-C8	ALL D	E1, E3, E4	F2-F4	H2, G2, G5 **Appendix 6**	H2, H3, H8

RESPONSIBILITIES TO/REGARDING:

1.3 Be honest and trustworthy.	A; A1-A8, A11-A12.	B1-B5	C; C1, C2, C3	ALL D	E2, E3, E4	F4	H2, G2, G5 Appendix 1,2, 4-12	ALL H
1.4 Be fair and take action not to discriminate.	A2-A4, A5-A6, A8, A9, A12	B1-B4, B6-B7	C4, C7	D5-D6, D10-D11	E1	F1-F3	H2- G4 Appendix 1, 7, 10	ALL H
1.5 Respect the work required to produce new ideas, inventions, creative works, and computing artifacts.	A5					F2	G4 Appendix 5, 10	
1.6 Respect privacy.		B4	C7	D4, D5, D7, D9-D11				
1.7 Honor confidentiality.		B4	C7	D4, D5, D7, D9-D11				H4-H6

2. PROFESSIONAL RESPONSIBILITIES.
A computing professional should…

2.1 Strive to achieve high quality in both the process and products of professional work.	ALL A		ALL C	D10	E2, E3, E4	ALL F	ALL G / Appendix 1, 2, 4, 8-10, 12	
2.2 Maintain high standards of professional competence, conduct, and ethical practice.	ALL A	ALL B	ALL C	ALL D	E2, E3, E4	ALL F	ALL G / ALL Appendix	ALL H
2.3 Know, respect, and apply existing rules pertaining to professional work.	A1, A4-A8, A11-A12	B1, B4, B5	C2, C4-C8	D1-D5, D7, D9-D11	E2-E4	F4, F5	H2, G2 / Appendix 1, 4, 7, 8-11	ALL H
2.4 Accept and provide appropriate professional review.	A1-A4	B5, B6		D1, D11	E3	F2	G2, G5 / Appendix 3, 7	H2

2. PROFESSIONAL RESPONSIBILITIES.
A computing professional should…

2.5 Give comprehensive and thorough evaluations of computer systems and their impacts, including analysis of possible risks.	(A1, A4)	(B1, B2, B6)	(C4, C5)	(D6, D9-D11)	(E4)		(G5) (Appendix 6)	
2.6 Have the necessary expertise, or the ability to obtain that expertise, for completing a work assignment before accepting it. Once accepted, that commitment should be honored.	A1, A2, A5, A10	B2	C1, C3	D1 (D9, D10)		F3	Appendix 1, 2	(H1)

2. PROFESSIONAL RESPONSIBILITIES.
A computing professional should…

2.7 Improve public awareness and understanding of computing, related technologies, and their consequences.	(A12)	(B1-B3, B5)	(C6)		(E3)	(F4, F5)	(H2, G5) (Appendix 9)	
2.8 Access computing and communication resources only when authorized to do so.		(B4)	(C7)	(D4, D5, D7, D11)	(E4)			(H5, H6)
2.9 Design and implement systems that are robustly and usably secure.	(A2, A4)	(B6)	(C2, C7)	(D4, D9, D10)	(E4)		(G5) (Appendix 1, 4)	(H5)

3. PROFESSIONAL LEADERSHIP PRINCIPLES.

In this section, "leader" means any member of an organization or group who has influence, educational responsibilities, or managerial responsibilities. These principles generally apply to organizations and groups, as well as their leaders.

A computing professional should…

3.1 Ensure that the public good is the central concern during all professional computing work.	(A2-A4, A8)	(B1, B4, B5)	(C2, C6)	(D1, D2, D6; D10, D11)	(E3, E4)	(F2, F3)	(H2, G2, G5) Appendix 4, 6	
3.2 Articulate, encourage acceptance of, and evaluate fulfillment of the social responsibilities of members of an organization or group.	(A9)		(C6, C8)	(D11)		(F3)	(G2, G5) Appendix 4, 6	(H2, H3)
3.3 Manage personnel and resources to enhance the quality of working life.	(A4, A8, A9, A12)	(B6)	(C2)			F3	ALL G Appendix 3, 10, 12	ALL H

3. PROFESSIONAL LEADERSHIP PRINCIPLES.

In this section, "leader" means any member of an organization or group who has influence, educational responsibilities, or managerial responsibilities. These principles generally apply to organizations and groups, as well as their leaders.
A computing professional should….

	(A4, A8, A9, A12)	(ALL B)	(C2, C4, C5, C8)	(ALL D)	(E2 – E4)	(F2-F5)	(ALL G)	(ALL H)
3.4 Articulate, apply, and support policies and processes that reflect the principles in the Code.							Appendix 1, 4, 6, 8-12	
3.5 Create opportunities for members of the organization or group to learn and be accountable for the scope, functions, limitations, and impacts of systems.	(A1)	(B6)	(C1)			(F2, F5)	G4 Appendix 3	
3.6 Retire legacy systems with care.		(B6)						

3. PROFESSIONAL LEADERSHIP PRINCIPLES.

In this section, "leader" means any member of an organization or group who has influence, educational responsibilities, or managerial responsibilities. These principles generally apply to organizations and groups, as well as their leaders.
A computing professional should...

3.7 Recognize when a computer system is becoming integrated into the infrastructure of society, and adopt an appropriate standard of care for that system and its users.			

4. COMPLIANCE WITH THE CODE.
A computing professional should...
(* suggested in the ASA preamble; included by implication only (that "the ethical statistical practitioner..." follows the GLs))

4.1 Uphold, promote, and respect the principles of the Code.	A9, A12	E2, E3	H2
4.2 Treat violations of the Code as inconsistent with membership in the ACM.			Appendix 1, 8, 12

This table also appears in Tractenberg, 2022.

Table 1.6.1. *Association of Computing Machinery Ethics (2018) Code v3 vs American Statistical Association Ethical Guidelines (2022)*

Table 1.6.1 shows a high level of *thematic concordance* between the two practice standards. That is, while the wording (and some of the manifestations of the specific feature of ethical practice) is different, both organizations specify similar features as foundational in ethical practice with the methods and materials of their respective disciplines. Several ACM Code elements are aligned with **every** ASA principle; these are highlighted in light grey.

Of the 25 ACM Code elements, two of them are not aligned with any aspect of the ASA Guidelines (highlighted in dark grey in Table 1), but every ASA Guideline has at least some thematic alignment with all of the other ACM elements. While ACM Principle 4, "compliance with the code" has matches to the ASA GLs the ACM Code goes further, specifically stating that violations of their Code should be considered "inconsistent with membership in the ACM" (ACM 4.2). The ACM created a specific method for members to report violations of their Code of Ethics, and an adjudication process to deal with them. This means that 4.2 can be followed by ACM members and dealt with in a standardized, objective, manner. The ASA does not (as of October 2022) have such a mechanism in place, which explains why there is no element of the ASA GLs that matches, even thematically, ACM 4.2.

The ASA and ACM clearly offer guidance for ethical practice in statistics and data science, and overlap in many of the aspects on which the guidance is offered, even though the practice standards arose wholly independently of one another, and they tend to represent very different professions. Given the alignment of these two sets of practice standards, they represent an excellent – coherent – resource for beginning to describe what constitutes ethical data science. In addition to ACM 4.2, there is one other ACM item, from Professional Leadership (3.7), "Recognize when a computer system is becoming integrated into the infrastructure of society, and adopt an appropriate standard of care for that system and its users." Because it is so focused on computing, it makes sense that there is no parallel item on the ASA GLs. Apart from these two (of 25) ACM elements, these standards overlap considerably.

Other sources of prerequisite knowledge to ensure ethical practice in quantitative fields could include the Statement on Ethics of the American Mathematical Society (http://www.ams.org/about-us/governance/policy-statements/sec-ethics), updated in January 2019, and the Code of Conduct of the Royal Statistical Society (http://www.rss.org.uk/Images/PDF/join-us/RSS-Code-of-Conduct-2014.pdf) which was most recently updated in 2014. Tractenberg (2020) reviews the alignment of the ASA guidance (from 2018) with these two resources. In this chapter we explore the similarities and

differences in terms of the ethical guidance offered from the four sources we
have discussed at length. Recall the Venn diagram (Figure 1) from Chapter 1.4:

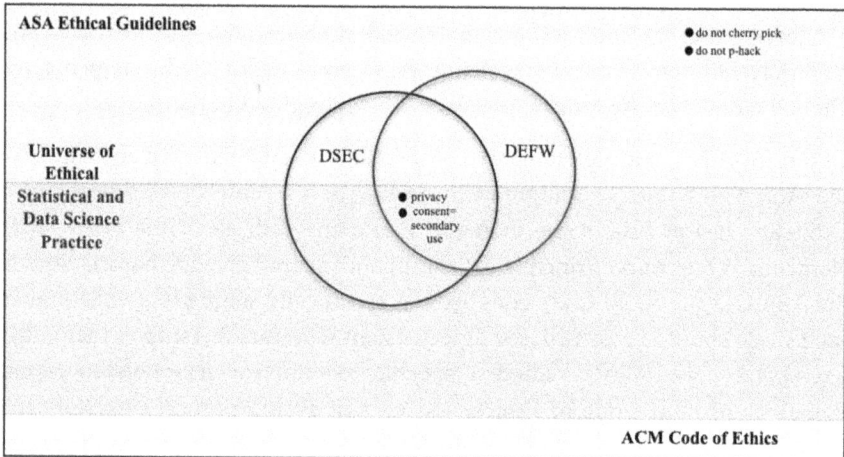

Figure 1.6.1. *Unions and disjunctions: ASA, ACM, DSEC, DEFW*

We saw in the last chapter, as well as from the figure originally, that the
checklists, DSEC and DEFW, do overlap some, but not fully. The discussions of
the DSEC and DEFW should also have reinforced this point.

We noted that the ASA and ACM standards do capture the full universe of
ethical practice in their respective fields – with only two elements of the 25
ACM CE falling completely outside the ethical practice standards of the ASA.
The challenge for statistics and data science is to understand how these
standards interact, overlap, and complement each other. Like the ASA GLs, the
ACM CE is comprehensive. However, just like the ASA GLs, it is difficult to
read them all and keep all in mind as you go through your working life.
Moreover, it is not plausible to expect that any person who sees or has access
to these practice standards would 'automatically' be able to utilize them
(Tractenberg, 2022; footnote 5; and McNamara, Smith & Murphy-Hill, 2018).

Ethical practice guidelines help individual practitioners to align their
professional identity with the characteristics that the wider, experienced
practice community has deemed descriptive and representative of "the ethical
practitioner" of their field, other guidance documents exist. We saw two
examples of additional reference materials in the Data Science Ethics Checklist
(DSEC), http://deon.drivendata.org/, which provides a list of things to consider
as a project goes on/in the practice of data science, and the UK government's
Data Ethics Framework and Workbook (DEFW), http://api.ethicscodes-
collection.org/files/57adc8fa-072a-4d43-8296-00e1ad5d4923.pdf. The DSEC

was designed to support decision making (i.e., it was intended to be "actionable"), but was developed from the perspective that "the primary benefit of a checklist is ensuring that we don't overlook important work". By contrast, the DEFW was developed (and is being maintained), to "set out clear principles for how data should be used in the public sector. It will help us maximize the value of data whilst also setting the highest standards for transparency and accountability when building or buying new data technology."

There are clear differences between the DEFW and the DSEC, which is reflected in the only slight overlap of their domains, as shown in the figure. Note that both references include both privacy and secondary uses of data. Importantly, DSEC and DEFW are described by their developers as focused on *data science ethics* and *data ethics*, respectively. When considering the extent of the "Universe of Ethical Statistical and Data Science Practice" shown in the figure that falls *outside* of the DSEC and DEFW, hopefully the reader recognizes the utility of the ASA and ACM practice standards. Specifically, not all of the quantitative practitioner's practice is specifically relating to data, but *all of the practice* needs to be ethical. The four specific – diverse - types of ethical challenges shown in the figure (privacy, secondary use of data, cherry-picking, and p-hacking) are all addressed by the ASA practice standards. The ethical practice of statistics and data science requires attention to all four of these issues – and more.

The overlap and gaps are summarized and discussed below, organized according to the 2018 DEFW Principles.

Principle 1 - Start with clear user need and public benefit

Describe the user need.

The starting point of any project, consideration of the need for whatever it is that data are needed to help achieve, is missing from DSEC. It is important to note that the ACM CE specifies that all computing needs to have the public good at the core/as a focus. The ASA GLs do not address the public benefit, unless the public is funding the work (in which case the ethical statistical practitioner has responsibilities to the funder and the public). The reason why the DEFW requires the starting point to be a clear user need is because, if data will be collected – particularly without specific consent, but the need and public benefit must be evaluable against any kind of risk to the data contributor once the collector has amassed that data. If the need and benefit cannot be concretely described, then any risks to the data contributor are clearly not going to be offset – meaning, that project should not be done.

Principle 2 - Be aware of relevant legislation and codes of practice

List the pieces of legislation, codes of practice and guidance that apply to your project. While both the ASA & ACM mention this, the DSEC does not.

Principle 3 - Use data that is proportionate to the user need

Describe how the data being used is proportionate to the user need.

Consistent with ACM but missing from that CE as an explicit item; however, this principle – also missing from DSEC – is an important aspect of ensuring that data are obtained with consent (included in ASA). Again, because the use of the data must be offset against any risks to the data contributors, documenting that you are literally justified in risking any harms to data contributors is a priority.

Principle 4 - Understand the limitations of the data

Identify the potential limitations of the data source(s) and how they are being mitigated.

A specific responsibility identified in both the ACM (with respect to "systems" rather than the data they utilize) and the ASA (Principle B). Also mentioned in DSEC.

Principle 5 - Ensure robust practices and work within your skillset

Explain the relevant expertise and approaches that are being employed to maximise the efficacy of the project. Describe the disciplines involved and why.

A specific responsibility identified in both the ACM (Principle 2) and the ASA (Principles A & B).

Principle 6 - Make your work transparent and be accountable

Describe how you have considered making your work transparent and your team accountable.

A specific responsibility identified in both the ACM (Principle 2) and the ASA (Principles A, B, C, and D).

Principle 7 - Embed data use responsibly

Describe the steps taken to ensure any insight is managed responsibly.

A specific responsibility identified in the ASA GLs as a function of supporting reproducible and interpretable (valid) results (ASA A), and addressed in ASA Principle D.

	Principle 1. Start with clear user need and public benefit	Principle 2. Be aware of relevant legislation and codes of practice	Principle 3. Use data that is proportionate to the user need	Principle 4. Understand the limitations of the data	Principle 5. Ensure robust practices and work within your skillset	Principle 6. Make your work transparent and be accountable	Principle 7. Embed data use responsibly
ASA PRINCIPLES HAVING ELEMENTS IN ALIGNMENT WITH DEFW							
ASA A	A2, A3, A4	A11	A2	A2	A1, A2	A2, A5, A6	A2, A4
ASA B	B1	B4	B3, B6	B1-B4, B6	B1, B2, B4, B6	B1-B4, B6	B1-B4, B6
ASA C	C1	C7	C4	C3	C1, C3	C1, C2, C4, C5	C1-C4, C7, C8
ASA D	D5	D1, D4, D9	D2, D3, D7, D8	D4, D5, D10		D6	D1, D2, D3, D11
ASA E		E1			E4	E3, E4	E4
ASA G						G5	G5
ASA F							F5
ASA H		H1					H2
ACM PRINCIPLES HAVING ELEMENTS IN ALIGNMENT WITH DEFW							
ACM 1				1.3	1.3	1.2	
ACM 2		2.3		2.5	2.5, 2.6	2.1, 2.4	
ACM 3							3.6, 3.7

Table. 1.6.2. *Alignment of DEFW Principles with ASA and ACM Principles.*

[9]In this table, the DEFW principles (columns) are shown to align with *all but one* of the eight ASA Principles. In fact, every DEFW Principle is aligned with ASA Principles A, B, and C, and all but one DEFW Principle is aligned with ASA Principle D (Responsibilities to research subjects, data subjects, or those directly affected by statistical practices). Given the top priority/first DEFW Principle ("Start with clear user need and public benefit"), the total alignment with ASA C (responsibilities to stakeholders) is not surprising. Perhaps also unsurprisingly is that there is no alignment with ASA F. Responsibilities to fellow statistical practitioners and the Profession. The total alignment with D, and lack of alignment with F, make sense because the DEFW is not specific to statistical practice but instead, is specific to "data ethics". Beyond the alignment with much of the ASA GLs, there is also alignment with three of the four ACM Principles (all but compliance with the code, ACM 4 – also not surprising). The table shows that every DEFW principle has at least one ASA or ACM principle that also addresses it. This table is more fully explored in Chapter 2.1.

Each cell of this table shows the specific element of the Guidance Principle (ASA or ACM) that can help practitioners to answer the questions that are asked in the DEFW worksheets. There are 72 different elements in the ASA GLs; 35 of these are common with DEFW, meaning that the ASA Guidelines address 72-35= 37 *other* aspects of ethical practice that the DEFW does not address. As noted, the DEFW does not address ASA **F**. Responsibilities to fellow statistics practitioners or the Profession, but it also has minimal alignment with ASA **G**. Responsibilities of leaders, supervisors, and mentors in statistical practice and ASA **H**, Responsibilities regarding potential misconduct.

Note also that DEFW Principle 7 (embed data use responsibly) is aligned with at least one element from every ASA GL Principle except F. This is noteworthy because a reader might imagine that the ASA Ethical Guidelines for Statistical Practice are more concerned with statistical analysis than with "data ethics", and yet every ASA Principle except for F has guidance for the embeddedness of 'data use' – more so than DEFW Principle 5 (ensure robust practices and work within your skillset).

Similarly, there are 25 ACM elements, and the DEFW items are aligned with 10 of these, leaving 15 aspects of ethical computing that are not considered in the DEFW worksheets. These gaps are, perhaps, not surprising because the DEFW

---

[9] This material is adapted from Tractenberg RE. (2020, February 19). Concordance of professional ethical practice standards for the domain of Data Science: A white paper. Published in the *Open Archive of the Social Sciences* (SocArXiv), 10.31235/osf.io/p7rj2

is a checklist specifically intended to support project-specific ethical collection, use, and storage of *data* – and to support conversations about what is ethical about a given project, rather than provide guidance about ethical practice overall. Another important thing about this DEFW-based table is the level of overlap between the ASA and ACM Principles; they both include relevant material to support all of the DEFW principles except DEFW Principle 1, which is only addressed by the ASA Guidelines. Also important to notice is that when considering DEFW Principle 7, "embed data use responsibly", the strongest alignment with ACM is in Principle 3, Leadership -i.e., the ACM perspective treats the responsibility around responsible data use as a *leadership* consideration. By contrast, all ASA GL principles *except* responsibilities of leaders, supervisors, and mentors (ASA G) and responsibilities to other practitioners and the profession (ASA F) are aligned with DEFW Principle 7. The ASA GLs charge each practitioner (not just those in leadership roles) with the responsibilities around responsible embedding of data use.

The strongest alignment between ASA GLs and DEFW is with ASA D. There is alignment between DEFW Principles and ACM Principles 1, 2 and 3. Since ACM Principle 4 is about compliance with their specific Code of Ethics, it is not surprising that there is no association between DEFW and ACM 4. Of the alignment between ASA, ACM, and DEFW, not all elements in the guideline contribute to the observed alignment. Thus, this table shows the areas of overlap – as well disjunction – between ASA, ACM, and DEFW when you refer back to the Venn diagram from Chapter 1.4.

Continuing the exploration of alignment between checklists and professional practice standards, the Table below examines the alignment of the DSEC checklist with the ASA GL Principles. Recall the overall GL Principle areas are:

Principle A. Professional integrity and accountability
Principle B. Integrity of data and methods
Principle C. Responsibilities to stakeholders
Principle E. Responsibilities to members of multidisciplinary teams
Principle F. Responsibilities to fellow statistical practitioners and the Profession
*Principle G. Responsibilities of leaders, supervisors, and mentors in statistical practice
Principle H. Responsibilities regarding potential misconduct
*Appendix. Responsibilities of organizations/institutions

An asterisk (*) appear above on Principle G (Responsibilities of leaders, supervisor, and mentors in statistical practice) and the Appendix because there is no intention of the DSEC to relate differently to people in leadership roles, nor to organizations or institutions. Therefore, these two are not included in the alignment tables below exploring ASA GLs (except not G or the Appendix) and the five DSEC domains. The final column gives the number (out of the seven we consider) of ASA Principles are aligned with each of the DSEC items.

ASA GL Principles[10]:	A Prof integrity	B Data/ Method Integrity	C Resp to stake-holder	D Resps to research/ data subjects	E Resps to team	F Resps to other Practiti-oners, profession	H Resps re: mis-conduct	Align-Ment (of 7)
Data Science Ethics Checklist:								
A. Data Collection								
A.1 Informed consent: If there are human subjects, have they given informed consent, where subjects affirmatively opt-in and have a clear understanding of the data uses to which they consent?	A11	B1, B4	C7	D; D1, D5, D6, D10, D11	E4	F5	H2‡	7‡ *G5
A.2 Collection bias: Have we considered sources of bias that could be introduced during data collection and survey design and taken steps to mitigate those?	A2, A4	B; B1, B2, B3, B6	C1, C2, C3, C4, C8	D6	E4	F5		6

ASA GL Principles[10]:	A Prof integrity	B Data/ Method Integrity	C Resp to stake-holder	D Resps to research/ data subjects	E Resps to team	F Resps to other Practitioners, profession	H Resps re: Miscon-duct	Align-ment (of 7)
A.3 Limit PII exposure: Have we considered ways to minimize exposure of personally identifiable information (PII) for example through anonymization or not collecting information that isn't relevant for analysis?			C7	D1, D4, D7, D9				2
B. Data Storage								
B.1 Data security: Do we have a plan to protect and secure data (e.g., encryption at rest and in transit, access controls on internal users and third parties, access logs, and up-to-date software)?		B1, B4, B6	C7	D4, D9, D10, D11				3
B.2 Right to be forgotten: Do we have a mechanism through which an individual can request their personal information be removed?		B4‡, B6‡	C7	D1‡, D4‡, D5‡, D7‡ D11				3‡

ASA GL Principles[10].	A Prof integrity	B Data/ Method Integrity	C Resp to stake-holder	D Resps to research/ data subjects	E Resps to team	F Resps to other Practitioners, profession	H Resps re: Miscon-duct	Align-ment (of 7)
B.3 Data retention plan: Is there a schedule or plan to delete the data after it is no longer needed?				D1‡, D4‡, D5‡, D7‡ D11				1‡
C. Analysis								
C.1 Missing perspectives: Have we sought to address blind spots in the analysis through engagement with relevant stakeholders (e.g., checking assumptions and discussing imply-cations with affected communities and subject matter experts)?	A2, A4	B1, B2	C1	D, D1, D6	E3	F4		6 *G5
C.2 Dataset bias: Have we examined the data for possible sources of bias and taken steps to mitigate or address these biases (e.g., stereotype perpetuation, confirmation bias, imbalanced classes, or omitted confounding variables)?	A2, A3, A4	B1, B2	C2	D1, D2, D3, D5, D6	E4			5

ASA GL Principles[10]:	A Prof integrity	B Data/ Method Integrity	C Resp to stake-holder	D Resps to research/ data subjects	E Resps to team	F Resps to other Practitioners, profession	H Resps re: Misconduct	Align-ment (of 7)
C.3 Honest representation: Are our visualizations, summary statistics, and reports designed to honestly represent the underlying data?	A3, A5, A7, A8, A11	B1, B2, B3, B5, B6	C2, C3, C5	D5, D6, D7, D10, D11	E4	F4		6
C.4 Privacy in analysis: Have we ensured that data with PII are not used or displayed unless necessary for the analysis?		B4	C4, C7	D4, D9		F4		4
C.5 Auditability: Is the process of generating the analysis well documented and reproducible if we discover issues in the future?	A5, A11	B4, B5		D6, D8		F2		4
D. Modeling								
D.1 Proxy discrimination: Have we ensured that the model does not rely on variables or proxies for variables that are unfairly discriminatory?	A2, A3, A4	B1, B2, B5, B6		D6	E4			4

ASA GL Principles[10]:	A Prof integrity	B Data/ Method Integrity	C Resp to stake-holder	D Resps to research/ data subjects	E Resps to team	F Resps to other Practitioners, profession	H Resps re: Miscon-duct	Align-ment (of 7)
D.2 Fairness across groups: Have we tested model results for fairness with respect to different affected groups (e.g., tested for disparate error rates)?	A2, A3, A4	B1, B2, B3, B7		D2, D5, D6, D10, D11	E4			4
D.3 Metric selection: Have we considered the effects of optimizing for our defined metrics and considered additional metrics?	A2	B1, B2, B3,	C1	D8				4
D.4 Explainability: Can we explain in understandable terms a decision the model made in cases where a justification is needed?	A1, A4	B2, B3, B4	C3, C4, C6	D10				4
D.5 Communicate bias: Have we communicated the shortcomings, limitations, and biases of the model to relevant stakeholders in ways that can be generally understood?	A2	B1, B2, B3	C2	D6, D10	E4	F2		5

ASA GL Principles[10]:	A Prof integrity	B Data/ Method Integrity	C Resp to stake-holder	D Resps to research/ data subjects	E Resps to team	F Resps to other Practitioners, profession	H Resps re: Miscon-duct	Align-ment (of 7)
E. Deployment								
E.1 Redress: Have we discussed with our organization a plan for response if users are harmed by the results (e.g., how does the data science team evaluate these cases and update analysis and models to prevent future harm)?	A2, A3, A4, A9	B5	C2, C4	D1, D4, D5, D6, D7, D8				4
E.2 Roll back: Is there a way to turn off or roll back the model in production if necessary?	A3, A4, A9	B5, B6	C2		E4	F2	H2	4*G5
E.3 Concept drift: Do we test and monitor for concept drift to ensure the model remains fair over time?	A2, A3	B1, B2, B5, B6						2*G5

ASA GL Principles[10]:	A Prof integrity	B Data/ Method Integrity	C Resp to stake-holder	D Resps to research/ data subjects	E Resps to team	F Resps to other Practitioners, profession	H Resps re: Miscon-duct	Align-ment (of 7)
E.4 Unintended use: Have we taken steps to identify and prevent unintended uses and abuse of the model and do we have a plan to monitor these once the model is deployed?	A2, A3, A4, A9	B3, B5, B6	C7	D; D4, D5, D6, D7, D10, D11	E4		H2‡	6‡ *G5

Notes: The greyed-out row have *complete* alignment: every ASA GL Principle has at least one element aligned with A1.
* Means there is a match between the language or the intention of one item (G5) from Principle G (which is otherwise excluded from the table) and the DSEC question.
‡ Means there is a match between the intention (and not the language) of at least one ASA GL element or the overall Principle and the DSEC question.

[1] The Appendix is left out of this table because these are not relevant to the DSEC project checklist items. Any items from Principle G are indicated in the final column.

TABLE 1.6.3. *DSEC x ASA GL Alignment*

Since the ASA GLs are intended to reflect the intentions of the profession, it is not surprising that DSEC and ASA GLs are concerned with different features of professional practice in statistics and data science. As you may have expected, every ASA GL Principle is aligned with DSEC A1 (informed consent) but there is little to no alignment between DSEC items and responsibilities owing to individuals on the team (ASA Principle E), other practitioners (ASA F), nor with much to do with misconduct (professional, statistical, or scientific, ASA H). Although it was omitted from the table due to general expectation of lack of alignment between DSEC items and ASA Principle G (responsibilities for leadership roles), there is alignment between DSEC A1, C1, and E2, E3 and E4 with one element in G (G5, "Establish a culture that values validation of assumptions, and assessment of model/algorithm performance over time and across relevant subgroups, as needed. Communicate with relevant stakeholders regarding model or algorithm maintenance, failure, or actual or proposed modifications"). In general, DSEC items are not aligned with leadership/supervisor/mentor roles as discussed in ASA Principle G.

It should not be surprising that the column showing overlap between the DSEC and ASA GL Principle H, "Responsibilities regarding potential misconduct", is *blank* except for two instances where H2 ("Avoids condoning or appearing to condone statistical, scientific, or professional misconduct. Encourages other practitioners to avoid misconduct or the appearance of misconduct.") is identified. The first instance is in informed consent (DSEC A1) and the other is unintended use (DSEC E4). The alignment of H2 with DSEC A1 and E4 has the ‡ notation that means the alignment with the identified ASA GL element is one of intention rather than explicit language match. The DSEC developers were not concerned with the concept of professional or scientific misconduct but instead focused on existing checklists and promoting ethical practices around data science projects – apparently which did not prioritize misconduct and avoiding that. In terms of DSEC E1, the ethical statistical practitioner notifies stakeholders of errors they identify (ASA B5); while this is not the same as "redress" (DSEC E1), if bias in an analysis, or arising from an analysis system, is identified – whether or not it causes harm – the ethical statistical practitioner has an obligation to disseminate the identification of the error (and any fix). Note also that the ASA GLs include an entire principle (ASA C) outlining responsibilities to all stakeholders, and another with responsibilities for data donors and those who are affected by statistical practices (ASA D). The intentional, but not language-matching, alignment between data security (DSEC B) and elements of ASA Principles C and D (indicated with the ‡ notation) reflects quite a lot of alignment in spite of the absence of matching verbiage.

The DSEC areas and content overlap most naturally with ASA Guideline Principles B (integrity of data and methods) and D (Responsibilities to research subjects, data subjects, or those directly affected by statistical practices), with minimal alignment with A. It is worth pointing out that ASA Principle A is "professional integrity and accountability" – it is not the purpose or intention of the DSEC to help you to be stewardly in your quantitative practice, whereas ASA Principle A was essentially created to do just that.

The DSEC questions are, in some sense, rhetorical: the "checklist" format means you tick the box when you have done the <verb> the questions ask if you've done. As noted, understanding the contents of the DSEC doesn't help you practice ethically, and answering the questions with "yes" or "no" is essentially insufficient. The table above shows considerable alignment between the DSEC questions and the ASA GLs: that is, ethical practice guidance can help answer most of the DSEC questions. Thus, instead of answering just "yes" or "no" to DSEC items, the identified ASA GL elements can help ensure that the way the "yes" or "no" is obtained follows these practice standards and norms for ethical quantitative practice – this is explored task by task in Section 2.

The alignment between the DSEC and ACM Principles can also be instructive:

Data Science Ethics Checklist:	ACM CE Principles:			
	1 General ethical principles	2 Professional responsibilities	3 Professional leadership principles	4 Compliance with the code
				Alignment
A. Data Collection				
A.1 Informed consent: If there are human subjects, have they given informed consent, where subjects affirmatively opt-in and have a clear understanding of the data uses to which they consent?	1.6	2.3, 2.9		2
A.2 Collection bias: Have we considered sources of bias that could be introduced during data collection and survey design and taken steps to mitigate those?	1.4	2.5		2
A.3 Limit PII exposure: Have we considered ways to minimize exposure of personally identifiable information (PII) for example through anonymization or not collecting information that isn't relevant for analysis?	1.2, 1.6	2.5, 2.8, 2.9		2

ACM CE Principles: Data Science Ethics Checklist:	1 General ethical principles	2 Professional responsibilities	3 Professional leadership principles	4 Compliance with the code	Alignment
B. Data Storage					
B.1 Data security: Do we have a plan to protect and secure data (e.g., encryption at rest and in transit, access controls on internal users and third parties, access logs, and up-to-date software)?	1.2, 1.6	2.5, 2.8, 2.9	3.7		3
B.2 Right to be forgotten: Do we have a mechanism through which an individual can request their personal information be removed?		2.8, 2.9	3.7		2
B.3 Data retention plan: Is there a schedule or plan to delete the data after it is no longer needed?	1.6				1

ACM CE Principles: Data Science Ethics Checklist:	1 General ethical principles	2 Professional responsibilities	3 Professional leadership principles	4 Compliance with the code	Alignment
C. Analysis					
C.1 Missing perspectives: Have we sought to address blind spots in the analysis through engagement with relevant stakeholders (e.g., checking assumptions and discussing implications with affected communities and subject matter experts)?	1.4				1
C.2 Dataset bias: Have we examined the data for possible sources of bias and taken steps to mitigate or address these biases (e.g., stereotype perpetuation, confirmation bias, imbalanced classes, or omitted confounding variables)?	1.4				1
C.3 Honest representation: Are our visualizations, summary statistics, and reports designed to honestly represent the underlying data?	1.4‡	2.1‡			2‡
C.4 Privacy in analysis: Have we ensured that data with PII are not used or displayed unless necessary for the analysis?	1.4, 1.6‡				1‡
C.5 Auditability: Is the process of generating the analysis well documented and reproducible if we discover issues in the future?		2.4‡			1‡

ACM CE Principles: Data Science Ethics Checklist:	1 General ethical principles	2 Professional responsibilities	3 Professional leadership principles	4 Compliance with the code	Alignment
D. Modeling					
D.1 Proxy discrimination: Have we ensured that the model does not rely on variables or proxies for variables that are unfairly discriminatory?	1.4	2.4, 2.5			2
D.2 Fairness across groups: Have we tested model results for fairness with respect to different affected groups (e.g., tested for disparate error rates)?	1.2, 1.4				1
D.3 Metric selection: Have we considered the effects of optimizing for our defined metrics and considered additional metrics?					0
D.4 Explainability: Can we explain in understandable terms a decision the model made in cases where a justification is needed?		2.5			1
D.5 Communicate bias: Have we communicated the shortcomings, limitations, and biases of the model to relevant stakeholders in ways that can be generally understood?		2.5			1

ACM CE Principles: Data Science Ethics Checklist:	1 General ethical principles	2 Professional responsibilities	3 Professional leadership principles	4 Compliance with the code	Alignment
E. Deployment					
E.1 Redress: Have we discussed with our organization a plan for response if users are harmed by the results (e.g., how does the data science team evaluate these cases and update analysis and models to prevent future harm)?	1.2, 1.4‡	2.5			1‡
E.2 Roll back: Is there a way to turn off or roll back the model in production if necessary?		2.5‡, 2.9‡			1‡
E.3 Concept drift: Do we test and monitor for concept drift to ensure the model remains fair over time?		2.5			1
E.4 Unintended use: Have we taken steps to identify and prevent unintended uses and abuse of the model and do we have a plan to monitor these once the model is deployed?	1.2, 1.4	2.5, 2.9			2

Notes: The greyed-out row reflect zero alignment between ACM CE elements and DSEC questions.
‡ Means there is a match between the intention (and not the language) of the ACM CE and the DSEC question.

Table 1.6.4. *DSEC x ACM CE Alignment table*

There is fairly limited alignment between DSEC and ACM CE Principles 1 and 2, reinforcing the point that the ethical practitioner needs more than the checklists like the DSEC for professional behavior when it comes to computing in particular. Unsurprisingly, there is very little alignment between ACM Principle 3 (Professional Leadership Principles) and none at all with Principle 4 (Compliance with the Code). The lack of alignment between DSEC D3 (metric selection) and DSEC E3 (concept drift) is not surprising given the orientation of the ACM CE to computing specifically and that of the DSEC to data science specifically, where metrics and drift are critical components of modeling and analysis. The fact that modeling and analysis are such essential components of data science (but not computing specifically) may also explain why the alignment between DSEC and the ACM CE is not better.

Summary

As noted in the table showing the DSEC alignment with the ACM and ASA standards, these two DSEC tables show that professional practice standards include much more than simply the answers to DSEC questions. As we saw in the Venn diagram (Chapter 1.4.2 and Figure 1.6.1), and as will become clearer in Sections 2 and 3, there is quite a wide area of quantitative practice that requires attention for ethical practice and/or stewardship of the profession, and for which DSEC and DEFW provide little to no guidance.

Checklists can be useful in documenting the features of a project and whether ethical features were considered at each stage, but the checklist items do not refer to professional practice standards, so cannot help to orient you towards the professional identity the way that practice standards do. However, every item on this worksheet (DEFW 2018) and checklists (DSEC), and possibly others does represent an opportunity to utilize the ASA GLs, and in some cases also the ACM CE, to address aspects of ethical practice with data (DEFW) or practice in data science (DSEC).

Checklists like these naturally do not speak to aspects of ethical professional practice *outside* of a given project. There may be key features of professional practice that are discussed in professional practice standards, but fall outside of any given "project" that a checklist must necessarily focus your attention on. For example, the ACM CE includes a set of Leadership Principles (with 7 items), which entails different decisions specific to the role; the ASA GLs also include a set of responsibilities of those with leadership/supervisor/mentor roles. These relate more specifically to professional conduct, and less specifically to any particular project like the DSEC and DEFW items do. Tools like the DSEC and DEFW can be useful, but exactly how to address the DSEC and DEFW items, and so much more about ethical professional practice, are contained in the practice standards. Thus, checklist type tools (particularly if they are described as shortcuts) should not supplant a commitment to ethical practice of statistics and data science. Moreover, such tools can interfere with professional identity development. "I use a checklist!" does not reflect a commitment to ethical practice, whereas even for individuals who are not members of organizations like ASA or ACM can commit to ethical practice simply by knowing, understanding, and applying their ethical practice standards.

Chapter 1.7

Augmenting your "Prerequisite Knowledge": Stakeholder Analysis[11].

Note that ACM and DEFW focus on the public good (DEFW) and avoidance of harm (ACM) as the keystone for all decision making. If not the driver, the public good is at least the focal consideration when using those reference documents. Among other differences, outlined in the previous chapters, ASA and DSEC do not have this specific focus.

The ASA, while not mentioning harms at all in its ethical practice standards, does emphasize avoiding specific harms like knowingly exploiting vulnerable populations or creating or perpetuating unfair outcomes (A3), misleading stakeholders (C2), violating subject rights (D11), and succumbing to pressures to use inappropriate methods (A4, C2, E4). Also, as noted, specific ethical violations like cherry-picking and p-hacking are not named, but are subsumed under multiple ASA GL elements like A2 ("Uses methodology and data that are valid, relevant, and appropriate, without favoritism or prejudice, and in a manner intended to produce valid, interpretable, and reproducible results."), B2 ("Is transparent about assumptions made in the execution and interpretation of statistical practices including methods used, limitations, possible sources of error, and algorithmic biases. Conveys results or applications of statistical practices in ways that are honest and meaningful.") and C2 ("Regardless of personal or institutional interests or external pressures, does not use statistical practices to mislead any stakeholder.").

This chapter introduces the concept of a stakeholder analysis to demonstrate that harms and benefits are not exchangeable, and to provide structure in how the ethical practitioner can approach their understanding, and documentation, of what harms and benefits may accrue whenever decisions are made by the statistician or data scientist. DSEC also never mentions harms, nor any specific negative outcomes (like ASA GLs do). This may imply that harms and benefits are *exchangeable*: i.e., it doesn't matter if there are more, or worse, harms than benefits resulting from any decision on the DSEC that is made. In addition to

---

[11] This chapter contains material originally published in the Socarxiv preprint archive as Section 3 of Tractenberg, RE. (2019-c, April 23). Teaching and learning about ethical practice: The case analysis. https://doi.org/10.31235/osf.io/58umw

considering the variety of harms the ASA GLs identify, as outlined below, even more general harms cannot be considered exchangeable.

Stakeholder is defined "one who is involved in or affected by a course of action" (Merriam-Webster https://www.merriam-webster.com/dictionary/stakeholder); in the context of ethical case analysis, the stakeholder is simply an individual, or group, that might be affected by the outcome of the case. Clearly, identifying *who* might be affected, and the nature of that effect, are essential in understanding risks and benefits that might be associated with any decision or activity. Ethical case analysis implicitly requires that stakeholders are identified; however, this is not often a focus of instruction or practice. Because statistics and data science can have far-reaching implications, consideration of stakeholders warrants more attention than is typical; so to facilitate the learning goals articulated in the previous section, the stakeholder analysis template was created and appears in Table 1.7.1. Notes on the Table follow the discussion of cells, starting on p. 97.

Potential result: Stakeholder[1]:	HARM[5]	BENEFIT[5]	UNKNOWN[4]	UNKNOWABLE[3]
YOU[2,3]				
Your boss/client				
Unknown individuals[2]				
Employer				
Colleagues				
Profession				
Public/public trust				

Table 1.7.1. *Stakeholder Analysis template*

There are two dimensions to the Stakeholder Analysis Template. The first dimension, captured in the columns, represents *"Potential Results"*. These capture those effects of a decision or action, summarizing them according to whether or not they may represent net negatives. Potential negative results of any action or decision, or *harms*, include costing money, time, effort; negatively impacting reputations or persons; and other types of conceptual (intangible) or actual (tangible) damage. Like harms, potential positive results, or *benefits*, could be tangible or intangible – and they can have immediate or delayed effects. Benefits could include earning or gaining money; the removal of a harm; saving time or effort; improving reputation; and demonstrating expertise or superiority; among other things.

Since the effects of any action or decision may be negative for one entity, person, or group while positive for another, the potential result must be considered with respect to each *"Potential Stakeholder"* (the second dimension in the template). It may be surprising to realize that one of the potential stakeholders is *you*, the person making the decision. As described earlier, harms (costs in time, effort, and reputation) are easily recognized to yourself, as are benefits. These should be considered first because understanding the potential harms and benefits to yourself can also help you to recognize what they may be for the next stakeholder, your boss or client. If a decision costs *you* time (a harm), this could be a harm to your immediate *boss or cli*ent as well. By contrast, deciding to formally identify a data breach at work could be perceived as a harm to you (you could be punished if your boss does not want others in the company to find out) and a benefit to you (you would be forewarned that maybe your employer will soon be out of business or targeted by authorities). Only when the results of decisions or actions – in terms of both harms and benefits – are recognized can they be balanced against each other to make a *justifiable* decision (Tractenberg & FitzGerald, 2012).

You and your boss/client are fairly clear, recognizable, stakeholders. By contrast, *"unknown individuals"* are not recognizable stakeholders per se, but if you make a programming decision in the creation of an algorithm, or a distributional assumption in an analysis, there could be predictable but not-specific results for these unknown individuals. A simple example is males versus females: prior to 2015 the National Institutes of Health did not require that genetic sex effects should be considered and potentially modeled separately (https://grants.nih.gov/grants/guide/notice-files/not-od-15-102.html) – making the scientific and statistical assumption that all human subjects (after controlling for age, height, and weight in many cases) are exchangeable. The decision to require specification of sex effect hypotheses in proposed research, or to justify not including such specification, represents an acknowledgement that failing to consider sex effects in biomedical science is not appropriate, and is insufficiently rigorous. In terms of harms or benefits, failing to consider that heart attack symptoms differ for men and women (Coventry et al. 2011) for example, and making assumptions in analyses or algorithms that ignore the specific symptoms that women experience, will end up potentially harming "unknown women". The benefit to "unknown men" might be "more is understood about the symptoms prior to a heart attack". Medical professionals may never warn women about what might in fact occur to *them*. Thus, the assumption that "human subjects are exchangeable" has great potential to do tangible and lasting harm to those –unknown individuals - about whom such an assumption is wrong. Even when we know that "women experience heart attack symptoms differently from men", we will not know the specific women for whom the decision to model all humans the same will have this harm, thus the category of stakeholder is "unknown individuals". Unknown individuals also comprise customers – e.g., all those customers who are known to have data/records associated with them, even though the specific ones whose data will be breached is unknown.

Identifying your *"employer"* as a stakeholder may seem more straightforward, but "your employer" could be a person or the corporate entity that is your company's name. Harms and benefits can accrue to both; making your decision about notifying the authorities about a data breach could affect the corporate "employer" without affecting the CEO/owner. The more important aspect of this stakeholder is that harms and benefits could accrue to them based on *your* performance or decision. If you are self-employed, you are the face of your company; so while a benefit may accrue to *you* when you fail to notify the authorities about a data breach "at your company" (e.g., save yourself a lot of paperwork; limit the likelihood that a reporter will publicize the breach and name you specifically, etc.), this would actually constitute a harm to your employer (your personal brand) – demonstrating that your company cannot be trusted by a potential customer.

Few practitioners in statistics and data science work alone, so recognizing that decisions we make about ethical problems can affect our *"colleagues"* should not be surprising. Some statisticians and data scientists are the only professional in that field on a team or in a working group – meaning that our colleagues in those situations are unaware of the ethical practice guidelines we are obliged to follow. If you fail to act on an ethical guideline principle, non-statisticians may never find out; but if they do, their trust in you would be diminished (a harm to you). You might, however, inadvertently create a situation where they might also be faced with ethical dilemmas caused by your action or decision; complicating their jobs creates a harm to *them*.

Professional Guidelines are developed by those in practice, in part, to delineate the qualifications to practice and to encourage public trust in *"the profession"* (Tractenberg et al. 2015). Some fields (e.g., medicine, law) have methods for controlling licenses to practice; all those who violate professional or ethical guidelines harm the profession – by suggesting that these controls do not act to keep "bad actors" out of practice. Those who ignore ethical challenges do not strengthen the profession – even if ignoring data breaches seems to benefit the individual ("keep your head down"; "go along to get along"), they harm the profession overall. When an individual, acting in a professional capacity, behaves unethically or ignores unethical behavior, the profession is impacted negatively.

The final stakeholder to consider is the *public/public trust*. When a representative of a profession is unethical, not only is that a harm to their profession but it also diminishes the public's trust in the profession, how that profession is regulated, and whether or not federal funds should be allocated to support or otherwise engage with that profession. The public – people who are not (yet) your customers, and people who are definitely "unknown individuals", but whose stake in the ethical practice of statistics and data science is very real – represent the cultural context in which we are educated, trained, and employed. The public also influences legislation –for or against our profession. Western culture tends to believe itself to be evidence-based: decisions are supposed to be supported by evidence and data (although see Pencheva, 2019 and McGoughy 2019, both discussing the campaign for Britain to leave the European Union ("Brexit"). Part of that evidence is "evidence that practitioners of statistics and data science can be trusted" – harms accrue to the public trust when, e.g., data breaches happen – the public may seek to require stricter legal controls on data collection (e.g., the General Data Protection Regulation, (EU) 2016/679; see https://en.wikipedia.org/wiki/General_ Data_Protection_Regulation). Public sentiment towards federally funded research can also become negative if the public, or the public trust, are harmed.

It is also important for the person completing the stakeholder analysis template to recognize two types of unknown information, which are the final two columns in the table:

- *Unknown*. It is possible for a decision to be required early in a project (for example), before an effect for a given stakeholder can be established as a "harm" or a "benefit". Thought experiments, in which these effects are imagined - rather than observed or remembered from personal experience - (https://plato.stanford.edu/entries/thought-experiment/) or simulations can help to determine which of these is more likely; and an important aspect of the stakeholder analysis template is to document where more information is needed. Additionally, whether a potential stakeholder is – or will become – an actual stakeholder may also need to be determined.

- *Unknowable*. Both early and late in a project, it may be simply impossible to determine whether the effects for a particular stakeholder will be positive (benefit) or negative (harm). Whenever something appears in the "unknowable" column, it suggests that whatever decision is taken currently may need to be revisited in the future. Recognizing something as "unknowable" does not mean it should be ignored, but rather suggests that more thinking or more specification is required –possibly both.

Now that we have discussed every cell (row x column intersection) of the stakeholder analysis template, we can consider what information it might hold as a whole.

1. Knowing to whom harms may accrue can guide you to where the professional guidelines can assist in decision making. This is the topic of the next section.

2. Articulating the harms that may accrue to YOU is essential for you to "treat others' data as you would your own" (Loukides et al., 2018: Chapter 3). You need to recognize the harms that can accrue *to you* before you can compare those to you and those to others. Moreover, "others' data" could relate to your boss/client, your employer, unknowable others, or the public. Recognizing whether or not benefits or harms accrue to these different types of "others" is the only way for you to make a decision about how you want other people to treat *your* data: in someone else's table, *you* are the client, an unknowable other, or part of the public.

3. If there are no recognizable harms, and plausibly no "unknowable" harms <for which your decision would be responsible>, then there can be no conflict. It is really important to recognize whether something truly is unknowable or if it is actually something that can be known – but you just don't know it. The key words here are "recognizable" and "plausible" – your failure to recognize something doesn't mean it does not exist. And beware of straw man[12] or red herring[13] harms! Keeping in mind that not all harms are exchangeable, invented harms (straw man) and distraction harms (red herring) will be recognized for what they are – and the weight they should actually be given in your consideration (i.e., *none* for straw man, and *only what is actually appropriate* for red herring harms).

4. If there are plausible harms (or benefits) that you cannot identify, but you believe/suspect may exist, then there is insufficient information for you to make a decision and you need more information. Recognizing

---

[12] "Straw Man": defined as "an argument, claim or opponent that is invented in order to win or create an argument", Cambridge English Dictionary.

[13] "Red Herring": defined as "something that takes attention away from a more important subject", Cambridge English Dictionary.

this – instead of making an uninformed decision – is currently *not part of the norm*. Learning how to use this table and complete a case analysis is essential for enabling *informed decisions about ethical challenges* for *current and* future practitioners.

5. All harms are not the same; all the benefits are not the same; and harms and benefits are not exchangeable.

Discussion about stakeholders and harms/benefits that may result from any of the standard activities in statistics and data science can strengthen the learners' engagement with GLs, and also encourage consideration –and acceptance - of the responsibilities to practice ethically that the GLs describe (lest any of the harms befall any of the stakeholders).

Section 2. Stewardship of the profession: prerequisite knowledge

Chapter 2.1 Introduction to Section 2

Chapter 2.2 Planning/Designing

Chapter 2.3 Data collection/munging/wrangling

Chapter 2.4 Analysis (perform or program to perform)

Chapter 2.5 Interpretation

Chapter 2.6 Documenting your work

Chapter 2.7 Reporting your results/communication

Chapter 2.8 Engaging in team science/ teamwork

Chapter 2.9 Summary of Section 2

Chapter 2.1
Introduction to Section 2

Each chapter in this section presents fairly targeted ASA and ACM input on each of the seven tasks. Full discussion of ASA and ACM guidance on each task is provided in Tractenberg (2022). Of course, the reader should carefully consider if there are other elements of ASA and ACM guidance that can be brought to bear in any of the tasks and cases in this book. Throughout Section 2, we focus our attention and discussion on the DSEC and DEFW questions, and where in the ASA GLs and ACM CE answers to those questions can be found. We follow the outlines for each task (and DSEC/DEFW questions that are relevant on each task) from Chapter 1.5.

The ASA and ACM acknowledge that there cannot be a rule or algorithm for every possible situation, whereas the existence or use of the DSEC and DEFW suggest that, in any situation there is a concise list of questions that, once answered, create an 'ethical' project. By contrast, the construct of ethical reasoning (ER) suggests that there is a core of "prerequisite knowledge" that can guide all of the decisions to be made in the course of practice (ER KSA 1); and also, that a decision-making framework can further structure those decisions (ER KSA 2). As discussed in Chapter 1.1, the two frameworks we focus on in this book are the **virtue approach**, where behaviors are determined by – and decisions made in alignment with – what "the ethical practitioner" would do; and the **utilitarian approach**, where behaviors and decisions are based on the action(s) that cause the least harm/maximize benefits.

Recall that the DSEC and DEFW lists also only encourage practitioners to consider each item, they offer no guidance about how to answer the questions they raise or pose. We saw in the Venn diagram (above, and repeated at Figure 2.1.1, below) and more specifically in the tables in Chapter 1.6 that while there is some alignment, DSEC does not overlap with much of the DEFW, the ASA GLs, nor with the ACM CE. However, we can still use the DSEC and DEFW questions to help us think about how to use and apply the ASA and ACM guidance. That is the purpose of this chapter (and section).

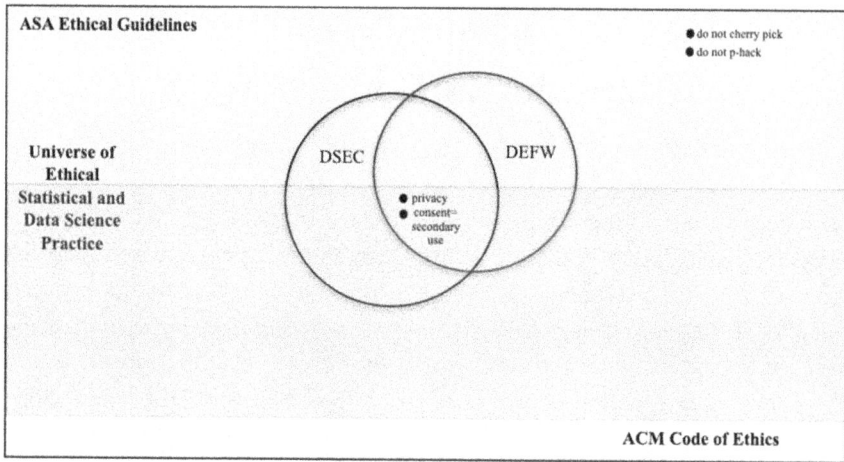

Figure 2.1.1. *Unions and disjunctions: ASA GLs, ACM CE, DSEC, DEFW*

In this chapter we explore the overlap more fully, so that throughout Section 2, we can understand and recognize where to find answers to DSEC and DEFW questions using the ASA and ACM practice standards. After this orienting chapter, each of the subsequent Section 2 chapters will specify which DSEC and DEFW questions – together with relevant ASA and ACM guidance – are applicable in each of the seven tasks in the statistics and data science pipeline. The features of the GLs and CE that are not addressed in the DSEC or DEFW questions will also be highlighted, in order to provide a fuller picture of what constitutes ethical execution of each task. Section 2 is intended to support ethical practice generally – focusing on ER KSAs 1-2, your prerequisite knowledge of the ethical practice standards. Implicit in this exploration is building a better understanding of the virtue (ASA) and utilitarian (ACM) perspectives on common activities in daily practice through the seven common tasks of statistics and data science.

Structure of this chapter: how ASA and ACM facilitate responses to each DSEC and DEFW question.

The Venn Diagram (above) and tables we have seen so far represents the overlap - and alignment – between DSEC and DEFW and the ASA and ACM practice standards. As explained above, this chapter explores the alignment between DSEC and the ASA and ACM practice standards and that between DEFW and these standards in greater detail. The discussion is organized around the 20 DSEC questions in each of its five areas (A. collection; B. data security; C. analysis; D. modeling; and E. deployment).

Two important caveats:

1. This Chapter captures a fairly superficial "matchup" of the DSEC questions with ASA and ACM practice standards. There are five tables that summarize this matching, followed by more detail about some – but not all – of the ASA and ACM guidance. In "matching" the ASA and ACM practice standards to each DSEC question, the reader will quickly see that many of the same elements of the practice standards are relevant for addressing the DSEC questions; therefore, repetition is minimized in favor of pointing out noteworthy features of either the guidance from a practice standard, or the ways in which the ASA and ACM guidance differ or are the same.

2. Other readers may find alignment or answers in other ASA or ACM content, or even in other resources such as policy documents from their workplace. The reader is invited to explore both the ASA and ACM guidance documents fully so as to confirm the observations here, but also to consider alternatives to the matches identified here. These alternatives may come from workplace policy documents, ethical practice standards from other fields, or other relevant resources. As noted earlier, the practice standards do not offer - or propose to offer- any strict rules or algorithms that will always lead to the "right" or "ethical" choice. However, decisions that are made based on the ASA or ACM practice standards will always be consistent with professional, ethical, behavior.

By considering the four resources (DSEC, DEFW, ASA GL, ACM CE) and what they can offer as the practitioner engages in each task, you will build familiarity with the material, as well as with how they reinforce each other and can be brought to bear across the entirety of quantitative practice. As this knowledge deepens, readers will be able to practice ethically and also think about their role as a steward of the discipline or profession, i.e., about being *recognized* and *recognizable* as one to whom "we can entrust the vigor, quality, and integrity of the field" (Golde & Walker 2006: p. 5).

DSEC A	A.1 Informed consent: If there are human subjects, have they given informed consent, where subjects affirmatively opt-in and have a clear understanding of the data uses to which they consent?	A.2 Collection bias: Have we considered sources of bias that could be introduced during data collection and survey design and taken steps to mitigate those?	A.3 Limit PII exposure: Have we considered ways to minimize exposure of personally identifiable information (PII) for example through anonymization or not collecting information that isn't relevant for analysis?
ASA A	A11	A2, A4	
ASA B	B1, B4	B; B1, B2, B3, B6	
ASA C	C7	C1, C2, C3, C4, C8	C7
ASA D	D; D1, D5, D6, D10, D11	D6	D1, D4, D7, D9
ASA E	E4	E4	
ASA F	F5	F5	
ASA G	G5		
ASA H	H2		
ACM 1	1.6	1.4	1.2, 1.6
ACM 2	2.3, 2.9	2.5	2.5, 2.8. 2.9

Table 2.1.1. *DSEC A. Data Collection*

Observations about Table 2.1.1. As expected, ASA Principles C (Responsibilities to stakeholders) and D (Responsibilities to research subjects, data subjects, or those directly affected by statistical practices) offer multiple elements to ensure ethical data collection -with supplements from A (Integrity and accountability) and B (integrity of the data and methods). Perhaps surprisingly, E, F, G and H also offer specific guidance, namely, about ensuring resisting pressure to engage in unethical activities (E, G, H) and to encourage ethical practices around data collection (F). Also, guidance is offered from ACM Principles 1 (general ethical principles) and 2 (professional responsibilities). Even though the ACM practitioner ("computing professional") is not traditionally focused on collecting data, the ACM CE is clearly relevant for ethical behavior in the data-centered world we now live in and in the context of data science and computationally intensive statistical practice.

DSEC A.1 Informed consent

If there are human subjects, have they given informed consent, where subjects affirmatively opt-in and have a clear understanding of the data uses to which they consent?

ASA

Elements of every ASA Principle pertain to ensuring that informed consent is obtained whenever data are collected. The greatest number of ASA elements that are relevant come, not surprisingly, from Principle D. **Responsibilities to research subjects, data subjects, or those directly affected by statistical practices.** The relevant D elements appear below:

> The ethical statistical practitioner does not misuse or condone the misuse of data. They protect and respect the rights and interests of human and animal subjects. These responsibilities extend to those who will be directly affected by statistical practices.

The ethical statistical practitioner:

D1   Keeps informed about and adheres to applicable rules, approvals, and guidelines for the protection and welfare of human and animal subjects. Knows when work requires ethical review and oversight.

D5 Uses data only as permitted by data subjects' consent when applicable or considering their interests and welfare when consent is not required. This includes primary and secondary uses, use of repurposed data, sharing data, and linking data with additional data sets.

D6 Considers the impact of statistical practice on society, groups, and individuals. Recognizes that statistical practice could adversely affect groups or the public perception of groups, including marginalized groups. Considers approaches to minimize negative impacts in applications or in framing results in reporting.

D10 Understands the provenance of the data, including origins, revisions, and any restrictions on usage, and fitness for use prior to conducting statistical practices.

D11 Does not conduct statistical practice that could reasonably be interpreted by subjects as sanctioning a violation of their rights. Seeks to use statistical practices to promote the just and impartial treatment of all individuals.

ACM

Guidance is also available from ACM Principles 1 (General ethical principles, 1.6) and 2 (Professional responsibilities, 2.5, 2.9):

1.6: Computing professionals should only use personal information for legitimate ends and without violating the rights of individuals and groups. This requires taking precautions to prevent re- identification of anonymized data or unauthorized data collection, ensuring the accuracy of data, understanding the provenance of the data, and protecting it from unauthorized access and accidental disclosure. Computing professionals should establish transparent policies and procedures that allow individuals to understand what data is being collected and how it is being used, to give informed consent for automatic data collection, and to review, obtain, correct inaccuracies in, and delete their personal data.

Only the minimum amount of personal information necessary should be collected in a system. The retention and disposal periods for that information should be clearly defined, enforced, and communicated to data subjects. Personal information gathered for a specific purpose should not be used for other purposes without the person's consent. Merged data collections can compromise privacy features present in the original collections. Therefore, computing professionals should take special care for privacy when merging data collections.

2.5 Give comprehensive and thorough evaluations of computer systems and their impacts, including analysis of possible risks. Computing professionals are in a position of trust, and therefore have a special responsibility to provide objective, credible evaluations and testimony to employers, employees, clients, users, and the public. Computing professionals should strive to be perceptive, thorough, and objective when evaluating, recommending, and presenting system descriptions and alternatives. Extraordinary care should be taken to identify and mitigate potential risks in machine learning systems. A system for which future risks cannot be reliably predicted requires frequent reassessment of risk as the system evolves in use, or it should not be deployed. Any issues that might result in major risk must be reported to appropriate parties.

2.9 Design and implement systems that are robustly and usably secure. Breaches of computer security cause harm. Robust security should be a primary consideration when designing and implementing systems. Computing professionals should perform due diligence to ensure the system functions as intended, and take appropriate action to secure resources against accidental and intentional misuse, modification, and denial of service. As threats can arise and change after a system is deployed, computing professionals should integrate mitigation techniques and policies, such as monitoring, patching, and vulnerability reporting. Computing professionals should also take steps to ensure parties affected by data breaches are notified in a timely and clear manner, providing appropriate guidance and remediation. To ensure the system achieves its intended purpose, security features should be designed to be as intuitive and easy to use as possible. Computing professionals should discourage security precautions that are too confusing, are situationally inappropriate, or otherwise inhibit legitimate use. In cases where misuse or harm are predictable or unavoidable, the best option may be to not implement the system.

A.2 Collection bias

Have we considered sources of bias that could be introduced during data collection and survey design and taken steps to mitigate those?

ASA

Elements of ASA Principles A, B, C, D, E and F are relevant for understanding and limiting bias in data collection. Importantly, elements of A (A4) and E (E4) specifically highlight responsibilities to avoid bias that may arise from pressures on the practitioner. Principle C offers multiple elements of importance:

The ethical statistical practitioner:

C1 Seeks to establish what stakeholders hope to obtain from any specific project. Strives to obtain sufficient subject-matter knowledge to conduct meaningful and relevant statistical practice.

C2 Regardless of personal or institutional interests or external pressures, does not use statistical practices to mislead any stakeholder.

C3 Uses practices appropriate to exploratory and confirmatory phases of a project, differentiating findings from each so the stakeholders can understand and apply the results.

C4 Informs stakeholders of the potential limitations on use and re-use of statistical practices in different contexts and offers guidance and alternatives, where appropriate, about scope, cost, and precision considerations that affect the utility of the statistical practice.

C8 Prioritizes both scientific integrity and the principles outlined in these Guidelines when interests are in conflict.

ACM

Guidance from ACM on how to address DSEC A2 includes:

1.4 Computing professionals should foster fair participation of all people, including those of underrepresented groups. The use of information and technology may cause new, or enhance existing, inequities. Technologies and practices should be as inclusive and accessible as possible and computing professionals should take action to avoid creating systems or technologies that disenfranchise or oppress people. Failure to design for inclusiveness and accessibility may constitute unfair discrimination.

2.5 Computing professionals should strive to be perceptive, thorough, and objective when evaluating, recommending, and presenting system descriptions and alternatives. Extraordinary care should be taken to identify and mitigate potential risks in machine learning systems. A system for which future risks cannot be reliably predicted requires frequent reassessment of risk as the system evolves in use, or it should not be deployed. Any issues that might result in major risk must be reported to appropriate parties.

A.3 Limit PII exposure

Have we considered ways to minimize exposure of personally identifiable information (PII) for example through anonymization or not collecting information that isn't relevant for analysis?

ASA

Most of the guidance from the ASA GLs on DSEC A3 is found in Principle D.

The ethical statistical practitioner:

D1 Keeps informed about and adheres to applicable rules, approvals, and guidelines for the protection and welfare of human and animal subjects. Knows when work requires ethical review and oversight.

D4 Protects people's privacy and the confidentiality of data concerning them, whether obtained from the individuals directly, other persons, or existing records. Knows and adheres to applicable rules, consents, and guidelines to protect private information.

D7 Refrains from collecting or using more data than is necessary. Uses confidential information only when permitted and only to the extent necessary. Seeks to minimize the risk of re-identification when sharing de-identified data or results where there is an expectation of confidentiality. Explains any impact of de-identification on accuracy of results.

D9 Knows the legal limitations on privacy and confidentiality assurances and does not over-promise or assume legal privacy and confidentiality protections where they may not apply.

ACM

ACM Principle 2 offers three elements to address DSEC A3.

2.5 Give comprehensive and thorough evaluations of computer systems and their impacts, including analysis of possible risks. Computing professionals are in a position of trust, and therefore have a special responsibility to provide objective, credible evaluations and testimony to employers, employees, clients, users, and the public. Computing professionals should strive to be perceptive, thorough, and objective when evaluating, recommending, and presenting system descriptions and alternatives. Extraordinary care should be taken to identify and mitigate potential risks in machine learning systems. A system for which future risks cannot be reliably predicted requires frequent reassessment of risk as the system evolves in use, or it should not be deployed. Any issues that might result in major risk must be reported to appropriate parties.

2.8 Access computing and communication resources only when authorized or when compelled by the public good. Individuals and organizations have the right to restrict access to their systems and data so long as the restrictions are consistent with other principles in the Code. Consequently, computing professionals should not access another's computer system, software, or data without a reasonable belief that such an action would be authorized or a compelling belief that it is consistent with the public good. A system being publicly accessible is not sufficient grounds on its own to imply authorization. Under exceptional circumstances a computing professional may use unauthorized access to disrupt or inhibit the functioning of malicious systems; extraordinary precautions must be taken in these instances to avoid harm to others.

2.9 Design and implement systems that are robustly and usably secure. Breaches of computer security cause harm. Robust security should be a primary consideration when designing and implementing systems. Computing professionals should perform due diligence to ensure the system functions as intended, and take appropriate action to secure resources against accidental and intentional misuse, modification, and denial of service. As threats can arise and change after a system is deployed, computing professionals should integrate mitigation techniques and policies, such as monitoring, patching, and vulnerability reporting. Computing professionals should also take steps to ensure parties affected by data breaches are notified in a timely and clear manner, providing appropriate guidance and remediation. To ensure the system achieves its intended purpose, security features should be designed to be as intuitive and easy to use as possible. Computing professionals should discourage security precautions that are too confusing, are situationally inappropriate, or otherwise inhibit legitimate use. In cases where misuse or harm are predictable or unavoidable, the best option may be to not implement the system.

	B.1 Data security: Do we have a plan to protect and secure data (e.g., encryption at rest and in transit, access controls on internal users and third parties, access logs, and up-to-date software)?	**B.2 Right to be forgotten:** Do we have a mechanism through which an individual can request their personal information be removed?	**B.3 Data retention plan**: Is there a schedule or plan to delete the data after it is no longer needed?
ASA B	B1, B4, B6	B4, B6	B4, B6
ASA C	C7	C7	C4, C7
ASA D	D4, D9, D10, D11	D1, D4, D5, D7, D11	D1, D4, D5, D7, D11
ASA F	F4		
ACM 1	1.2, 1.6		1.6
ACM 2	2.5, 2.8	2.8, 2.9	
ACM 3	3.7	3.7	

Table 2.1.2. *DSEC B. Data Storage*

Observations about Table 2.1.2. The majority of guidance about data storage comes from ASA Principle D (Responsibilities to research subjects, data subjects, or those directly affected by statistical practices). The elements identified in the table represent a need to balance obligations to share data – and store securely for secure and appropriate sharing – with obligations to the data donors and those affected by statistical practice. Guidance is also offered from ACM Principles 1 (general ethical principles) and 2 (professional responsibilities).

B.1 Data security

Do we have a plan to protect and secure data (e.g., encryption at rest and in transit, access controls on internal users and third parties, access logs, and up-to-date software)?

ASA

Data security is not specifically addressed in the ASA Ethical Guidelines, but Principles B, C, and D offer guidance. Principle F, discussing responsibilities to other practitioners and the Profession, also highlights that, while sharing data and methods, the responsible practitioner "Promotes reproducibility and replication, whether results are "significant" or not, by sharing data, methods, and documentation to the extent possible." (F4) Additional support comes from Principle B:

> The ethical statistical practitioner:

> B1 Communicates data sources and fitness for use, including data generation and collection processes and known biases. Discloses and manages any conflicts of interest relating to the data sources. Communicates data processing and transformation procedures, including missing data handling.

> B4 Meets obligations to share the data used in the statistical practices, for example, for peer review and replication, as allowable. Respects expectations of data contributors when using or sharing data. Exercises due caution to protect proprietary and confidential data, including all data that might inappropriately harm data subjects.

> B6 For models and algorithms designed to inform or implement decisions repeatedly, develops and/or implements plans to validate assumptions and assess performance over time, as needed. Considers criteria and mitigation plans for model or algorithm failure and retirement.

ACM

ACM Principles 1 (1.2, 1.6), 2 (2.5, 2.8) and 3 (3.7) each offer useful guidance about data security, although not explicitly. For example,

> **1.2 Avoid harm.** In this document, "harm" means negative consequences, especially when those consequences are significant and unjust. Examples of harm include unjustified physical or mental injury, unjustified destruction or disclosure of information, and unjustified damage to property, reputation, and the environment. This list is not exhaustive. Well-intended

actions, including those that accomplish assigned duties, may lead to harm. When that harm is unintended, those responsible are obliged to undo or mitigate the harm as much as possible. Avoiding harm begins with careful consideration of potential impacts on all those affected by decisions. When harm is an intentional part of the system, those responsible are obligated to ensure that the harm is ethically justified. In either case, ensure that all harm is minimized. To minimize the possibility of indirectly or unintentionally harming others, computing professionals should follow generally accepted best practices unless there is a compelling ethical reason to do otherwise. Additionally, the consequences of data aggregation and emergent properties of systems should be carefully analyzed. Those involved with pervasive or infrastructure systems should also consider Principle 3.7.

3.7 Recognize and take special care of systems that become integrated into the infrastructure of society. Even the simplest computer systems have the potential to impact all aspects of society when integrated with everyday activities such as commerce, travel, government, healthcare, and education. When organizations and groups develop systems that become an important part of the infrastructure of society, their leaders have an added responsibility to be good stewards of these systems. Part of that stewardship requires establishing policies for fair system access, including for those who may have been excluded. That stewardship also requires that computing professionals monitor the level of integration of their systems into the infrastructure of society. As the level of adoption changes, the ethical responsibilities of the organization or group are likely to change as well. Continual monitoring of how society is using a system will allow the organization or group to remain consistent with their ethical obligations outlined in the Code. When appropriate standards of care do not exist, computing professionals have a duty to ensure they are developed.

B.2 Right to be forgotten

Do we have a mechanism through which an individual can request their personal information be removed?

ASA

Again, guidance from the ASA on data security features is not specific. However, Principle D offers multiple recommendations.

The ethical statistical practitioner:

D1 Keeps informed about and adheres to applicable rules, approvals, and guidelines for the protection and welfare of human and animal subjects. Knows when work requires ethical review and oversight.

D4 Protects people's privacy and the confidentiality of data concerning them, whether obtained from the individuals directly, other persons, or existing records. Knows and adheres to applicable rules, consents, and guidelines to protect private information.

D5 Uses data only as permitted by data subjects' consent when applicable or considering their interests and welfare when consent is not required. This includes primary and secondary uses, use of repurposed data, sharing data, and linking data with additional data sets.

D7 Refrains from collecting or using more data than is necessary. Uses confidential information only when permitted and only to the extent necessary. Seeks to minimize the risk of re-identification when sharing de-identified data or results where there is an expectation of confidentiality. Explains any impact of de-identification on accuracy of results.

D11 Does not conduct statistical practice that could reasonably be interpreted by subjects as sanctioning a violation of their rights. Seeks to use statistical practices to promote the just and impartial treatment of all individuals.

(NB: in some cases, ethical statistical practice does not include a need for "forgetting" because data are collected with consent for specific purposes – about which data contributors are informed.)

ACM

Computing professionals are encouraged to consider the individual whose data has been obtained (2.8. 2.9) although forgetting and individual removal are not specifically addressed. For example:

2.8 Access computing and communication resources only when authorized or when compelled by the public good. Individuals and organizations have the right to restrict access to their systems and data so long as the restrictions are consistent with other principles in the Code. Consequently, computing professionals should not access another's computer system, software, or data without a reasonable belief that such an action would be authorized or a compelling belief that it is consistent with the public good. A system being publicly accessible is not sufficient grounds on its own to imply authorization. Under exceptional circumstances a computing professional may use unauthorized access to disrupt or inhibit the functioning of malicious systems; extraordinary precautions must be taken in these instances to avoid harm to others.

2.9 Design and implement systems that are robustly and usably secure. Breaches of computer security cause harm. Robust security should be a primary consideration when designing and implementing systems. Computing professionals should perform due diligence to ensure the system functions as intended, and take appropriate action to secure resources against accidental and intentional misuse, modification, and denial of service. As threats can arise and change after a system is deployed, computing professionals should integrate mitigation techniques and policies, such as monitoring, patching, and vulnerability reporting. Computing professionals should also take steps to ensure parties affected by data breaches are notified in a timely and clear manner, providing appropriate guidance and remediation. To ensure the system achieves its intended purpose, security features should be designed to be as intuitive and easy to use as possible. Computing professionals should discourage security precautions that are too confusing, are situationally inappropriate, or otherwise inhibit legitimate use.

B.3 Data retention plan

Is there a schedule or plan to delete the data after it is no longer needed?

NB: when data are collected using federal funds in the United States (e.g., for NIH, NSF, or CDC funded research), there are increasingly considerations being given to ensuring that these data – the collection of which is publicly funded – should be shared with "qualified" researchers who make a legitimate request. However, data disposal is also a required component in NSF data management plans (as of October 2022), and as data management and storage are becoming increasingly resource-intensive with exponential growth in data collecting and analysis capabilities, considerations about data management –to include disposal – are becoming increasingly important in every research-data collection enterprise. This is one area where what constitutes ethical practice in the context of business, government, and academic applications may vary widely. A key to understanding "what to do" when it comes to data retention is in the nature of the data and its collection: when data are obtained in most experimental, federal (descriptive samples), or strictly scientific settings, plans for deleting the data may not be necessary (and may be contrary to the funder's objectives). In business settings, there is likely to be a need for a plan to delete the data.

ASA

Because the ASA Ethical Guidelines are intended for use by statistics practitioners across many settings, the caveat above is integrated in the support offered by Principle D (as highlighted throughout this section on DSEC B). Two elements each from Principles B and C are relevant:

The ethical statistical practitioner:

B4 Meets obligations to share the data used in the statistical practices, for example, for peer review and replication, as allowable. Respects expectations of data contributors when using or sharing data. Exercises due caution to protect proprietary and confidential data, including all data that might inappropriately harm data subjects.

B6 For models and algorithms designed to inform or implement decisions repeatedly, develops and/or implements plans to validate assumptions and assess performance over time, as needed. Considers criteria and mitigation plans for model or algorithm failure and retirement.

C4 Informs stakeholders of the potential limitations on use and re-use of statistical practices in different contexts and offers guidance and alternatives, where appropriate, about scope, cost, and precision considerations that affect the utility of the statistical practice.

C7 Understands and conforms to confidentiality requirements for data collection, release, and dissemination and any restrictions on its use established by the data provider (to the extent legally required). Protects the use and disclosure of data accordingly. Safeguards privileged information of the employer, client, or funder.

ACM

The caveats above are also relevant for ACM CE guidance on data retention. When contemplating what to do about the storage of data, the following is relevant:

1.6 Computing professionals should only use personal information for legitimate ends and without violating the rights of individuals and groups. This requires taking precautions to prevent re- identification of anonymized data or unauthorized data collection, ensuring the accuracy of data, understanding the provenance of the data, and protecting it from unauthorized access and accidental disclosure. Computing professionals should establish transparent policies and procedures that allow individuals to understand what data is being collected and how it is being used, to give informed consent for automatic data collection, and to review, obtain, correct inaccuracies in, and delete their personal data.

	C.1 Missing perspectives: Have we sought to address blind spots in the analysis through engagement with relevant stakeholders (e.g., checking assumptions and discussing implications with affected communities and subject matter experts)?	C.2 Dataset bias: Have we examined the data for possible sources of bias and taken steps to mitigate or address these biases (e.g., stereotype perpetuation, confirmation bias, imbalanced classes, or omitted confounding variables)?	C.3 Honest representation: Are our visualizations, summary statistics, and reports designed to honestly represent the underlying data?	C.4 Privacy in analysis: Have we ensured that data with PII are not used or displayed unless necessary for the analysis?	C.5 Auditability: Is the process of generating the analysis well documented and reproducible if we discover issues in the future?
ASA A	A2, A4	A2, A3, A4	A3, A4, A5, A7, A8, A11		A5, A11
ASA B	B1, B2	B1, B2	B1, B2, B3, B5, B6	B4	B4, B5
ASA C	C1, C2	C2	C2, C3, C5	C4, C7	
ASA D	D, D1, D6	D1, D2, D3, D5, D6	D5, D6, D7, D10, D11	D4, D5, D9	D6, D8
ASA E	E3	E4	E4		
ASA F	F4		F4	F4	F2
ACM 1	1.4	1.4		1.6	
ACM 2			2.1		(2.4)

Table 2.1.3. *DSEC C. Analysis*

Observations about Table 2.1.3. Readers might be surprised by the fact that six of the eight ASA GL Principles all have recommendations to offer for ethical "analysis". Clearly, professional integrity and accountability (ASA Principle A) in analysis includes responsibilities about the integrity of the data and methods (Principle B), as well as responsibilities to stakeholders (Principle C) and data donors and those affected by statistical practice (Principle D). Since "analysis" is not typically part of the computing professional's role, it is not surprising that minimal guidance is offered from ACM Principles 1 (general ethical principles) and 2 (professional responsibilities).

C.1 Missing perspectives

Have we sought to address blind spots in the analysis through engagement with relevant stakeholders (e.g., checking assumptions and discussing implications with affected communities and subject matter experts)?

ASA

Many of the ASA Principles are informative about DSEC C (Analysis), not surprisingly. Consideration of guidance from Principle A is important because it offers two different perspectives on addressing blind spots. The ethical statistical practitioner must balance:

A2. Uses methodology and data that are valid, relevant, and appropriate, without favoritism or prejudice, and in a manner intended to produce valid, interpretable, and reproducible results.

A4. Opposes efforts to predetermine or influence the results of statistical practices and resists pressure to selectively interpret data.

Guidance for ethical analysis from ASA Principle A describes two general aspects of ethical statistical practice – including but beyond "analysis". Namely, without allowing efforts to influence the results or selective interpretation of data (A4), the ethical practitioner uses methodology and data that are relevant for the specific objectives (A2). This is not just a feature of the ethical practitioner's integrity and accountability (Principle A), but is also a responsibility owed to stakeholders, as outlined in C1 and C2:

The ethical statistical practitioner:

C1 Seeks to establish what stakeholders hope to obtain from any specific project. Strives to obtain sufficient subject-matter knowledge to conduct meaningful and relevant statistical practice.

C2 Regardless of personal or institutional interests or external pressures, does not use statistical practices to mislead any stakeholder.

Principle D also offers guidance, including ensuring that studies involving animal subjects are also not subject to bias or blind spots (D3):

The ethical statistical practitioner:

> D1 Keeps informed about and adheres to applicable rules, approvals, and guidelines for the protection and welfare of human and animal subjects. Knows when work requires ethical review and oversight.
>
> D6 Considers the impact of statistical practice on society, groups, and individuals. Recognizes that statistical practice could adversely affect groups or the public perception of groups, including marginalized groups. Considers approaches to minimize negative impacts in applications or in framing results in reporting.

ACM

ACM guidance is more general than that of the ASA when it comes to avoiding bias in data that are collected (thereby reducing bias in analysis -which itself is not explicitly treated in the ACM CE).

> **1.4 Computing professionals should foster fair participation of all people, including those of underrepresented groups.** The use of information and technology may cause new, or enhance existing, inequities. Technologies and practices should be as inclusive and accessible as possible and computing professionals should take action to avoid creating systems or technologies that disenfranchise or oppress people. Failure to design for inclusiveness and accessibility may constitute unfair discrimination.

C.2 Dataset bias

Have we examined the data for possible sources of bias and taken steps to mitigate or address these biases (e.g., stereotype perpetuation, confirmation bias, imbalanced classes, or omitted confounding variables)?

ASA

Adding to what was considered above, Principle A specifically charges the ethical practitioner to avoid bias:

The ethical statistical practitioner:

> A3 Does not knowingly conduct statistical practices that exploit vulnerable populations or create or perpetuate unfair outcomes.

Principle B is also relevant for much of DSEC C. For DSEC C1, C2 and C3 two elements are particularly relevant:

The ethical statistical practitioner:

B1 Communicates data sources and fitness for use, including data generation and collection processes and known biases. Discloses and manages any conflicts of interest relating to the data sources. Communicates data processing and transformation procedures, including missing data handling.

B2 Is transparent about assumptions made in the execution and interpretation of statistical practices including methods used, limitations, possible sources of error, and algorithmic biases. Conveys results or applications of statistical practices in ways that are honest and meaningful.

Principle D also offers guidance, including ensuring that studies involving animal subjects are also not subject to bias or blind spots (D3):

The ethical statistical practitioner:

D1 Keeps informed about and adheres to applicable rules, approvals, and guidelines for the protection and welfare of human and animal subjects. Knows when work requires ethical review and oversight.

D2 Makes informed recommendations for sample size and statistical practice methodology in order to avoid the use of excessive or inadequate numbers of subjects and excessive risk to subjects.

D3 For animal studies, seeks to leverage statistical practice to reduce the number of animals used, refine experiments to increase the humane treatment of animals, and replace animal use where possible.

D5 Uses data only as permitted by data subjects' consent when applicable or considering their interests and welfare when consent is not required. This includes primary and secondary uses, use of repurposed data, sharing data, and linking data with additional data sets.

D6 Considers the impact of statistical practice on society, groups, and individuals. Recognizes that statistical practice could adversely affect groups or the public perception of groups, including marginalized groups. Considers approaches to minimize negative impacts in applications or in framing results in reporting.

Finally, the ASA GLs include important guidance for bias that *may arise from their team* (Principle E) rather than the data itself:

The ethical statistical practitioner:

E4 Avoids compromising validity for expediency. Regardless of pressure on or within the team, does not use inappropriate statistical practices

ACM

The ACM CE guidance relates to both the workplace, those impacted by computing, and those who may be impacted by analysis that is enabled by computing practices.

1.4 Computing professionals should foster fair participation of all people, including those of underrepresented groups. The use of information and technology may cause new, or enhance existing, inequities. Technologies and practices should be as inclusive and accessible as possible and computing professionals should take action to avoid creating systems or technologies that disenfranchise or oppress people. Failure to design for inclusiveness and accessibility may constitute unfair discrimination.

C.3 Honest representation

Are our visualizations, summary statistics, and reports designed to honestly represent the underlying data?

ASA

ASA Principle A is about integrity and accountability, but offers several elements supporting honesty in representation of analyses. Accordingly,

The ethical statistical practitioner:

> A3 Does not knowingly conduct statistical practices that exploit vulnerable populations or create or perpetuate unfair outcomes.
>
> A4 Opposes efforts to predetermine or influence the results of statistical practices and resists pressure to selectively interpret data.
>
> A5 Accepts full responsibility for their own work, does not take credit for the work of others, and gives credit to those who contribute. Respects and acknowledges the intellectual property of others.
>
> A7 Discloses conflicts of interest, financial and otherwise, and manages or resolves them according to established policies, regulations, and laws.
>
> A8 Promotes the dignity and fair treatment of all people. Neither engages in nor condones discrimination based on personal characteristics. Respects personal boundaries in interactions and avoids harassment, including sexual harassment, bullying, and other abuses of power or authority.
>
> A11 Follows applicable policies, regulations, and laws relating to their professional work, unless there is a compelling ethical justification to do otherwise.

Relating to the integrity of data and methods, Principle B also has multiple elements that are relevant:

The ethical statistical practitioner:

B1 Communicates data sources and fitness for use, including data genera-tion and collection processes and known biases. Discloses and manages any conflicts of interest relating to the data sources. Communicates data processing and transformation procedures, including missing data handling.

B2 Is transparent about assumptions made in the execution and interpretation of statistical practices including methods used, limitations, possible sources of error, and algorithmic biases. Conveys results or applications of statistical practices in ways that are honest and meaningful.

B3 Communicates the stated purpose and the intended use of statistical practices. Is transparent regarding a priori versus post hoc objectives and planned versus unplanned statistical practices. Discloses when multiple comparisons are conducted, and any relevant adjustments.

B5 Strives to promptly correct substantive errors discovered after publication or implementation. As appropriate, disseminates the correction publicly and/or to others relying on the results.

B6 For models and algorithms designed to inform or implement decisions repeatedly, develops and/or implements plans to validate assumptions and assess performance over time, as needed. Considers criteria and mitigation plans for model or algorithm failure and retirement.

One additional element each from Principles E and F are relevant, underscoring the range of stakeholders to whom DSEC C3 (Honest representation) should be considered to apply.

The ethical statistical practitioner:

E4 Avoids compromising validity for expediency. Regardless of pressure on or within the team, does not use inappropriate statistical practices.

F4 Promotes reproducibility and replication, whether results are "significant" or not, by sharing data, methods, and documentation to the extent possible.

ACM

2.1 Strive to achieve high quality in both the processes and products of professional work. Computing professionals should insist on and support high-quality work from themselves and from colleagues. The dignity of employers, employees, colleagues, clients, users, and anyone else affected either directly or indirectly by the work should be respected throughout the process. Computing professionals should respect the right of those involved to transparent communication about the project. Professionals should be cognizant of any serious negative consequences affecting any stakeholder that may result from poor quality work and should resist inducements to neglect this responsibility.

C.4 Privacy in analysis

Have we ensured that data with PII are not used or displayed unless necessary for the analysis?

ASA

ASA Guidelines charge the ethical practitioner with responsibilities relating to both stakeholders (Principle C) and data donors or those who might be affected by statistical practices (Principle D). The elements below offer a few dimensions that must be balanced in terms of conforming to applicable rules and laws (Principle C) and the intentions of data donors (Principle D).

The ethical statistical practitioner:

> C4 Informs stakeholders of the potential limitations on use and re-use of statistical practices in different contexts and offers guidance and alternatives, where appropriate, about scope, cost, and precision considerations that affect the utility of the statistical practice.
> C7 Understands and conforms to confidentiality requirements for data collection, release, and dissemination and any restrictions on its use established by the data provider (to the extent legally required). Protects the use and disclosure of data accordingly. Safeguards privileged information of the employer, client, or funder.

> D4 Protects people's privacy and the confidentiality of data concerning them, whether obtained from the individuals directly, other persons, or existing records. Knows and adheres to applicable rules, consents, and guidelines to protect private information.
> D5 Uses data only as permitted by data subjects' consent when applicable or considering their interests and welfare when consent is not required. This includes primary and secondary uses, use of repurposed data, sharing data, and linking data with additional data sets.

D9 Knows the legal limitations on privacy and confidentiality assurances and does not over-promise or assume legal privacy and confidentiality protections where they may not apply.

Note that ASA C4 specifically outlines the responsibility to notify stakeholders about limitations on use *and re-use* (emphasis added) of data. When engaging with stakeholders in discussions about the data to be used, the ethical practitioner should ensure that all recognize the limitations on permissions as well as the extent of PII in the dataset.

ACM

Again, rather than specific elements relating to analysis, the ACM CE is instead more general about the data to be used (in analyses or other activities).

1.6: Computing professionals should only use personal information for legitimate ends and without violating the rights of individuals and groups. This requires taking precautions to prevent re- identification of anonymized data or unauthorized data collection, ensuring the accuracy of data, understanding the provenance of the data, and protecting it from unauthorized access and accidental disclosure. Computing professionals should establish transparent policies and procedures that allow individuals to understand what data is being collected and how it is being used, to give informed consent for automatic data collection, and to review, obtain, correct inaccuracies in, and delete their personal data.

Only the minimum amount of personal information necessary should be collected in a system. The retention and disposal periods for that information should be clearly defined, enforced, and communicated to data subjects. Personal information gathered for a specific purpose should not be used for other purposes without the person's consent. Merged data collections can compromise privacy features present in the original collections. Therefore, computing professionals should take special care for privacy when merging data collections.

C.5 Auditability

Is the process of generating the analysis well documented and reproducible if we discover issues in the future?

ASA

The alignment table shows how considerations about data will naturally lead to protections relating to auditable practices. Among other elements, specific guidance from the ASA can also be found in Principles D and F.

The ethical statistical practitioner:

> D6 Considers the impact of statistical practice on society, groups, and individuals. Recognizes that statistical practice could adversely affect groups or the public perception of groups, including marginalized groups. Considers approaches to minimize negative impacts in applications or in framing results in reporting.
> D7 Refrains from collecting or using more data than is necessary. Uses confidential information only when permitted and only to the extent necessary. Seeks to minimize the risk of re-identification when sharing de-identified data or results where there is an expectation of confidentiality. Explains any impact of de-identification on accuracy of results.
> D8 To maximize contributions of data subjects, considers how best to use available data sources for exploration, training, testing, validation, or replication as needed for the application. The ethical statistical practitioner appropriately discloses how the data is used for these purposes and any limitations.

> F2 Helps strengthen, and does not undermine, the work of others through appropriate peer review or consultation. Provides feedback or advice that is impartial, constructive, and objective.

Note that while the elements from Principle D are intended to ensure that the impact of an auditable process on data donors is appropriate, element F2 is a responsibility that relates to ensuring that the *work of others* is ethical. This creates both direct and indirect effects of the ASA GLs on stakeholders – because one's own work should be done ethically, but also one has a responsibility to strengthen the work of others. This responsibility (from Principle F) applies even if one is not (yet) at a career stage where Principle G (Responsibilities of leaders, supervisors, and mentors in statistical practice) is relevant.

ACM

While auditability – of analyses – is not an explicit part of the computing professionals' CE, ACM 2.4 notes the need to "accept and provide appropriate professional review." Thus, like the direct and indirect effects of the ASA GLs on practice, the ACM CE has a similar two-dimensional influence on ethical computing practice through 2.4 (among other elements).

	D.1 Proxy discrimination: Have we ensured that the model does not rely on variables or proxies for variables that are unfairly discriminatory?	D.2 Fairness across groups: Have we tested model results for fairness with respect to different affected groups (e.g., tested for disparate error rates)?	D.3 Metric selection: Have we considered the effects of optimizing for our defined metrics and considered additional metrics?	D.4 Explainability: Can we explain in understandable terms a decision the model made in cases where a justification is needed?	D.5 Communicate bias: Have we communicated the shortcomings, limitations, and biases of the model to relevant stakeholders in ways that can be generally understood?
ASA A	A2, A3, A4	A2, A3, A4	A2	A1, A4	A2
ASA B	B1, B2, B5, B6	B1, B2, B3, B7	B1, B2, B3	B2, B3, B4	B1, B2, B3
ASA C			C1	C3, C4, C6	C2
ASA D	D6	D2, D5, D6, D10, D11	D8	D10	D6, D10
ASA E	E4	E4			E4
ASA F					F2
ACM 1	1.4	1.2, 1.4			
ACM 2	2.4, 2.5			2.5, 2.9	2.5

Table 2.1.4. *DSEC D. Modeling*

Observations about Table 2.1.4. Like "analysis" (DSEC area C), "modeling" may be considered more the purview of statistics than computing. Thus, the guidance offered on modeling by the ASA GLs emphasizes responsibilities about data and methods (ASA Principle B) and about data donors and those affected by statistical practice (ASA Principle D). Note that, while ACM elements offer guidance on four of five of DSEC area D, there is no consideration for metric selection (DSEC D3) in the ACM CE. Both ASA and ACM describe the provision of "appropriate peer review or consultation" (ASA F2) and "appropriate professional review" (ACM 2.4). While these are certainly not limited to the communication of bias about modeling (DSEC D5), it is worth pointing out that communication about the results of (or designs of) modeling may first occur with other practitioners. Communications about the bias that may be inherent in modeling methods or results should begin at the actual modeling stage, and the ethical practitioner does not wait until the modeling is completed and shared with stakeholders (or implemented) to begin contemplating the potential for bias.

D.1 Proxy discrimination

Have we ensured that the model does not rely on variables or proxies for variables that are unfairly discriminatory?

ASA

Much of statistical practice involves modeling. More specifically, building, testing, and interpreting statistical models requires a great many decisions: for example, about distributions to utilize, which metrics to use, and even how to know when the model iterations have been optimized (i.e., "stopping rules"). Understanding the variables ("proxies") that are used in the analysis or model are critical to making justifiable assumptions and defensible interpretations. To address the very wide range of applications of modeling across statistical practice, there are a few specific elements of the ASA GLs addressing the obligation to guard against the potential for bias arising from decisions about variables or proxies that are used.

The ethical statistical practitioner:

> A2 Uses methodology and data that are valid, relevant, and appropriate, without favoritism or prejudice, and in a manner intended to produce valid, interpretable, and reproducible results.
> A3 Does not knowingly conduct statistical practices that exploit vulnerable populations or create or perpetuate unfair outcomes.

A4 Opposes efforts to predetermine or influence the results of statistical practices and resists pressure to selectively interpret data.

D6 Considers the impact of statistical practice on society, groups, and individuals. Recognizes that statistical practice could adversely affect groups or the public perception of groups, including marginalized groups. Considers approaches to minimize negative impacts in applications or in framing results in reporting.

E4 Avoids compromising validity for expediency. Regardless of pressure on or within the team, does not use inappropriate statistical practices

ACM

While modeling, and particularly the choice of variables to include in models, may not typically be considered in computing, DSEC D.1 is addressed by three elements of the ACM CE:

1.4 Computing professionals should foster fair participation of all people, including those of underrepresented groups. The use of information and technology may cause new, or enhance existing, inequities. Technologies and practices should be as inclusive and accessible as possible and computing professionals should take action to avoid creating systems or technologies that disenfranchise or oppress people. Failure to design for inclusiveness and accessibility may constitute unfair discrimination.

2.4 Accept and provide appropriate professional review. High-quality professional work in computing depends on professional review at all stages. Whenever appropriate, computing professionals should seek and utilize peer and stakeholder review. Computing professionals should also provide constructive, critical reviews of others' work.

2.5 Give comprehensive and thorough evaluations of computer systems and their impacts, including analysis of possible risks. Computing professionals are in a position of trust, and therefore have a special responsibility to provide objective, credible evaluations and testimony to employers, employees, clients, users, and the public. Computing professionals should strive to be perceptive, thorough, and objective when evaluating, recommending, and presenting system descriptions and alternatives. Extraordinary care should be taken to identify and mitigate potential risks in machine learning systems. A system for which future risks cannot be reliably predicted requires frequent reassessment of risk as the system evolves in use, or it should not be deployed. Any issues that might result in major risk must be reported to appropriate parties.

D.2 Fairness across groups

Have we tested model results for fairness with respect to different affected groups (e.g., tested for disparate error rates)?

ASA

Given the importance of modeling to statistical practice, multiple elements across Principles A, B, and D are relevant for ensuring fairness across groups based on the results of modeling. For example,

The ethical statistical practitioner:

> A2 Uses methodology and data that are valid, relevant, and appropriate, without favoritism or prejudice, and in a manner intended to produce valid, interpretable, and reproducible results.
> A3 Does not knowingly conduct statistical practices that exploit vulnerable populations or create or perpetuate unfair outcomes.
> A4 Opposes efforts to predetermine or influence the results of statistical practices and resists pressure to selectively interpret data.

> B1 Communicates data sources and fitness for use, including data generation and collection processes and known biases. Discloses and manages any conflicts of interest relating to the data sources. Communicates data processing and transformation procedures, including missing data handling.
> B2 Is transparent about assumptions made in the execution and interpretation of statistical practices including methods used, limitations, possible sources of error, and algorithmic biases. Conveys results or applications of statistical practices in ways that are honest and meaningful.
> B3 Communicates the stated purpose and the intended use of statistical practices. Is transparent regarding a priori versus post hoc objectives and planned versus unplanned statistical practices. Discloses when multiple comparisons are conducted, and any relevant adjustments.
> B7 Explores and describes the effect of variation in human characteristics and groups on statistical practice when feasible and relevant.

> D2 Makes informed recommendations for sample size and statistical practice methodology in order to avoid the use of excessive or inadequate numbers of subjects and excessive risk to subjects.
> D5 Uses data only as permitted by data subjects' consent when applicable or considering their interests and welfare when consent is not required. This includes primary and secondary uses, use of repurposed data, sharing data, and linking data with additional data sets.

D6 Considers the impact of statistical practice on society, groups, and individuals. Recognizes that statistical practice could adversely affect groups or the public perception of groups, including marginalized groups. Considers approaches to minimize negative impacts in applications or in framing results in reporting.

D10 Understands the provenance of the data, including origins, revisions, and any restrictions on usage, and fitness for use prior to conducting statistical practices.

D11 Does not conduct statistical practice that could reasonably be interpreted by subjects as sanctioning a violation of their rights. Seeks to use statistical practices to promote the just and impartial treatment of all individuals.

ACM

Two key elements of ACM CE Principle 1 are highly relevant for DSEC D.2:

1.2 Avoid harm. In this document, "harm" means negative consequences, especially when those consequences are significant and unjust. Examples of harm include unjustified physical or mental injury, unjustified destruction or disclosure of information, and unjustified damage to property, reputation, and the environment. This list is not exhaustive. Well-intended actions, including those that accomplish assigned duties, may lead to harm. When that harm is unintended, those responsible are obliged to undo or mitigate the harm as much as possible. Avoiding harm begins with careful consideration of potential impacts on all those affected by decisions. When harm is an intentional part of the system, those responsible are obligated to ensure that the harm is ethically justified. In either case, ensure that all harm is minimized. To minimize the possibility of indirectly or unintentionally harming others, computing professionals should follow generally accepted best practices unless there is a compelling ethical reason to do otherwise. Additionally, the consequences of data aggregation and emergent properties of systems should be carefully analyzed. Those involved with pervasive or infrastructure systems should also consider Principle 3.7.

1.4 Computing professionals should foster fair participation of all people, including those of underrepresented groups. The use of information and technology may cause new, or enhance existing, inequities. Technologies and practices should be as inclusive and accessible as possible and computing professionals should take action to avoid creating systems or technologies that disenfranchise or oppress people. Failure to design for inclusiveness and accessibility may constitute unfair discrimination.

D.3 Metric selection

Have we considered the effects of optimizing for our defined metrics and considered additional metrics?

ASA

Guidance about metric selection, while not explicit in the ASA GLs, comes from Principles A, B, C, and D. That is, metric selection is a function of professional integrity (ASA A), methods and data (ASA B), stakeholders (ASA C), and data donors and those affected by statistical practice (ASA D).

The ethical statistical practitioner:

A2 Uses methodology and data that are valid, relevant, and appropriate, without favoritism or prejudice, and in a manner intended to produce valid, interpretable, and reproducible results.

B1 Communicates data sources and fitness for use, including data generation and collection processes and known biases. Discloses and manages any conflicts of interest relating to the data sources. Communicates data processing and transformation procedures, including missing data handling.
B2 Is transparent about assumptions made in the execution and interpretation of statistical practices including methods used, limitations, possible sources of error, and algorithmic biases. Conveys results or applications of statistical practices in ways that are honest and meaningful.
B3 Communicates the stated purpose and the intended use of statistical practices. Is transparent regarding a priori versus post hoc objectives and planned versus unplanned statistical practices. Discloses when multiple comparisons are conducted, and any relevant adjustments.

C1 Seeks to establish what stakeholders hope to obtain from any specific project. Strives to obtain sufficient subject-matter knowledge to conduct meaningful and relevant statistical practice.

D8 To maximize contributions of data subjects, considers how best to use available data sources for exploration, training, testing, validation, or replication as needed for the application. The ethical statistical practitioner appropriately discloses how the data is used for these purposes and any limitations.

ACM

As noted in the previous chapter, the selection of metrics in modeling may be more specifically statistical, or more part of data scientist work, than for computation. The table shows the DSEC D.3 x ACM cell greyed out because there is no specific or even indirect guidance from ACM on metric selection. NB: if the reader was a computing professional, utilizing DSEC and the ACM CE for guidance on a project, and realized that there was nothing in the ACM CE to enable an ethical choice with respect to model selection (DSEC D3), this would constitute a "gap" – the ethical computing professional understands "that the public good is always the primary consideration" (ACM CE Preamble). Further, "the Code is not an algorithm for solving ethical problems; rather it serves as a basis for ethical decision-making." That is, even if the ACM CE does not contain specific guidance, the ethical practitioner can look elsewhere for guidance but should not conclude that, because the ACM CE is not specific about metric selection (DSEC D. 3), then nothing should be done about this key aspect of modeling.

D.4 Explainability

Can we explain in understandable terms a decision the model made in cases where a justification is needed?

ASA

Guidance on this, like for others of the DSEC D questions, can be found in ASA Principles A (A1 A4) and B (B2, B3, B4). Additional responsibilities relating to stakeholders (Principle C) and data donors and those affected by statistical practice (Principle D) include the following.

The ethical statistical practitioner:

> C3 Uses practices appropriate to exploratory and confirmatory phases of a project, differentiating findings from each so the stakeholders can understand and apply the results.
> C4 Informs stakeholders of the potential limitations on use and re-use of statistical practices in different contexts and offers guidance and alternatives, where appropriate, about scope, cost, and precision considerations that affect the utility of the statistical practice.
> C6 Strives to make new methodological knowledge widely available to provide benefits to society at large. Presents relevant findings, when possible, to advance public knowledge.

D6 Considers the impact of statistical practice on society, groups, and individuals. Recognizes that statistical practice could adversely affect groups or the public perception of groups, including marginalized groups. Considers approaches to minimize negative impacts in applications or in framing results in reporting

Many of the ASA GL elements describe ethical responsibilities relating to communication. These highlight the relationship between communication and explainability. Simply reporting results may not be sufficient (e.g., as outlined in ASA D6). The impact of the communication must also be considered by the ethical practitioner.

ACM

2.5 Give comprehensive and thorough evaluations of computer systems and their impacts, including analysis of possible risks. Computing professionals are in a position of trust, and therefore have a special responsibility to provide objective, credible evaluations and testimony to employers, employees, clients, users, and the public. Computing professionals should strive to be perceptive, thorough, and objective when evaluating, recommending, and presenting system descriptions and alternatives. Extraordinary care should be taken to identify and mitigate potential risks in machine learning systems. A system for which future risks cannot be reliably predicted requires frequent reassessment of risk as the system evolves in use, or it should not be deployed. Any issues that might result in major risk must be reported to appropriate parties.

D.5 Communicate bias

Have we communicated the shortcomings, limitations, and biases of the model to relevant stakeholders in ways that can be generally understood?

ASA

As noted in the discussion around the ASA GLs that support answering DSEC D4, communication is an important consideration for the ethical statistical practitioner. The communication of bias is a component of many of the ASA GL Principles, with responsibilities that include the following.

The ethical statistical practitioner:

A2 Uses methodology and data that are valid, relevant, and appropriate, without favoritism or prejudice, and in a manner intended to produce valid, interpretable, and reproducible results.

B1 Communicates data sources and fitness for use, including data generation and collection processes and known biases. Discloses and manages any conflicts of interest relating to the data sources. Communicates data processing and transformation procedures, including missing data handling.

B2 Is transparent about assumptions made in the execution and interpretation of statistical practices including methods used, limitations, possible sources of error, and algorithmic biases. Conveys results or applications of statistical practices in ways that are honest and meaningful.

B3 Communicates the stated purpose and the intended use of statistical practices. Is transparent regarding a priori versus post hoc objectives and planned versus unplanned statistical practices. Discloses when multiple comparisons are conducted, and any relevant adjustments.

C2 Regardless of personal or institutional interests or external pressures, does not use statistical practices to mislead any stakeholder.

D6 Considers the impact of statistical practice on society, groups, and individuals. Recognizes that statistical practice could adversely affect groups or the public perception of groups, including marginalized groups. Considers approaches to minimize negative impacts in applications or in framing results in reporting.

D10 Understands the provenance of the data, including origins, revisions, and any restrictions on usage, and fitness for use prior to conducting statistical practices.

E4 Avoids compromising validity for expediency. Regardless of pressure on or within the team, does not use inappropriate statistical practices

Note that the responsibilities relating to communication of bias include responsibilities to resist pressure (C2, E4) to either use inappropriate methods (or methods that result in undue bias), or to fail to communicate the bias inherent in a model clearly, and to all relevant stakeholders.

ACM

Again, since modeling may not seem a core part of the computing professional's work, practitioners may not consider the importance of effectively communicating the bias that does or might result from the application or deployment of a model. The ACM CE guidance is focused on communication "to employers, employees, clients, users, and the public", as outlined in ACM CE 2.5:

2.5 Give comprehensive and thorough evaluations of computer systems and their impacts, including analysis of possible risks. Computing professionals are in a position of trust, and therefore have a special responsibility to provide objective, credible evaluations and testimony to employers, employees, clients, users, and the public. Computing professionals should strive to be perceptive, thorough, and objective when evaluating, recommending, and presenting system descriptions and alternatives. Extraordinary care should be taken to identify and mitigate potential risks in machine learning systems. A system for which future risks cannot be reliably predicted requires frequent reassessment of risk as the system evolves in use, or it should not be deployed. Any issues that might result in major risk must be reported to appropriate parties.

	E.1 Redress: Have we discussed with our organization a plan for response if users are harmed by the results (e.g., how does the data science team evaluate these cases and update analysis and models to prevent future harm)?	E.2 Roll back: Is there a way to turn off or roll back the model in production if necessary?	E.3 Concept drift: Do we test and monitor for concept drift to ensure the model remains fair over time?	E.4 Unintended use: Have we taken steps to identify and prevent unintended uses and abuse of the model and do we have a plan to monitor these once the model is deployed?
ASA A	A2, A3, A4, A9	A3, A4, A9	A2, A3	A2, A3, A4, A9
ASA B	B5	B5, B6	B1, B2, B5, B6	B3, B5, B6
ASA C	C2, C4	C2		C7
ASA D	D1, D4, D5, D6, D7, D8			D; D4, D5, D6, D7, D10, D11
ASA E		E4		E4
ASA F		F2		
ASA G		G5	G5	G5
ASA H		H2		H2
ACM 2	2.5	2.5, 2.9	2.5	2.5, 2.9

Table 2.1.5. *DSEC E. Deployment*

Observations about Table 2.1.5. Like DSEC Area A (data collection), all eight of the ASA Ethical Guideline Principles are relevant for addressing the four questions in DSEC Area E. In fact, all eight Principles offer guidance specifically with respect to DSEC E.4, "unintended use". The only other DSEC item with this same level of coverage by the ASA GLs is DSEC A.1, "informed consent". Also noteworthy is that all the ACM guidance comes from Principle 2 (professional responsibilities), mostly with respect to ACM CE 2.5 ("2.5 Give comprehensive and thorough evaluations of computer systems and their impacts, including analysis of possible risks"), but also with contributions from ACM CE 2.9 ("2.9 Design and implement systems that are robustly and usably secure. Breaches of computer security cause harm").

E.1 Redress

Have we discussed with our organization a plan for response if users are harmed by the results (e.g., how does the data science team evaluate these cases and update analysis and models to prevent future harm)?

ASA

In many cases, once a statistical analysis is complete, the results are then transmitted or shared. The "user" of the results of statistical work may be different from the user of the results of data science work, and the user of the computing professional's work. Moreover, whenever quantitative work is utilized for scientific research, the results are published (often) – greatly multiplying the "users" of that work. It can be very difficult to identify exactly who the "user" is when it comes to the "harm" referred to in DSEC E.1. However, there are aspects of the ASA Ethical Guidelines that implicitly refer to limiting the potential harms to users of statistical and data science practice. For example,

The ethical statistical practitioner:

A2 Uses methodology and data that are valid, relevant, and appropriate, without favoritism or prejudice, and in a manner intended to produce valid, interpretable, and reproducible results.

A3 Does not knowingly conduct statistical practices that exploit vulnerable populations or create or perpetuate unfair outcomes.

A4 Opposes efforts to predetermine or influence the results of statistical practices and resists pressure to selectively interpret data.

A9 Takes appropriate action when aware of deviations from these guidelines by others.

Note that ASA A2 requires that the end user's purposes – whatever they may be – will only be based on valid, relevant, and appropriate methodology and will be applied "in a manner intended to produce valid, interpretable, and reproducible results." If someone takes those ethically generated outputs and uses them to harm others, this is beyond the scope of the ethical statistical practitioner's responsibility. However, ASA A3 charges the ethical practitioner with the responsibility of not knowingly engaging in practice that creates or perpetuates unfair outcomes, and ASA A9 creates the responsibility to take some action when the ethical practitioner becomes aware of deviations from these guidelines – by other statistical practitioners or by users.

Additional ASA Guidance includes elements from Principles B, C and D. ASA B and C elements relate to what to do after the work is done, so that users will not be misled by the work (C2), or use it without knowing there might be (C4) or actually was (B5) a substantive error. By contrast, ASA D is concerned with responsibilities to assure that harms do not accrue to those whose data are then utilized.

The ethical statistical practitioner:

B5 Strives to promptly correct substantive errors discovered after publication or implementation. As appropriate, disseminates the correction publicly and/or to others relying on the results.

C2 Regardless of personal or institutional interests or external pressures, does not use statistical practices to mislead any stakeholder.

C4 Informs stakeholders of the potential limitations on use and re-use of statistical practices in different contexts and offers guidance and alternatives, where appropriate, about scope, cost, and precision considerations that affect the utility of the statistical practice.

D1 Keeps informed about and adheres to applicable rules, approvals, and guidelines for the protection and welfare of human and animal subjects. Knows when work requires ethical review and oversight.

D4 Protects people's privacy and the confidentiality of data concerning them, whether obtained from the individuals directly, other persons, or existing records. Knows and adheres to applicable rules, consents, and guidelines to protect private information.

D5 Uses data only as permitted by data subjects' consent when applicable or considering their interests and welfare when consent is not required. This includes primary and secondary uses, use of repurposed data, sharing data, and linking data with additional data sets.

D7 Refrains from collecting or using more data than is necessary. Uses confidential information only when permitted and only to the extent necessary. Seeks to minimize the risk of re-identification when sharing de-identified data or results where there is an expectation of confidentiality. Explains any impact of de-identification on accuracy of results.

D8 To maximize contributions of data subjects, considers how best to use available data sources for exploration, training, testing, validation, or replication as needed for the application. The ethical statistical practitioner appropriately discloses how the data is used for these purposes and any limitations.

ACM

Like the ASA Guidance, the ACM guidance charges the ethical practitioner with looking ahead (to predict future risks) and to thoroughly and objectively evaluate systems once created to identify and mitigate risks of harm. ACM 2.5 goes further than any of the ASA elements by suggesting that, if future risks cannot be predicted or controlled via frequent monitoring, then the system should not be deployed.

> **2.5 Give comprehensive and thorough evaluations of computer systems and their impacts, including analysis of possible risks**. Computing professionals are in a position of trust, and therefore have a special responsibility to provide objective, credible evaluations and testimony to employers, employees, clients, users, and the public. Computing professionals should strive to be perceptive, thorough, and objective when evaluating, recommending, and presenting system descriptions and alternatives. Extraordinary care should be taken to identify and mitigate potential risks in machine learning systems. A system for which future risks cannot be reliably predicted requires frequent reassessment of risk as the system evolves in use, or it should not be deployed. Any issues that might result in major risk must be reported to appropriate parties.

E.2 Roll back

Is there a way to turn off or roll back the model in production if necessary?

ASA

Again, once an analysis is completed, it or its results may become part of the scientific record, making it impossible to "roll back". Another consideration for the ethical statistical practitioner and data scientist is that sometimes "users" want the results to turn out a specific way, or support a specific outcome. This is blatantly unethical and is addressed by A4, C2, E4, and H2 (i.e., across many of the ASA Principles). Thus, the conditions under which it might be, become, or only *seem* to be necessary to "turn off or roll back" a model need to be carefully considered, in order to determine whether there is an actual and legitimate need to turn it off or roll it back. Recognizing that this determination is essential to acting on DSEC E.2, the ASA Ethical Guidelines have several relevant elements to consider.

The ethical statistical practitioner:

A3 Does not knowingly conduct statistical practices that exploit vulnerable populations or create or perpetuate unfair outcomes.

A4 Opposes efforts to predetermine or influence the results of statistical practices and resists pressure to selectively interpret data.

A9 Takes appropriate action when aware of deviations from these guidelines by others.

C2 Regardless of personal or institutional interests or external pressures, does not use statistical practices to mislead any stakeholder

H2 Avoids condoning or appearing to condone statistical, scientific, or professional misconduct. Encourages other practitioners to avoid misconduct or the appearance of misconduct.

Statistical practitioners create models and methods that may be used in inappropriate and unintended ways (see NASEM 2017). In the event that a methodological error occurs (this would definitely qualify as a situation where roll back or stopping a model is "necessary"):

B5. The ethical statistical practitioner strives to promptly correct any errors discovered while producing the final report or after publication. As appropriate, disseminates the correction publicly or to others relying on the results.

Furthermore, Principle B describes full and transparent reporting about data and methods to promote valid and reliable interpretations. Again, it is unethical to withhold, or fail to report results, because they are inconvenient or even may be costly or cause costly decisions to be faced (see, e.g., A4 and E4). However, in the normal course of work, there is a clear need for evaluating a system, model, or analysis that is utilized repeatedly. Specifically,

The ethical statistical practitioner:

B6 For models and algorithms designed to inform or implement decisions repeatedly, develops and/or implements plans to validate assumptions and assess performance over time, as needed.

Additional relevant guidance comes from ASA GL Principles relating to team members (ASA Principle E), other practitioners and the Profession (Principle F), and to those in leadership roles (Principle G).

The ethical statistical practitioner:

E4 Avoids compromising validity for expediency. Regardless of pressure on or within the team, does not use inappropriate statistical practices.

F2 Helps strengthen, and does not undermine, the work of others through appropriate peer review or consultation. Provides feedback or advice that is impartial, constructive, and objective.

G5 Establish a culture that values validation of assumptions, and assessment of model/algorithm performance over time and across relevant subgroups, as needed. Communicate with relevant stakeholders regarding model or algorithm maintenance, failure, or actual or proposed modifications.

ACM

As we have seen, ACM Professional Responsibilities (Principle 2) charges the ethical practitioner with thinking ahead to risks of harm with any system. This arises in evaluating systems as well as designing them.

2.5 Give comprehensive and thorough evaluations of computer systems and their impacts, including analysis of possible risks. Computing professionals are in a position of trust, and therefore have a special responsibility to provide objective, credible evaluations and testimony to employers, employees, clients, users, and the public. Computing professionals should strive to be perceptive, thorough, and objective when evaluating, recommending, and presenting system descriptions and alternatives. Extraordinary care should be taken to identify and mitigate potential risks in machine learning systems. A system for which future risks cannot be reliably predicted requires frequent reassessment of risk as the system evolves in use, or it should not be deployed. Any issues that might result in major risk must be reported to appropriate parties.

2.9 Design and implement systems that are robustly and usably secure. Breaches of computer security cause harm. Robust security should be a primary consideration when designing and implementing systems.

Computing professionals should perform due diligence to ensure the system functions as intended, and take appropriate action to secure resources against accidental and intentional misuse, modification, and denial of service. As threats can arise and change after a system is deployed, computing professionals should integrate mitigation techniques and policies, such as monitoring, patching, and vulnerability reporting. Computing professionals should also take steps to ensure parties affected by data breaches are notified in a timely and clear manner, providing appropriate guidance and remediation. To ensure the system achieves its intended purpose, security features should be designed to be as intuitive and easy to use as possible. Computing professionals should discourage security precautions that are too confusing, are situationally inappropriate, or otherwise inhibit legitimate use.

In cases where misuse or harm are predictable or unavoidable, the best option may be to not implement the system.

Both ACM 2.5 and ACM 2.9 specify that, if the risks of harm cannot be reasonably predicted (sufficiently to mitigate them), then the system should not be implemented. This is not a "roll back", as suggested in DSEC E.2 but a preemptive non-deployment instead. Thus, the DSEC suggestion is to plan for a potential roll back, while the ACM guidance suggests planning for non-deployment (making a roll back unnecessary).

E.3 Concept drift

Do we test and monitor for concept drift to ensure the model remains fair over time?

ASA

An important aspect of DSEC E.3 is "concept drift", which generally means that the idea – typically contained within the target or to-be-predicted variable – may change over time. In machine learning and computing applications, the definition of drift is based on the utility of the target variable or what is being predicted or optimized. When that changes, the model or method built to predict, target, optimize or otherwise function with respect to that target variable will be or become incorrect. An example is the use of particular words: if an artificial intelligence system was developed to count instances in natural language of the word "nice" in the 1200s, that system might have been built to capture insults: in the 1200s the word "nice" meant *ignorant*. By the mid-1700s the word had move away from negative connotations and by the mid-1800s it was a compliment.[14] The system would still work to identify every use of the word "nice", but the original purpose of the model would no longer be served. This sort of socially driven drift is slow. Other types of drift can be abrupt, for example, when how a construct may actually change meaning because new information emerges. For example, on 24 August 2006, the International Astronomical Union announced that what had been "known" to be the furthest-out planet in Earth's solar system, Pluto, since 1930, had been re-classified as a "dwarf planet" (not a 'real' planet'). As of that date, all of the facts about the solar system changed (one fewer planet, a new dwarf planet member, etc.)[15] Note that these types of drift make the model less useful – and so ongoing assessment is always needed to assure that the model is doing what it was designed to do.

When a statistical analysis or model is planned or executed, as long as appropriate methodology and data fit for purpose are utilized (i.e., following ASA Ethical Guideline Principle A and A2), concept drift is only relevant in the sense that the data, and analysis of that data at that time, are fixed in time and this needs to be recognized by later users of the data/analysis. Following ASA Principles A and B is typically sufficient in many cases. In particular,

The ethical statistical practitioner:

A2 Uses methodology and data that are valid, relevant, and appropriate, without favoritism or prejudice, and in a manner intended to produce valid, interpretable, and reproducible results.
A3 Does not knowingly conduct statistical practices that exploit vulnerable populations or create or perpetuate unfair outcomes.

B1 Communicates data sources and fitness for use, including data generation and collection processes and known biases. Discloses and manages any conflicts of interest relating to the data sources. Communicates data processing and transformation procedures, including missing data handling.
B2 Is transparent about assumptions made in the execution and interpretation of statistical practices including methods used, limitations, possible sources of error, and algorithmic biases. Conveys results or applications of statistical practices in ways that are honest and meaningful.

---

[14] For the history of the word "nice", see https://www.etymonline.com/word/nice
[15] 15-year anniversary article about Pluto's demotion: https://earthsky.org/human-world/pluto-dwarf-planet-august-24-2006/

However, ASA B5 and B6 offer guidance in terms of responsibilities to follow up on that originally ethical work, particularly when it comes to concept drift that causes an erroneous output or unfair/biased results:

The ethical statistical practitioner:

> B5 Strives to promptly correct substantive errors discovered after publication or implementation. As appropriate, disseminates the correction publicly and/or to others relying on the results.
>
> B6 For models and algorithms designed to inform or implement decisions repeatedly, develops and/or implements plans to validate assumptions and assess performance over time, as needed. Considers criteria and mitigation plans for model or algorithm failure and retirement.

ACM

As we have seen the ACM guidance on DSEC E.3 is mainly with respect to evaluation of systems; the potential for drift should be considered because the user of the system will need it to stay current for useable outputs. The potential for the system to become biased or unfair over time due to concept drift must also be considered a risk to be mitigated or else the drift/potential for drift might lead to the decision not to deploy the system if this risk is too great or cannot be mitigated satisfactorily.

> **2.5 Give comprehensive and thorough evaluations of computer systems and their impacts, including analysis of possible risks**. Specifically, "Computing professionals should strive to be perceptive, thorough, and objective when evaluating, recommending, and presenting system descriptions and alternatives. A system for which future risks cannot be reliably predicted requires frequent reassessment of risk as the system evolves in use, or it should not be deployed. Any issues that might result in major risk must be reported to appropriate parties."

When these evaluations are thorough, models and systems that are used repeatedly over time will be assessed for their utility and fitness for purpose over time, as well as for risks of bias and unfair results.

E.4 Unintended use

Have we taken steps to identify and prevent unintended uses and abuse of the model and do we have a plan to monitor these once the model is deployed?

NOTE: It is imperative for readers (and practitioners) to recognize that the way the ASA and ACM respond to a need to "prevent unintended uses and abuses of the model" arises from the obligation to assure that data are only used in ways that the donor intended, that models are only used in ways that are appropriate, fair, and unbiased, and that unauthorized uses of either data or models (certainly both) are viewed as abuses and violations of ethical practice standards.

ASA

As discussed with others of the DSEC Area E considerations, statistical practitioners create models and methods that may be used in inappropriate and unintended ways (see NASEM 2017) after the statistical practitioner is not, or no longer, involved. As noted earlier, every one of the ASA Ethical Guideline Principles has relevant guidance for addressing DSEC E.4, but this is because use or abuse of what was originally ethical statistical practice is so damaging to statistics, data science, the public trust, and also to the trust of data donors.

The ethical statistical practitioner:

A2 Uses methodology and data that are valid, relevant, and appropriate, without favoritism or prejudice, and in a manner intended to produce valid, interpretable, and reproducible results.

A3 Does not knowingly conduct statistical practices that exploit vulnerable populations or create or perpetuate unfair outcomes.

A4 Opposes efforts to predetermine or influence the results of statistical practices and resists pressure to selectively interpret data.

A9 Takes appropriate action when aware of deviations from these guidelines by others.

B3 Communicates the stated purpose and the intended use of statistical practices. Is transparent regarding a priori versus post hoc objectives and planned versus unplanned statistical practices. Discloses when multiple comparisons are conducted, and any relevant adjustments.

B5 Strives to promptly correct substantive errors discovered after publication or implementation. As appropriate, disseminates the correction publicly and/or to others relying on the results.

B6 For models and algorithms designed to inform or implement decisions repeatedly, develops and/or implements plans to validate assumptions and assess performance over time, as needed. Considers criteria and mitigation plans for model or algorithm failure and retirement.

C7 Understands and conforms to confidentiality requirements for data collection, release, and dissemination and any restrictions on its use established by the data provider (to the extent legally required). Protects the use and disclosure of data accordingly. Safeguards privileged information of the employer, client, or funder.

In addition to the elements of ASA Principles listed above, responsibilities to research subjects, data subjects, or those directly affected by statistical practices (ASA Principle D) reflects the ethical obligation of the practitioner to prevent uses of data that are unintended by the data donor. Uses or reports of the results of statistical practices that are unintended by the statistical practitioner specifically represent "detrimental practices" that the NASEM 2017 report warns against. Similarly, uses of data that the data donor did not intend are unethical.

> Principle D: The ethical statistical practitioner does not misuse or condone the misuse of data. They protect and respect the rights and interests of human and animal subjects. These responsibilities extend to those who will be directly affected by statistical practices.

The ethical statistical practitioner:

> D4 Protects people's privacy and the confidentiality of data concerning them, whether obtained from the individuals directly, other persons, or existing records. Knows and adheres to applicable rules, consents, and guidelines to protect private information.
>
> D5 Uses data only as permitted by data subjects' consent when applicable or considering their interests and welfare when consent is not required. This includes primary and secondary uses, use of repurposed data, sharing data, and linking data with additional data sets.
>
> D6 Considers the impact of statistical practice on society, groups, and individuals. Recognizes that statistical practice could adversely affect groups or the public perception of groups, including marginalized groups. Considers approaches to minimize negative impacts in applications or in framing results in reporting.
>
> D7 Refrains from collecting or using more data than is necessary. Uses confidential information only when permitted and only to the extent necessary. Seeks to minimize the risk of re-identification when sharing de-identified data or results where there is an expectation of confidentiality. Explains any impact of de-identification on accuracy of results.
>
> D10 Understands the provenance of the data, including origins, revisions, and any restrictions on usage, and fitness for use prior to conducting statistical practices.
>
> D11 Does not conduct statistical practice that could reasonably be interpreted by subjects as sanctioning a violation of their rights. Seeks to use statistical practices to promote the just and impartial treatment of all individuals.

E4 Avoids compromising validity for expediency. Regardless of pressure on or within the team, does not use inappropriate statistical practices

H2 Avoids condoning or appearing to condone statistical, scientific, or professional misconduct. Encourages other practitioners to avoid misconduct or the appearance of misconduct.

Additionally, to prevent statisticians from working on data that others may want (or try to bring pressure to bear) to apply unintended methods to data or models, another element from Principle G, Responsibilities of leaders, supervisors, and mentors in statistical practice, is relevant for DSEC E.4:

Those leading, supervising, or mentoring statistical practitioners are expected to:

G5 Establish a culture that values validation of assumptions, and assessment of model/algorithm performance over time and across relevant subgroups, as needed. Communicate with relevant stakeholders regarding model or algorithm maintenance, failure, or actual or proposed modifications. This is an analog to ASA A9 ("Takes appropriate action when aware of deviations from these guidelines by others").

ACM

The main guiding principle of the ACM CE is to avoid harm. Ethical computing professionals follow all the CE Principles (ACM 4.1), and when they violate any of the CE elements. Any computing professional who is asked to use or misuse a system (or data) should follow ACM 2.5 to determine whether or not there are any risks – including, the risk that their involvement on a project is actually an unintended use (i.e., misuse and potentially an abuse).

2.5 Give comprehensive and thorough evaluations of computer systems and their impacts, including analysis of possible risks. Computing professionals are in a position of trust, and therefore have a special responsibility to provide objective, credible evaluations and testimony to employers, employees, clients, users, and the public. Computing professionals should strive to be perceptive, thorough, and objective when evaluating, recommending, and presenting system descriptions and alternatives. Extraordinary care should be taken to identify and mitigate potential risks in machine learning systems. A system for which future risks cannot be reliably predicted requires frequent reassessment of risk as the system evolves in use, or it should not be deployed. Any issues that might result in major risk must be reported to appropriate parties.

2.9 Design and implement systems that are robustly and usably secure. Breaches of computer security cause harm. Robust security should be a primary consideration when designing and implementing systems. Computing professionals should perform due diligence to ensure the system functions as intended, and take appropriate action to secure resources against accidental and intentional misuse, modification, and denial of service. As threats can arise and change after a system is deployed, computing professionals should integrate mitigation techniques and policies, such as monitoring, patching, and vulnerability reporting. Computing professionals should also take steps to ensure parties affected by data breaches are notified in a timely and clear manner, providing appropriate guidance and remediation. To ensure the system achieves its intended purpose, security features should be designed to be as intuitive and easy to use as possible. Computing professionals should discourage security precautions that are too confusing, are situationally inappropriate, or otherwise inhibit legitimate use.

In cases where misuse or harm are predictable or unavoidable, the best option may be to not implement the system.

Notes

Important note about ACM & DSEC: in several Principles, ACM CE states that, in the event that a system cannot be deployed in a manner that avoids harms that can be predicted, then that system should not be deployed or developed. However, the DSEC questions suggest that, even if harms can be predicted, plans for remediation or rolling back are sufficient – but that those models should in fact *be deployed*, or that their deployment should not be objected to/protested or stopped. This is another example of how project-specific lists like the DSEC and DEFW differ from the guidance that comes from the professional practice standards.

Important note about ASA & DSEC: statistical practitioners and data scientists may create statistical/analytic models and methods, and ASA GLs encourage the full and transparent reporting and sharing of these innovations. However, statistics in particular is a field from which excellent methods are taken and misused (Stark & Saltelli, 2018) by others. Even when misuse is not purposefully deceptive, the damage and harms that detrimental methods create are real, and accrue to science, the scientific community, and to the public trust, rather than to specific groups or populations of individuals. Because of the near-ubiquity of both statistics, statistical, and computational methods, the monitoring of "unintended uses and abuse" of these innovations is essentially impossible. While it is wholly predictable that "miscreant individuals" (NASEM, 2017 p. 176) will purposefully misuse and abuse both statistical and computational methods, by the time the miscreant individuals act to misuse a system, model, or data, the ethical practitioner might be unable to follow the ACM admonition that "In cases where misuse or harm are predictable or unavoidable, the best option may be to not implement the system." However, what every practitioner *can* do – and is in fact ethically obliged to do - is ensure that they do not contribute to any misuse or abuse of any system or model.

We have seen that DSEC areas and specific questions overlap with multiple elements of the ASA and ACM ethical practice standards. That is, if the practitioner needs to utilize – or answer questions from – the DSEC, they can use any/all of the ethical practice standard elements identified in order to do so. The reader may have identified other codes, or other aspects of the ASA and ACM reference documents, to utilize. The next section explores how these guidance documents can be similarly utilized to address the seven Principles outlined in the 2018 Data Ethics Framework (DEFW).

DEFW (2018) with ASA and ACM

Like the DSEC, the DEFW (2018) worksheet asks the practitioner to consider several questions that can enable them to follow each of the seven DEFW Principles. The reader may have other policy guidance, or other codes, to follow, but any statistical practitioner, data scientist, and computing professional can utilize the ASA and ACM ethical practice standards. We have seen the table below earlier, in Chapter 1.6. In this section we will explore the alignment more closely in order to better understand the DEFW Principles and how the ASA and ACM guidance resource can be used to answer the questions posed in the DEFW worksheet (2018).

	Principle 1. Start with clear user need and public benefit	Principle 2. Be aware of relevant legislation and codes of practice	Principle 3. Use data that is proportionate to the user need	Principle 4. Understand the limitations of the data	Principle 5. Ensure robust practices and work within your skillset	Principle 6. Make your work transparent and be accountable	Principle 7. Embed data use responsibly
			ASA PRINCIPLES HAVING ELEMENTS IN ALIGNMENT WITH DEFW				
ASA A	A2, A3, A4	A11	A2	A2	A1, A2	A2, A5, A6	A2, A4
ASA B	B1	B4	B3, B6	B1-B4, B6	B1, B2, B4, B6	B1-B4, B6	B1-B4, B6
ASA C	C1	C7	C4	C3	C1, C3	C1, C2, C4, C5	C1-C4, C7, C8
ASA D	D5	D1, D4, D9	D2, D3, D7, D8	D4, D5, D10		D6	D1, D2, D3, D11
ASA E		E1			E4	E3, E4	E4
ASA F							F5
ASA G						G5	G5
ASA H		H1					H2
			ACM PRINCIPLES HAVING ELEMENTS IN ALIGNMENT WITH DEFW				
ACM 1			1.6	1.3	1.3	1.2	
ACM 2		2.3		2.5	2.5, 2.6	2.1, 2.4	
ACM 3							3.6, 3.7

Table 2.1.6. *DEFW alignment with ASA and ACM*

Observations about Table 2.1.6. One Principle from the ASA GL and one from the ACM CE are not aligned with any of the DEFW principles. These relate to engagement with other practitioners (ASA Principle F) or the code of ethics (ACM Principle 4), although DEFW does include Principle 2, "Be aware of relevant legislation and codes of practice".

DEFW Principle 1 - Start with clear user need and public benefit

Describe the user need.

Does everyone in the team understand the user need?
How does this benefit the public?
What would be the harm in not using data science - what needs might not be met?
Do you have supporting evidence for the approach being likely to meet a user need or provide public benefit?

NOTE: Neither the ASA nor the ACM articulates any obligation for practitioners to consider "public benefit". Since the DEFW (2018) is the output of a government body (in the United Kingdom), their uses of data *must* be founded on public need or at least a benefit to the public that results from the work.

ASA

It is impossible for a statistical project to be planned/executed without a clear need articulated. However, the ASA GLs describe how the ethical statistical practitioner operates with this need in mind.

The ethical statistical practitioner:

A2 Uses methodology and data that are valid, relevant, and appropriate, without favoritism or prejudice, and in a manner intended to produce valid, interpretable, and reproducible results.
A3 Does not knowingly conduct statistical practices that exploit vulnerable populations or create or perpetuate unfair outcomes.
A4 Opposes efforts to predetermine or influence the results of statistical practices and resists pressure to selectively interpret data.

Note that ASA A3 implies a need to balance the "clear user need" and/or "public benefit" against the potential to "exploit vulnerable populations or create or perpetuate unfair outcomes."

B1 Communicates data sources and fitness for use, including data generation and collection processes and known biases. Discloses and manages any conflicts of interest relating to the data sources. Communicates data processing and transformation procedures, including missing data handling.

C1 Seeks to establish what stakeholders hope to obtain from any specific project. Strives to obtain sufficient subject-matter knowledge to conduct meaningful and relevant statistical practice.

D5 Uses data only as permitted by data subjects' consent when applicable or considering their interests and welfare when consent is not required. This includes primary and secondary uses, use of repurposed data, sharing data, and linking data with additional data sets.

ASA guidance assures that fitness for use (of the data, B1), recognition of what the user actually wants to achieve with the work (C1), and data donor knowledge and intent (D5) are recognized as components at the start of any project (as well as the need).

ACM

Just as it is for the statistical practitioner, it is unlikely that any computing project can be executed without a clear need, so the ACM CE does not (need to) refer to ensuring that user need is understood.

DEFW Principle 2 - Be aware of relevant legislation and codes of practice

List the pieces of legislation, codes of practice and guidance that apply to your project.

Do all team members understand how relevant laws apply to the project?
If necessary, have you consulted with relevant experts?
Have you spoken to your information assurance team?
If using personal data, do you understand your obligations under data protection legislation?
Do you have plans in place to handle any potential security breach?

ASA

Statistical practice that follows the ASA Ethical Guidelines will naturally represent these professional practice standards. However, in addition to the code of practice and guidance, DEFW (and the ASA & ACM) charge the practitioner with knowing local laws and regulations as well. Specifically,

The ethical statistical practitioner:

A11 Follows applicable policies, regulations, and laws relating to their professional work, unless there is a compelling ethical justification to do otherwise.

B4 Meets obligations to share the data used in the statistical practices, for example, for peer review and replication, as allowable. Respects expectations of data contributors when using or sharing data. Exercises due caution to protect proprietary and confidential data, including all data that might inappropriately harm data subjects.

C7 Understands and conforms to confidentiality requirements for data collection, release, and dissemination and any restrictions on its use established by the data provider (to the extent legally required). Protects the use and disclosure of data accordingly. Safeguards privileged information of the employer, client, or funder.

D1 Keeps informed about and adheres to applicable rules, approvals, and guidelines for the protection and welfare of human and animal subjects. Knows when work requires ethical review and oversight.

D4 Protects people's privacy and the confidentiality of data concerning them, whether obtained from the individuals directly, other persons, or existing records. Knows and adheres to applicable rules, consents, and guidelines to protect private information.

D9 Knows the legal limitations on privacy and confidentiality assurances and does not over-promise or assume legal privacy and confidentiality protections where they may not apply.

E1 Recognizes and respects that other professions may have different ethical standards and obligations. Dissonance in ethics may still arise even if all members feel that they are working towards the same goal. It is essential to have a respectful exchange of views.

H1 Knows the definitions of and procedures relating to misconduct in their institutional setting. Seeks to clarify facts and intent before alleging misconduct by others. Recognizes that differences of opinion and honest error do not constitute unethical behavior.

Note that, while the DEFW Principle is straightforward, the responsibilities in the ASA Ethical Guidelines span six different Principles, and describe ethical practice from diverse perspectives (from professional integrity to misconduct).

ACM

The ACM has similar guidance to the ASA and to DEFW:

2.3 Know and respect existing rules pertaining to professional work. "Rules" here include local, regional, national, and international laws and regulations, as well as any policies and procedures of the organizations to which the professional belongs. Computing professionals must abide by these rules unless there is a compelling ethical justification to do otherwise. Rules that are judged unethical should be challenged. A rule may be unethical when it has an inadequate moral basis or causes recognizable harm. A computing professional should consider challenging the rule through existing channels before violating the rule. A computing professional who decides to violate a rule because it is unethical, or for any other reason, must consider potential consequences and accept responsibility for that action.

DEFW Principle 3 - Use data that is proportionate to the user need

Describe how the data being used is proportionate to the user need.

Could you clearly explain why you need to use this data to members of the public?
Does this use of data interfere with the rights of individuals?
If yes, is there a less intrusive way of achieving the objective?
Is there a fair balance between the rights of individuals and the interests of the community?
Has the data you're using been specifically provided for your analysis?
By using data that the public has freely volunteered, would your project jeopardise people providing this again in the future?
How can you meet the project aim using the minimum personal data possible?
Is there a way to achieve the same aim with less identifiable data?
Can you use synthetic data?
If using personal data is unavoidable, have you answered the questions for determining proportionality?
If using personal data identifying individuals, what measures are in place to control access? How widely are you searching personal data?

ASA

The ethical statistical practitioner:

A2 Uses methodology and data that are valid, relevant, and appropriate, without favoritism or prejudice, and in a manner intended to produce valid, interpretable, and reproducible results.

B3 Communicates the stated purpose and the intended use of statistical practices. Is transparent regarding a priori versus post hoc objectives and planned versus unplanned statistical practices. Discloses when multiple comparisons are conducted, and any relevant adjustments.

B6 For models and algorithms designed to inform or implement decisions repeatedly, develops and/or implements plans to validate assumptions and assess performance over time, as needed.

C4 Informs stakeholders of the potential limitations on use and re-use of statistical practices in different contexts and offers guidance and alternatives, where appropriate, about scope, cost, and precision considerations that affect the utility of the statistical practice.

D2 Makes informed recommendations for sample size and statistical practice methodology in order to avoid the use of excessive or inadequate numbers of subjects and excessive risk to subjects.

D3 For animal studies, seeks to leverage statistical practice to reduce the number of animals used, refine experiments to increase the humane treatment of animals, and replace animal use where possible.

D7 Refrains from collecting or using more data than is necessary. Uses confidential information only when permitted and only to the extent necessary. Seeks to minimize the risk of re-identification when sharing de-identified data or results where there is an expectation of confidentiality. Explains any impact of de-identification on accuracy of results.

D8 To maximize contributions of data subjects, considers how best to use available data sources for exploration, training, testing, validation, or replication as needed for the application. The ethical statistical practitioner appropriately discloses how the data is used for these purposes and any limitations.

As expected, guidance on addressing DEFW Principle 3 comes from multiple ASA GL Principles.

ACM

Guidance from the ACM on addressing DEFW Principle 3 is focused:

1.6: Computing professionals should only use personal information for legitimate ends and without violating the rights of individuals and groups. This requires taking precautions to prevent re- identification of anonymized data or unauthorized data collection, ensuring the accuracy of data, understanding the provenance of the data, and protecting it from unauthorized access and accidental disclosure. Computing professionals should establish transparent policies and procedures that allow individuals to understand what data is being collected and how it is being used, to give informed consent for automatic data collection, and to review, obtain, correct inaccuracies in, and delete their personal data.

DEFW Principle 3 NB: "**proportionate to the user need**" requires that "user need" is a formal component of your process (it is Principle 1 for DEFW) – ACM *implies* this principle, while ASA (A, B) specify that methods used and the data must be appropriate and "Fit for purpose" fit the need (of the research). DEFW Principle 3 is consistent with ACM perspective, but this specific articulation ("proportionate to user need") is missing from the ACM CE. Keep in mind also that the "user" here is the person(s) who will use the data that is collected. To minimize risk to the public, and to optimize the public benefit, the user (who wants the public's data) must have a clearly established *need*. The intention of DEFW Principle 3 is for those contemplating, designing, or executing a project to consider the trade-off between the user's need and how that can still represent a net benefit to the public -without unfairly biasing or harming others.

This Principle is aligned with elements of the ASA and ACM practice standards that focus on ensuring that data are obtained with consent (included in ASA D5, D6 and ACM 1.6), because *informed* consent requires that contributors of data are told the purposes for which their data are being collected. That way, data contributors can determine whether their contributions of data satisfy a need on the part of the data collector that is worth (offset by – i.e., proportionate to) any risk of harm/potential harm to themselves.

DEFW Principle 4 - Understand the limitations of the data

Identify the potential limitations of the data source(s) and how they are being mitigated.

What data source(s) is being used?
Are all metadata and field names clearly understood?
What processes do you have in place to ensure and maintain data integrity?
Is there a plan in place to identify errors and biases?

What are the caveats?

How will the caveats be taken into account for any future policy or service which uses this work as an evidence base?

ASA

Since statistical practice is always concerned with data, DEFW Principle 4 is strongly addressed by multiple Guideline Principles. For example,

The ethical statistical practitioner:

A2 Uses methodology and data that are valid, relevant, and appropriate, without favoritism or prejudice, and in a manner intended to produce valid, interpretable, and reproducible results.

B1 Communicates data sources and fitness for use, including data generation and collection processes and known biases. Discloses and manages any conflicts of interest relating to the data sources. Communicates data processing and transformation procedures, including missing data handling.

B2 Is transparent about assumptions made in the execution and interpretation of statistical practices including methods used, limitations, possible sources of error, and algorithmic biases. Conveys results or applications of statistical practices in ways that are honest and meaningful.

B3 Communicates the stated purpose and the intended use of statistical practices. Is transparent regarding a priori versus post hoc objectives and planned versus unplanned statistical practices. Discloses when multiple comparisons are conducted, and any relevant adjustments.

B4 Meets obligations to share the data used in the statistical practices, for example, for peer review and replication, as allowable. Respects expectations of data contributors when using or sharing data. Exercises due caution to protect proprietary and confidential data, including all data that might inappropriately harm data subjects.

B6 For models and algorithms designed to inform or implement decisions repeatedly, develops and/or implements plans to validate assumptions and assess performance over time, as needed. Considers criteria and mitigation plans for model or algorithm failure and retirement.

C3 Uses practices appropriate to exploratory and confirmatory phases of a project, differentiating findings from each so the stakeholders can understand and apply the results.

D4 Protects people's privacy and the confidentiality of data concerning them, whether obtained from the individuals directly, other persons, or existing records. Knows and adheres to applicable rules, consents, and guidelines to protect private information.

D5 Uses data only as permitted by data subjects' consent when applicable or considering their interests and welfare when consent is not required. This includes primary and secondary uses, use of repurposed data, sharing data, and linking data with additional data sets.

D10 Understands the provenance of the data, including origins, revisions, and any restrictions on usage, and fitness for use prior to conducting statistical practices.

If all of the ASA Principles are followed in response to DEFW Principle 4, then other elements relating to resisting pressure to misuse data (e.g., ASA A4), failures to communicate limitations of data or interpretations (e.g. ASA C2), and misconduct (e.g., ASA E4 and H2) will not need to be brought to bear.

ACM

1.3 Be honest and trustworthy. Honesty is an essential component of trustworthiness. A computing professional should be transparent and provide full disclosure of all pertinent system capabilities, limitations, and potential problems to the appropriate parties. Making deliberately false or misleading claims, fabricating or falsifying data, offering or accepting bribes, and other dishonest conduct are violations of the Code. Computing professionals should be honest about their qualifications, and about any limitations in their competence to complete a task. **Computing professionals should be forthright about any circumstances that might lead to either real or perceived conflicts of interest or otherwise tend to undermine the independence of their judgment.** Furthermore, commitments should be honored. Computing professionals should not misrepresent an organization's policies or procedures, and should not speak on behalf of an organization unless authorized to do so.

2.5 Give comprehensive and thorough evaluations of computer systems and their impacts, including analysis of possible risks. Computing professionals are in a position of trust, and therefore have a special responsibility to provide objective, credible evaluations and testimony to employers, employees, clients, users, and the public. Computing professionals should strive to be perceptive, thorough, and objective when evaluating, recommending, and presenting system descriptions and alternatives. Extraordinary care should be taken to identify and mitigate potential risks in machine learning systems. A system for which future risks cannot be reliably predicted requires frequent reassessment of risk as the system evolves in use, or it should not be deployed. Any issues that might result in major risk must be reported to appropriate parties.

DEFW Principle 5 - Ensure robust practices and work within your skillset

Explain the relevant expertise and approaches that are being employed to maximise the efficacy of the project. Describe the disciplines involved and why.

Is there expertise that the project requires that you don't currently have?
Have you designed the approach with the policy team or a subject matter expert?
Has all subject matter context, from policy experts or otherwise, been taken into account when determining the appropriate loss function for the model?
If necessary, how can you (or external scrutiny) check that the algorithm is achieving the right output decision when new data is added?
How has reproducibility been ensured? Could another analyst repeat your procedure based on your documentation?
How confident are you that the algorithm is robust, and that any assumptions are met?
What is the quality of the model outputs, and how does this stack up against the project objectives?
If using data about people, is it possible that a data science technique is basing analysis on proxies for protected variables which could lead to a discriminatory policy decision?

ASA

DEFW Principle 5 is addressed clearly by ASA Principle A, but Principles about data and methods (ASA B) and stakeholders (ASA C) are also informative.

The ethical statistical practitioner:

A1 Takes responsibility for evaluating potential tasks, assessing whether they have (or can attain) sufficient competence to execute each task and that the work and timeline are feasible. Does not solicit or deliver work for which they are not qualified or that they would not be willing to have peer reviewed.

A2 Uses methodology and data that are valid, relevant, and appropriate, without favoritism or prejudice, and in a manner intended to produce valid, interpretable, and reproducible results.

B1 Communicates data sources and fitness for use, including data generation and collection processes and known biases. Discloses and manages any conflicts of interest relating to the data sources. Communicates data processing and transformation procedures, including missing data handling.

B2 Is transparent about assumptions made in the execution and interpretation of statistical practices including methods used, limitations, possible sources of error, and algorithmic biases. Conveys results or applications of statistical practices in ways that are honest and meaningful.

B4 Meets obligations to share the data used in the statistical practices, for example, for peer review and replication, as allowable. Respects expectations of data contributors when using or sharing data. Exercises due caution to protect proprietary and confidential data, including all data that might inappropriately harm data subjects.

B6 For models and algorithms designed to inform or implement decisions repeatedly, develops and/or implements plans to validate assumptions and assess performance over time, as needed. Considers criteria and mitigation plans for model or algorithm failure and retirement.

C1 Seeks to establish what stakeholders hope to obtain from any specific project. Strives to obtain sufficient subject-matter knowledge to conduct meaningful and relevant statistical practice.

C3 Uses practices appropriate to exploratory and confirmatory phases of a project, differentiating findings from each so the stakeholders can understand and apply the results.

E4 Avoids compromising validity for expediency. Regardless of pressure on or within the team, does not use inappropriate statistical practices

ACM

1.3 Computing professionals should be honest about their qualifications, and about any limitations in their competence to complete a task. Computing professionals should be forthright about any circumstances that might lead to either real or perceived conflicts of interest or otherwise tend to undermine the independence of their judgment. Furthermore, commitments should be honored.

2.5 Give comprehensive and thorough evaluations of computer systems and their impacts, including analysis of possible risks. Computing professionals are in a position of trust, and therefore have a special responsibility to provide objective, credible evaluations and testimony to employers, employees, clients, users, and the public. Computing professionals should strive to be perceptive, thorough, and objective when evaluating, recommending, and presenting system descriptions and alternatives. Extraordinary care should be taken to identify and mitigate potential risks in machine learning systems. A system for which future risks cannot be reliably predicted requires frequent reassessment of risk as the system evolves in use, or it should not be deployed. Any issues that might result in major risk must be reported to appropriate parties.

2.6 **Perform work only in areas of competence.** A computing professional is responsible for evaluating potential work assignments. This includes evaluating the work's feasibility and advisability, and making a judgment about whether the work assignment is within the professional's areas of competence. If at any time before or during the work assignment the professional identifies a lack of a necessary expertise, they must disclose this to the employer or client. The client or employer may decide to pursue the assignment with the professional after additional time to acquire the necessary competencies, to pursue the assignment with someone else who has the required expertise, or to forgo the assignment. A computing professional's ethical judgment should be the final guide in deciding whether to work on the assignment.

DEFW Principle 6 - Make your work transparent and be accountable

Describe how you have considered making your work transparent and your team accountable.

Have you spoken to your organisation to find out if you can speak about your project openly?
Have you considered how both internal and external engagement could benefit your project?

ASA:

Transparency and accountability represent core aspects of responsibility throughout the ASA Ethical Guidelines. For example,

The ethical statistical practitioner:

A2 Uses methodology and data that are valid, relevant, and appropriate, without favoritism or prejudice, and in a manner intended to produce valid, interpretable, and reproducible results.
A5 Accepts full responsibility for their own work, does not take credit for the work of others, and gives credit to those who contribute. Respects and acknowledges the intellectual property of others.

A6 Strives to follow, and encourages all collaborators to follow, an established protocol for authorship. Advocates for recognition commensurate with each person's contribution to the work. Recognizes that inclusion as an author does imply, while acknowledgement may imply, endorsement of the work.

B1 Communicates data sources and fitness for use, including data generation and collection processes and known biases. Discloses and manages any conflicts of interest relating to the data sources. Communicates data processing and transformation procedures, including missing data handling.

B2 Is transparent about assumptions made in the execution and interpretation of statistical practices including methods used, limitations, possible sources of error, and algorithmic biases. Conveys results or applications of statistical practices in ways that are honest and meaningful.

B3 Communicates the stated purpose and the intended use of statistical practices. Is transparent regarding a priori versus post hoc objectives and planned versus unplanned statistical practices. Discloses when multiple comparisons are conducted, and any relevant adjustments.

B4 Meets obligations to share the data used in the statistical practices, for example, for peer review and replication, as allowable. Respects expectations of data contributors when using or sharing data. Exercises due caution to protect proprietary and confidential data, including all data that might inappropriately harm data subjects.

B6 For models and algorithms designed to inform or implement decisions repeatedly, develops and/or implements plans to validate assumptions and assess performance over time, as needed. Considers criteria and mitigation plans for model or algorithm failure and retirement.

C1 Seeks to establish what stakeholders hope to obtain from any specific project. Strives to obtain sufficient subject-matter knowledge to conduct meaningful and relevant statistical practice.

C2 Regardless of personal or institutional interests or external pressures, does not use statistical practices to mislead any stakeholder.

C4 Informs stakeholders of the potential limitations on use and re-use of statistical practices in different contexts and offers guidance and alternatives, where appropriate, about scope, cost, and precision considerations that affect the utility of the statistical practice.

C5 Explains any expected adverse consequences from failing to follow through on an agreed-upon sampling or analytic plan.

D6 Considers the impact of statistical practice on society, groups, and individuals. Recognizes that statistical practice could adversely affect groups or the public perception of groups, including marginalized groups. Considers approaches to minimize negative impacts in applications or in framing results in reporting.

E3 Ensures that all communications regarding statistical practices are consistent with these Guidelines. Promotes transparency in all statistical practices.
E4 Avoids compromising validity for expediency. Regardless of pressure on or within the team, does not use inappropriate statistical practices

G5 Establish a culture that values validation of assumptions, and assessment of model/algorithm performance over time and across relevant subgroups, as needed. Communicate with relevant stakeholders regarding model or algorithm maintenance, failure, or actual or proposed modifications.

Note that, while Principle G is concerned with "Responsibilities of leaders, supervisors, and mentors in statistical practice", each practitioner can contribute to the establishment of a culture that supports ethical statistical practice and following the Guidelines (ASA G5).

ACM

The ACM CE addresses DEFW Principle 6 across two of its four Principles:

1.2 **Avoid harm.** A computing professional has an additional obligation to report any signs of system risks that might result in harm. If leaders do not act to curtail or mitigate such risks, it may be necessary to "blow the whistle" to reduce potential harm. However, capricious or misguided reporting of risks can itself be harmful. Before reporting risks, a computing professional should carefully assess relevant aspects of the situation.

2.1 **Strive to achieve high quality in both the processes and products of professional work.** Computing professionals should insist on and support high-quality work from themselves and from colleagues. The dignity of employers, employees, colleagues, clients, users, and anyone else affected either directly or indirectly by the work should be respected throughout the process. Computing professionals should respect the right of those involved to transparent communication about the project. Professionals should be cognizant of any serious negative consequences affecting any stakeholder that may result from poor quality work and should resist inducements to neglect this responsibility.

2.4 **Accept and provide appropriate professional review.** High quality professional work in computing depends on professional review at all stages. Whenever appropriate, computing professionals should seek and utilize peer and stakeholder review. Computing professionals should also provide constructive, critical reviews of others' work.

DEFW Principle 7 - Embed data use responsibly

Describe the steps taken to ensure any insight is managed responsibly.

How many people will be affected by the new model, insight or service?
Who are the users of the insight, model, or new service?
Do users have the appropriate support and training to maintain the new technology?
Have future events been planned for?
Is your implementation plan correlated with the impact of a particular model?
How often will you report on these plans to Senior Responsible Officers?

ASA

Every one of the ASA Guideline Principles offers support to address DEFW Principle 7. For example,

The ethical statistical practitioner:

A2 Uses methodology and data that are valid, relevant, and appropriate, without favoritism or prejudice, and in a manner intended to produce valid, interpretable, and reproducible results.
A4 Opposes efforts to predetermine or influence the results of statistical practices and resists pressure to selectively interpret data.

B1 Communicates data sources and fitness for use, including data generation and collection processes and known biases. Discloses and manages any conflicts of interest relating to the data sources. Communicates data processing and transformation procedures, including missing data handling.
B2 Is transparent about assumptions made in the execution and interpretation of statistical practices including methods used, limitations, possible sources of error, and algorithmic biases. Conveys results or applications of statistical practices in ways that are honest and meaningful.
B3 Communicates the stated purpose and the intended use of statistical practices. Is transparent regarding a priori versus post hoc objectives and planned versus unplanned statistical practices. Discloses when multiple comparisons are conducted, and any relevant adjustments.

B4 Meets obligations to share the data used in the statistical practices, for example, for peer review and replication, as allowable. Respects expectations of data contributors when using or sharing data. Exercises due caution to protect proprietary and confidential data, including all data that might inappropriately harm data subjects.

B6 For models and algorithms designed to inform or implement decisions repeatedly, develops and/or implements plans to validate assumptions and assess performance over time, as needed. Considers criteria and mitigation plans for model or algorithm failure and retirement.

C1 Seeks to establish what stakeholders hope to obtain from any specific project. Strives to obtain sufficient subject-matter knowledge to conduct meaningful and relevant statistical practice.

C2 Regardless of personal or institutional interests or external pressures, does not use statistical practices to mislead any stakeholder.

C3 Uses practices appropriate to exploratory and confirmatory phases of a project, differentiating findings from each so the stakeholders can understand and apply the results.

C4 Informs stakeholders of the potential limitations on use and re-use of statistical practices in different contexts and offers guidance and alternatives, where appropriate, about scope, cost, and precision considerations that affect the utility of the statistical practice.

C7 Understands and conforms to confidentiality requirements for data collection, release, and dissemination and any restrictions on its use established by the data provider (to the extent legally required). Protects the use and disclosure of data accordingly. Safeguards privileged information of the employer, client, or funder.

C8 Prioritizes both scientific integrity and the principles outlined in these Guidelines when interests are in conflict.

D1 Keeps informed about and adheres to applicable rules, approvals, and guidelines for the protection and welfare of human and animal subjects. Knows when work requires ethical review and oversight.

D2 Makes informed recommendations for sample size and statistical practice methodology in order to avoid the use of excessive or inadequate numbers of subjects and excessive risk to subjects.

D3 For animal studies, seeks to leverage statistical practice to reduce the number of animals used, refine experiments to increase the humane treatment of animals, and replace animal use where possible.

D11 Does not conduct statistical practice that could reasonably be interpreted by subjects as sanctioning a violation of their rights. Seeks to use statistical practices to promote the just and impartial treatment of all individuals.

E4 Avoids compromising validity for expediency. Regardless of pressure on or within the team, does not use inappropriate statistical practices

F5 Serves as an ambassador for statistical practice by promoting thoughtful choices about data acquisition, analytic procedures, and data structures among non-practitioners and students. Instills appreciation for the concepts and methods of statistical practice.

G5 Establish a culture that values validation of assumptions, and assessment of model/algorithm performance over time and across relevant subgroups, as needed. Communicate with relevant stakeholders regarding model or algorithm maintenance, failure, or actual or proposed modifications.

H2 Avoids condoning or appearing to condone statistical, scientific, or professional misconduct. Encourages other practitioners to avoid misconduct or the appearance of misconduct.

As the extensive reflection of DEFW Principle 7 throughout the ASA Ethical Guidelines suggests, "responsible data use" is a critical aspect of ethical statistical practice.

ACM

While "data use" is not specifically a component of ethical computational practice like it clearly is for ethical statistical practice, there are still important aspects of the ACM CE that are relevant for addressing DEFW Principle 7. For example,

3.6 Use care when modifying or retiring systems. Interface changes, the removal of features, and even software updates have an impact on the productivity of users and the quality of their work. Leaders should take care when changing or discontinuing support for system features on which people still depend. Leaders should thoroughly investigate viable alternatives to removing support for a legacy system. If these alternatives are unacceptably risky or impractical, the developer should assist stakeholders' graceful migration from the system to an alternative. Users should be notified of the risks of continued use of the unsupported system long before support ends. Computing professionals should assist system users in monitoring the operational viability of their computing systems, and help them understand that timely replacement of inappropriate or outdated features or entire systems may be needed.

3.7 Recognize and take special care of systems that become integrated into the infrastructure of society. Even the simplest computer systems have the potential to impact all aspects of society when integrated with everyday activities such as commerce, travel, government, healthcare, and education. When organizations and groups develop systems that become an important part of the infrastructure of society, their leaders have an added responsibility to be good stewards of these systems. Part of that stewardship requires establishing policies for fair system access, including for those who may have been excluded. That stewardship also requires that computing professionals monitor the level of integration of their systems into the infrastructure of society. As the level of adoption changes, the ethical responsibilities of the organization or group are likely to change as well. Continual monitoring of how society is using a system will allow the organization or group to remain consistent with their ethical obligations outlined in the Code. When appropriate standards of care do not exist, computing professionals have a duty to ensure they are developed

To summarize, this chapter has outlined examples of where ASA GLs and ACM CE principles specifically address the DSEC and DEFW questions, to the extent that the contents match. As can be seen in the tables, the DSEC or DEFW questions are clearly *not* the limit of "ethical practice". Much of the ASA and ACM ethical practice standards are useful in addressing the DSEC and DEFW questions. It should be reiterated that, as the ACM and ASA elements identified above have demonstrated, ethical practice of statistics and data science requires that the practice standards be considered before a project is completed, and some of the practice standard elements are relevant throughout a project.

The following chapters refer back to this chapter's alignments, to outline both how the DSEC and DEFW questions relate to our seven tasks, but also, how the ASA GLs and ACM CE go further, to promote ethical practice on all tasks.

Chapter 2.2
Planning/Designing

As noted earlier, all statistics and data science projects involve planning and design. Ethical planning and design obviously require focus at the start of any project, but to ensure that the decisions made in the planning and design phase are either adhered to, or changed in justifiable ways, there must also be ongoing monitoring of the plan and whether it is being followed. This is true whether you are the original planner/designer, or you received a plan with directions to execute it. Thus, although the most obvious point at which planning and designing require attention to ethical practice standards is the beginning of a project, that is by no means to only time that consideration – of the plan/design and of the relevant ethical considerations – is needed.

As we saw in Chapter 2.1, several sections of DSEC and Principles of DEFW (2018) are relevant in the initial, and the ongoing/monitoring stages with respect to planning or the plan that was created. Here we consider how the professional practice standards (ACM CE, ASA GLs) inform answers to the questions DSEC and DEFW ask you to consider as you contemplate this task. The tables below highlight which DSEC questions -in all areas- and which DEFW Principles are most relevant to this task, showing how the ASA and ACM ethical practice standards help address DSEC and DEFW (2018) questions specific to for Planning/Designing.

The tables re-iterate the ASA and ACM elements that support answers to DSEC/DEFW (2018) questions, but in these tables, only the questions that are relevant to the task (planning and designing) are shown.

	A.1 Informed consent: If there are human subjects, have they given informed consent, where subjects affirmatively opt-in and have a clear understanding of the data uses to which they consent?	**A.2 Collection bias**: Have we considered sources of bias that could be introduced during data collection and survey design and taken steps to mitigate those?	**A.3 Limit PII exposure**: Have we considered ways to minimize exposure of personally identifiable information (PII) for example through anonymization or not collecting information that isn't relevant for analysis?
ASA A	A11	A2, A4	
ASA B	B1, B4	B; B1, B2, B3, B6	
ASA C	C7	C1, C2, C3, C4, C8	C7
ASA D	D; D1, D5, D6, D10, D11	D6	D1, D4, D7, D9
ASA E	E4	E4	
ASA F	F5	F5	
ASA G	G5		
ASA H	H2		
ACM 1	1.6	1.4	1.2, 1.6
ACM 2	2.3, 2.9	2.5	2.5, 2.8. 2.9

Table 2.2.1. *ASA and ACM Alignment with DSEC A. Data Collection. Which DSEC Area A questions are relevant for Planning/Designing, and which ASA and ACM elements support addressing these DSEC questions?*

	B.1 Data security: Do we have a plan to protect and secure data (e.g., encryption at rest and in transit, access controls on internal users and third parties, access logs, and up-to-date software)?	**B.2 Right to be forgotten**: Do we have a mechanism through which an individual can request their personal information be removed?	**B.3 Data retention plan**: Is there a schedule or plan to delete the data after it is no longer needed?
ASA B	B1, B4, B6	B4, B6	B4, B6
ASA C	C7	C7	C4, C7
ASA D	D4, D9, D10, D11	D1, D4, D5, D7, D11	D1, D4, D5, D7, D11
ASA F	F4		
ACM 1	1.2, 1.6		1.6
ACM 2	2.5, 2.8	2,.8, 2.9	
ACM 3	3.7	3.7	

Table 2.2.2. *ASA and ACM Alignment with DSEC B. Data Storage. Which DSEC Area B questions are relevant for Planning/Designing, and which ASA and ACM elements support addressing these DSEC questions?*

	C.1 Missing perspectives: Have we sought to address blind spots in the analysis through engagement with relevant stakeholders (e.g., checking assumptions and discussing implications with affected communities and subject matter experts)?	C.2 Dataset bias: Have we examined the data for possible sources of bias and taken steps to mitigate or address these biases (e.g., stereotype perpetuation, confirmation bias, imbalanced classes, or omitted confounding variables)?	C.3 Honest representation: Are our visualizations, summary statistics, and reports designed to honestly represent the underlying data?	C.4 Privacy in analysis: Have we ensured that data with PII are not used or displayed unless necessary for the analysis?	C.5 Auditability: Is the process of generating the analysis well documented and reproducible if we discover issues in the future?
ASA A	A2, A4				A5, A11
ASA B	B1, B2				B4, B5
ASA C	C1, C2				
ASA D	D, D1, D6				D6, D8
ASA E	E3				
ASA F	F4				F2
ACM 1	1.4				
ACM 2					(2.4)

Table 2.2.3. ASA and ACM Alignment with DSEC C. Analysis. *Which DSEC Area C questions are relevant for Planning/Designing, and which ASA and ACM elements support addressing these DSEC questions?*

	D.1 Proxy discrimination: Have we ensured that the model does not rely on variables or proxies for variables that are unfairly discriminatory?	D.2 Fairness across groups: Have we tested model results for fairness with respect to different affected groups (e.g., tested for disparate error rates)?	D.3 Metric selection: Have we considered the effects of optimizing for our defined metrics and considered additional metrics?	D.4 Explainability: Can we explain in understandable terms a decision the model made in cases where a justification is needed?	D.5 Communicate bias: Have we communicated the shortcomings, limitations, and biases of the model to relevant stakeholders in ways that can be generally understood?
ASA A			A2	A1, A4	
ASA B			B1, B2, B3	B2, B3, B4	
ASA C			C1	C3, C4, C6	
ASA D			D8	D10	
ASA E					
ACM 1					
ACM 2				2.5, 2.9	

Table 2.2.4. ASA and ACM Alignment with DSEC D. Modeling. *Which DSEC Area D questions are relevant for Planning/Designing, and which ASA and ACM elements support addressing these DSEC questions?*

	E.1 Redress: Have we discussed with our organization a plan for response if users are harmed by the results (e.g., how does the data science team evaluate these cases and update analysis and models to prevent future harm)?	E.2 Roll back: Is there a way to turn off or roll back the model in production if necessary?	E.3 Concept drift: Do we test and monitor for concept drift to ensure the model remains fair over time?	E.4 Unintended use: Have we taken steps to identify and prevent unintended uses and abuse of the model and do we have a plan to monitor these once the model is deployed?
ASA A	A2, A3, A4, A9		A2, A3	A2, A3, A4, A9
ASA B	B5		B1, B2, B5, B6	B3, B5, B6
ASA C	C2, C4			C7
ASA D	D1, D4, D5, D6, D7, D8			D; D4, D5, D6, D7, D10, D11
ASA E				E4
ASA F				
ASA G			G5	G5
ASA H				H2
ACM 2	2.5		2.5	2.5, 2.9

Table 2.2.5. ASA and ACM Alignment with DSEC E. Deployment. *Which DSEC Area E questions are relevant for Planning/Designing, and which ASA and ACM elements support addressing these DSEC questions?* Note that not all of the DSEC area questions are relevant for Task 1 (plan/design). Similarly, not all DEFW Principles are relevant.

	Principle 1. Start with clear user need and public benefit	Principle 2. Be aware of relevant legislation and codes of practice	Principle 3. Use data that is proportionate to the user need	Principle 4. Understand the limitations of the data	Principle 5. Ensure robust practices and work within your skillset	Principle 6. Make your work transparent and be accountable	Principle 7. Embed data use responsibly
ASA PRINCIPLES HAVING ELEMENTS IN ALIGNMENT WITH DEFW on this task							
ASA A	A2, A3, A4	A11	A2				A2, A4
ASA B	B1	B4	B3, B6				B1-B4, B6
ASA C	C1	C7	C4				C1-C4, C7, C8
ASA D	D5	D1, D4, D9	D2, D3, D7, D8				D1, D2, D3, D11
ASA E		E1					E4
ASA F							F5
ASA G							G5
ASA H		H1					H2
ACM PRINCIPLES HAVING ELEMENTS IN ALIGNMENT WITH DEFW on this task							
ACM 1			1.6				
ACM 2		2.3					
ACM 3							3.6, 3.7

Table 2.2.6. ASA and ACM Alignment with DEFW: *Which DEFW (2018) Principles are relevant for Planning/Designing, and which ASA and ACM elements support addressing these DEFW questions?*

As you might imagine, ethical practice begins with the plan and design – so these excerpts from the full tables in the previous chapter show the importance of many ASA and ACM principles to guide this task. Beyond the DSEC and DEFW questions and the ACM CE- and ASA GL- suggested answers, there are other general practice standards that more fully support ethical planning and design. Below we outline the principal GL and ACM indicators of how to plan/design ethically projects ethically. Tractenberg (2022) gives the total elaboration of *everything* in the GL/CE that is relevant for the task.

ASA

The first two Principles of the ASA GLs outline exactly how the ethical practitioner operates – respectfully, transparently, and accountably – throughout practice. These attributes may be most important to be incorporated/demonstrated at the earliest stages (i.e., Planning and Design) of any project.

Many other aspects of the ASA Ethical Guidelines are important for ethical planning and design, and as is shown across the tables above, these responsibilities arise with respect to different stakeholders and others. That is, every ASA GL Principle is invoked in ethical planning and design.

ACM

The guiding principle for utilizing the ACM CE to plan or design a project is outlined in Principle 1.2 *Avoid harm*; however, this is not specifically invoked in any of the DSEC or DEFW questions. Also missing from DSEC and DEFW (2018) is any role for ACM CE Principle 3, Professional Leadership Principles in considerations of planning and design. This is most likely due to the combination of facts that the DSEC and DEFW (2018) questions were

a) designed to be answered for a specific project (that is already formed DSEC or is forming DEFW); and
b) drafted with the assumption that the answerer is not necessarily a team leader.

Comment:

Like ASA Principle G, ACM Principles 1.2 and 3 (3.1 – 3.7) are intended to support a professional context in which all work – including planning and design – will be done.

As can be seen, addressing any DSEC or DEFW (2018) questions – and planning and designing statistics and data science projects in general – require and benefit from use of the ASA GLs and ACM CE. Similarly, ASA Principles E (responsibilities to multidisciplinary team members) and F (responsibilities to fellow statistics practitioners and the Profession) are rarely invoked by DSEC or DEFW, but offer important guidance for working with others, as most statisticians and data scientists must do. Finally, ASA Principle G and ACM Principle 3 outline the responsibilities that are specific to those in leadership roles. These are not contemplated in DSEC or DEFW (2018), which is problematic because it can suggest that a practitioner's responsibilities never change through their career. The inclusion of Principle G in the ASA GLs was intended to both notify practitioners early in their careers of the qualities of ethical leaders, and to provide leaders with a list of ethical practice standards specific to later career roles. That is, just as ASA Principle G is useful in planning and designing projects in statistics and data science, it can also be useful in planning a career and choosing teams with ethical leadership.

Chapter 2.3
Data Collection/Munging/Wrangling

Data must be considered "tied" with planning and design for the position of most important component in data analysis and data science. The ways in which data are collected or synthesized in simulations, munged (transformed), and wrangled (manipulated, integrated), are all essential; justification of the need for data is foundational in all of the DEFW questions, and collection (DSEC A) and storage (DSEC B) are the first two areas of consideration in DSEC. As noted earlier, careful planning is essential for effective and ethical data collection, and failures to obtain appropriate permissions/consent will undermine munging and wrangling efforts. At the beginning of the project, data collection design and planning are essential. In some cases, data may be collected in an ongoing way; monitoring the collection process is critical to ensure sampling is balanced/following the design. Sensitivity checks may also be implemented to ensure that the ultimate results are robust, interpretable, and reproducible. Attention to interpretable, valid results from collected, munged, and wrangled data underlines the role for ASA and ACM practice standards in responses to the DSEC and DEFW questions relating to this task.

DSEC: As we saw in Chapter 2.1, the DSEC has an entire section about "data collection" (that is A), but storage and security (DSEC B) must also be considered in the collection and wrangling or munging of data. DSEC C2 relates to dataset bias, and this can arise in either collection or manipulation (munging/wrangling) of data. Because data collection requires consideration of security, DSEC E4 is also an important consideration for this task.

DEFW: Principles 2 and 3, and to some extent, 4 are important for data collection and any manipulation of that data (including security) going forward. Although the DEFW is about "data ethics", security of that data is implied, with only one question (plan for response to security breach) under Principle 2.

The following tables show how the ASA and ACM ethical practice standards help address DSEC and DEFW (2018) questions specific to Data collection/munging/wrangling.

DSEC A	A.1 Informed consent: If there are human subjects, have they given informed consent, where subjects affirmatively opt-in and have a clear understanding of the data uses to which they consent?	A.2 Collection bias: Have we considered sources of bias that could be introduced during data collection and survey design and taken steps to mitigate those?	A.3 Limit PII exposure: Have we considered ways to minimize exposure of personally identifiable information (PII) for example through anonymization or not collecting information that isn't relevant for analysis?
ASA A	A11	A2, A4	
ASA B	B1, B4	B; B1, B2, B3, B6	
ASA C	C7	C1, C2, C3, C4, C8	C7
ASA D	D; D1, D5, D6, D10, D11	D6	D1, D4, D7, D9
ASA E	E4	E4	
ASA F	F5	F5	
ASA G	G5		
ASA H	H2		
ACM 1	1.6	1.4	1.2, 1.6
ACM 2	2.3, 2.9	2.5	2.5, 2.8. 2.9

Table 2.3.1. ASA and ACM support for addressing DSEC A. Data Collection questions. *Which DSEC Area A questions are relevant for Data collection/munging/wrangling, and which ASA and ACM elements support addressing these DSEC questions?*

	B.1 Data security: Do we have a plan to protect and secure data (e.g., encryption at rest and in transit, access controls on internal users and third parties, access logs, and up-to-date software)?	B.2 Right to be forgotten: Do we have a mechanism through which an individual can request their personal information be removed?	B.3 Data retention plan: Is there a schedule or plan to delete the data after it is no longer needed?
ASA B	B1, B4, B6	B4, B6	B4, B6
ASA C	C7	C7	C4, C7
ASA D	D4, D9, D10, D11	D1, D4, D5, D7, D11	D1, D4, D5, D7, D11
ASA F	F4		
ACM 1	1.2, 1.6		1.6
ACM 2	2.5, 2.8	2,8, 2.9	
ACM 3	3.7	3.7	

Table 2.3.2. ASA and ACM support for addressing DSEC B. Data Storage questions. *Which DSEC Area B questions are relevant for Data collection/munging/wrangling, and which ASA and ACM elements support addressing these DSEC questions?*

	C.1 Missing perspectives: Have we sought to address blind spots in the analysis through engagement with relevant stakeholders (e.g., checking assumptions and discussing implications with affected communities and subject matter experts)?	C.2 Dataset bias: Have we examined the data for possible sources of bias and taken steps to mitigate or address these biases (e.g., stereo-type perpetuation, confirmation bias, imbalanced classes, or omitted confounding variables)?	C.3 Honest representtation: Are our visualizations, summary statistics, and reports designed to honestly represent the underlying data?	C.4 Privacy in analysis: Have we ensured that data with PII are not used or displayed unless necessary for the analysis?	C.5 Auditability: Is the process of generating the analysis well documented and reproducible if we discover issues in the future?
ASA A		A2, A3, A4			
ASA B		B1, B2			
ASA C		C2			
ASA D		D1, D2, D3, D5, D6			
ASA E		E4			
ACM 1		1.4			

Table 2.3.3. *ASA and ACM support for addressing DSEC C. Analysis questions. Which DSEC Area C questions are relevant for Data collection/munging/wrangling, and which ASA and ACM elements support addressing these DSEC questions?*

As noted, only some of the DSEC Area questions are relevant to the task of data collection/munging/wrangling, and of the DSEC areas that are relevant (DSEC A, B, C), not all of the questions within DSEC Area C are relevant to ethical data collection/munging/wrangling. Similarly, only three of seven DEFW (2018) principles address this task.

	Principle 1. Start with clear user need and public benefit	Principle 2. Be aware of relevant legislation and codes of practice	Principle 3. Use data that is proportionate to the user need	Principle 4. Understand the limitations of the data	Principle 5. Ensure robust practices and work within your skillset	Principle 6. Make your work transparent and be accountable	Principle 7. Embed data use responsibly
ASA A		A11	A2	A2			
ASA B		B4	B3, B6	B1-B4, B6			
ASA C		C7	C4	C3			
ASA D		D1, D4, D9	D2, D3, D7, D8	D4, D5, D10			
ASA E		E1					
ASA H		H1					
ACM 1			1.6	1.3			
ACM 2		2.3		2.5			

Table 2.3.4. ASA and ACM support for addressing DEFW questions: *Which DEFW (2018) Principles are relevant for Data collection/munging/wrangling, and which ASA and ACM elements support addressing these DEFW questions?*

ASA

It might surprise you to see how much of the ASA Guidelines are relevant for data collection – the tables here (and in the previous chapter) show that responsibilities accrue to the ethical statistical and data science practitioner from more ASA Guideline Principles than just B (integrity of data and methods) and D (Responsibilities to research subjects, data subjects, or those directly affected by statistical practices).

ACM

The support from ACM CE for addressing DSEC and DEFW questions is outlined in the tables. It might seem surprising to see the extent of support for ethical data collection, munging, and wrangling for computing professionals, but this highlights the role of computing in modern statistical and data science practice.

Comment:

While data collection and analysis must certainly be among the most important aspects of statistical practice, the ASA GLs offer guidance on several other key activities beyond what is discussed in DSEC and DEFW (2018). These tables underscore the breadth of what must be considered "ethical practice" in statistics and data science. There is more to "statistics and data science" than *data* – as suggested by the extent of the ASA GLs. Note in the DEFW table, Principle 3 mentions that the collected data must be "**proportionate to the user need**". This must have been a component in the plan and design of the project (as outlined in the previous chapter); user need must also be kept in mind as the data are collected and munged. Collecting data *beyond what is needed* is contrary to both the ACM CE and ASA GLs. If the data are collected "just in case", it means that consent for using that data in this case, *unspecified at the time of collecting*, cannot be obtained. Thus, these Principles must all be considered in both planning and in data collection. That is, ethical considerations are not specific to the beginning or end of a project, but must be prioritized throughout. Ostensibly these would be sequential phases of a project but filling out the DSEC or DEFW at the start, or at the end, or a specific project could end up missing the need to apply important aspects of ACM and ASA ethical practice standards.

Chapter 2.4
Analysis (perform or program to perform)

Analysis is an essential part of the statistician's job and, while the data scientist may consider the analysis to be the role of an algorithm, program, or system, the analysis is critical to the use, utility, and relevance of any system/program/algorithm. For computing professionals, "analysis" is less about statistics and more about an evaluation of the system itself. In either perspective, the analysis of a system or of data (or of a system that collects/analyses/uses data) is an *essential* function of the statistician and data scientist. For the statistical perspective, the analysis must be designed (early stage) but monitored to ensure that results are valid and interpretable. In experimental settings, this must also be planned and reported; in longitudinal contexts, analyses that are not *directly* relevant for the ultimate objectives of the collection/analysis may be programmed into a system, for example, to ensure that recruiting and retention goals are being met, that surveys and other collection mechanisms are functioning as intended. In many cases, the analysis is intended for pre-planned time points or milestones, or at the end of a project.

DSEC: DSEC A2, relating to bias in data collection, should be considered when carrying out the analysis task. Although DSEC Area C is labelled "Analysis", DSEC C1, C2, C4 are directly related to analysis while DSEC C3 (representation of results) and DSEC C5 (documentation of analysis) are *indirectly* related. Four of five questions in Area D (modeling) are relevant in analysis, as is DSEC E3 (concept drift) – another question that would be answered using (planned) analyses.

DEFW: Like DSEC, the DEFW Principles apply widely to, or are answered via, analyses. Addressing DEFW Principle 4 (understand the limitations of the data) begins with planning, but it is also relevant in analysis. Similarly, while DEFW Principle 5 (ensure robust practices) may seem most relevant in planning, it is also important for analysis.

DSEC A	**A.1 Informed consent**: If there are human subjects, have they given informed consent, where subjects affirmatively opt-in and have a clear understanding of the data uses to which they consent?	**A.2 Collection bias**: Have we considered sources of bias that could be introduced during data collection and survey design and taken steps to mitigate those?	**A.3 Limit PII exposure**: Have we considered ways to minimize exposure of personally identifiable information (PII) for example through anonymization or not collecting information that isn't relevant for analysis?
ASA A		A2, A4	
ASA B		B; B1, B2, B3, B6	
ASA C		C1, C2, C3, C4, C8	
ASA D		D6	
ASA E		E4	
ASA F		F5	
ACM 1		1.4	
ACM 2		2.5	

Table 2.4.1. *ASA and ACM support for addressing DSEC A. Data Collection questions. Which DSEC Area A questions are relevant for Analysis, and which ASA and ACM elements support addressing these DSEC questions?*

	C.1 Missing perspectives: Have we sought to address blind spots in the analysis through engagement with relevant stakeholders (e.g., checking assumptions and discussing implications with affected communities and subject matter experts)?	C.2 Dataset bias: Have we examined the data for possible sources of bias and taken steps to mitigate or address these biases (e.g., stereotype perpetuation, confirmation bias, imbalanced classes, or omitted confounding variables)?	C.3 Honest representation: Are our visualizations, summary statistics, and reports designed to honestly represent the underlying data?	C.4 Privacy in analysis: Have we ensured that data with PII are not used or displayed unless necessary for the analysis?	C.5 Auditability: Is the process of generating the analysis well documented and reproducible if we discover issues in the future?
ASA A	A2, A4	A2, A3, A4	A3, A4, A5, A7, A8, A11		A5, A11
ASA B	B1, B2	B1, B2	B1, B2, B3, B5, B6	B4	B4, B5
ASA C	C1, C2	C2	C2, C3, C5	C4, C7	
ASA D	D, D1, D6	D1, D2, D3, D5, D6	D5, D6, D7, D10, D11	D4, D5, D9	D6, D8
ASA E	E3	E4	E4		
ASA F	F4		F4	F4	F2
ACM 1	1.4	1.4		1.6	
ACM 2		2.1	2.1		(2.4)

Table 2.4.2. ASA and ACM support for addressing DSEC C. Analysis questions. *Which DSEC Area C questions are relevant for Analysis, and which ASA and ACM elements support addressing these DSEC questions?*

	D.1 Proxy discrimination: Have we ensured that the model does not rely on variables or proxies for variables that are unfairly discriminatory?	D.2 Fairness across groups: Have we tested model results for fairness with respect to different affected groups (e.g., tested for disparate error rates)?	D.3 Metric selection: Have we considered the effects of optimizing for our defined metrics and considered additional metrics?	D.4 Explainability: Can we explain in understandable terms a decision the model made in cases where a justification is needed?	D.5 Communi-cate bias: Have we communicated the shortcomings, limitations, and biases of the model to relevant stakeholders in ways that can be generally understood?
ASA A	A2, A3, A4	A2, A3, A4	A2	A1, A4	
ASA B	B1, B2, B5, B6	B1, B2, B3, B7	B1, B2, B3	B2, B3, B4	
ASA C			C1	C3, C4, C6	
ASA D	D6	D2, D5, D6, D10, D11	D8	D10	
ASA E	E4	E4			
ACM1	1.4	1.2, 1.4			
ACM2	2.4, 2.5			2.5, 2.9	

Table 2.4.3. ASA and ACM support for addressing DSEC D. Modeling questions. *Which DSEC Area D questions are relevant for Analysis, and which ASA and ACM elements support addressing these DSEC questions?*

	E.1 Redress: Have we discussed with our organization a plan for response if users are harmed by the results (e.g., how does the data science team evaluate these cases and update analysis and models to prevent future harm)?	E.2 Roll back: Is there a way to turn off or roll back the model in production if necessary?	E.3 Concept drift: Do we test and monitor for concept drift to ensure the model remains fair over time?	E.4 Unintended use: Have we taken steps to identify and prevent unintended uses and abuse of the model and do we have a plan to monitor these once the model is deployed?
ASA A			A2, A3	
ASA B			B1, B2, B5, B6	
ASA G			G5	
ACM 2			2.5	

Table 2.4.4. *ASA and ACM support for addressing DSEC E. Deployment questions. Which DSEC Area E questions are relevant for Analysis, and which ASA and ACM elements support addressing these DSEC questions?*

	Principle 1. Start with clear user need and public benefit	Principle 2. Be aware of relevant legislation and codes of practice	Principle 3. Use data that is proportionate to the user need	Principle 4. Understand the limitations of the data	Principle 5. Ensure robust practices and work within your skillset	Principle 6. Make your work transparent and be accountable	Principle 7. Embed data use responsibly
ASA A				A2	A1, A2		
ASA B				B1-B4, B6	B1, B2, B4, B6		
ASA C				C3	C1, C3		
ASA D				D4, D5, D10			
ASA E					E4		
ACM 1				1.3	1.3		
ACM 2				2.5	2.5, 2.6		

Table 2.4.5. ASA and ACM support for addressing DEFW questions: *Which DEFW (2018) Principles are relevant for analysis, and which ASA and ACM elements support addressing these DEFW questions?*

ASA

The task of analysis engages much of the ASA Guidelines. Even though "statistical analysis" might seem to be the only thing a statistics practitioner does, clearly the Guidelines a) do not have a Principle for "analysis" (like DSEC Area C), and b) support the diverse components of ethical statistical practice from a range of Principles. Many of the ASA Principles are important for ensuring that the analyses are planned and executed correctly and transparently. Also, influence from other team members (ASA E4, or ASA C3) should not be allowed to influence the analyses.

ACM

Much of the ACM CE Principle 2 focuses on the evaluation (analysis) of systems and their capabilities, limitations, and risks. This understanding of the term "analysis" is clear when the ASA and ACM practice standards are understood, but is lacking when a unidimensional conceptualization of the term (and construct, as in DSEC area C) is utilized. Solely focusing on DSEC or DEFW could lead to a lack of the appropriate professional review (2.4) or the comprehensive and thorough evaluations of systems (2.5) – leading to inadvertent but clear violations of the ACM CE (4.1).

Comment:

Although DSEC C (Analysis) is the most obviously relevant set of questions for the "analysis" task, it is by no means the only DSEC area that is relevant for this task. As noted in the previous chapter, there is more to data science than "just data" and there is more to data analysis than "just analysis", so considerations of the integrity of the data – and of its provenance – are important for the ethical practitioner.

Chapter 2.5
Interpretation

The interpretation of results from a system or an analysis is an essential feature of statistical practice. However, in the computing context, the "analysis" or evaluation includes the interpretation, which means that our discussion of this task (interpretation) will be limited to the specific context of interpreting the results of any given statistical or descriptive analysis. (See Tractenberg, 2022).

Even when a practitioner simply executes an analysis plan, or analysis results are "provided" to a requestor, there must be preliminary evaluation of the results by the analyst to ensure that assumptions were not violated, and to detail the limitations and/or bias that might be present. This is a fundamental part of the statistician's role when working with others in team science and teamwork; particularly when no one else on the team has statistical expertise. However, even when others on the team understand statistical analysis and the methods that were used, the ethical practitioner ensures that the analysis was done appropriately and that assumptions, methods, and results are all presented transparently and without bias by providing interpretation to at least that extent. Without even this superficial interpretation, the expert's analysis will unlikely be useful to a non-expert, while the expert's interpretation is a crucial part of communication to other experts in the field. Whenever analyses are executed, interpretation will be needed. This can occur in early system or algorithm checks, at any point to determine if there is concept drift or bias in results, and in the "final analysis". Thus, interpretation is a task that can occur throughout a project, and is also required in any peer review or other system evaluations that are performed outside of specific projects.

DSEC: The identification of bias in a data set (DSEC A2) derives from interpretation of analyses of that dataset as does the identification of missing perspectives (DSEC C1) and whether any personally identifying information (PII) is needed for an analysis is also based on interpretation. The determination that variables in an analysis are actually proxies that may lead to discrimination (DSEC D1), unfairness across groups (DSEC D2), or concept drift across time (DSEC E3) are all interpretive (based on analyses). Finally, the selection or evaluation of any metric used in analyses (DSEC D3) also requires interpretation.

DEFW: Appreciation of the limitations of data (Principle 4), and the identification of robust practices (Principle 5) require interpretation of the analysis or functioning of a model. Principle 7, Embed data use responsibly, also requires interpretation.

DSEC A	A.1 Informed consent: If there are human subjects, have they given informed consent, where subjects affirmatively opt-in and have a clear understanding of the data uses to which they consent?	A.2 Collection bias: Have we considered sources of bias that could be introduced during data collection and survey design and taken steps to mitigate those?	A.3 Limit PII exposure: Have we considered ways to minimize exposure of personally identifiable information (PII) for example through anonymization or not collecting information that isn't relevant for analysis?
ASA A		A2, A4	
ASA B		B; B1, B2, B3, B6	
ASA C		C1, C2, C3, C4, C8	
ASA D		D6	
ASA E		E4	
ASA F		F5	
ACM 1		1.4	
ACM 2		2.5	

Table 2.5.1. *ASA and ACM support for addressing DSEC A. Data Collection questions.* Which DSEC Area A questions are relevant for interpretation, and which ASA and ACM elements support addressing these DSEC questions?

	C.1 Missing perspectives: Have we sought to address blind spots in the analysis through engagement with relevant stakeholders (e.g., checking assumptions and discussing implications with affected communities and subject matter experts)?	C.2 Dataset bias: Have we examined the data for possible sources of bias and taken steps to mitigate or address these biases (e.g., stereotype perpetuation, confirmation bias, imbalanced classes, or omitted confounding variables)?	C.3 Honest representation: Are our visualizations, summary statistics, and reports designed to honestly represent the underlying data?	C.4 Privacy in analysis: Have we ensured that data with PII are not used or displayed unless necessary for the analysis?	C.5 Auditability: Is the process of generating the analysis well documented and reproducible if we discover issues in the future?
ASA A	A2, A4				
ASA B	B1, B2				
ASA C	C1, C2				
ASA D	D, D1, D6				
ASA E	E3				
ASA F	F4				
ACM 1	1.4				
ACM 2					

Table 2.5.2. ASA and ACM support for addressing DSEC C. Analysis questions. *Which DSEC Area C questions are relevant for interpretation, and which ASA and ACM elements support addressing these DSEC questions?*

	D.1 Proxy discrimination: Have we ensured that the model does not rely on variables or proxies for variables that are unfairly discriminatory?	D.2 Fairness across groups: Have we tested model results for fairness with respect to different affected groups (e.g., tested for disparate error rates)?	D.3 Metric selection: Have we considered the effects of optimizing for our defined metrics and considered additional metrics?	D.4 Explainability: Can we explain in understandable terms a decision the model made in cases where a justification is needed?	D.5 Communicate bias: Have we communicated the shortcomings, limitations, and biases of the model to relevant stakeholders in ways that can be generally understood?
ASA A	A2, A3, A4	A2, A3, A4	A2		
ASA B	B1, B2, B5, B6	B1, B2, B3, B7	B1, B2, B3		
ASA C			C1		
ASA D	D6	D2, D5, D6, D10, D11	D8		
ASA E	E4	E4			
ACM1	1.4	1.2, 1.4			
ACM2	2.4, 2.5				

Table 2.5.3. ASA and ACM support for addressing DSEC D. Modeling questions. *Which DSEC Area D questions are relevant for interpretation, and which ASA and ACM elements support addressing these DSEC questions?*

	E.1 Redress: Have we discussed with our organization a plan for response if users are harmed by the results (e.g., how does the data science team evaluate these cases and update analysis and models to prevent future harm)?	E.2 Roll back: Is there a way to turn off or roll back the model in production if necessary?	E.3 Concept drift: Do we test and monitor for concept drift to ensure the model remains fair over time?	E.4 Unintended use: Have we taken steps to identify and prevent unintended uses and abuse of the model and do we have a plan to monitor these once the model is deployed?
ASA A			A2, A3	
ASA B			B1, B2, B5, B6	
ASA G			G5	
ACM 2			2.5	

Table 2.5.4. ASA and ACM support for addressing DSEC E. Deployment questions. *Which DSEC Area E questions are relevant for interpretation, and which ASA and ACM elements support addressing these DSEC questions?*

	Principle 1. Start with clear user need and public benefit	Principle 2. Be aware of relevant legislation and codes of practice	Principle 3. Use data that is proportionate to the user need	Principle 4. Understand the limitations of the data	Principle 5. Ensure robust practices and work within your skillset	Principle 6. Make your work transparent and be accountable	Principle 7. Embed data use responsibly
ASA A				A2	A1, A2		A2, A4
ASA B				B1-B4, B6	B1, B2, B4, B6		B1-B4, B6
ASA C				C3	C1, C3		C1-C4, C7, C8
ASA D				D4, D5, D10			D1, D2, D3, D11
ASA E					E4		E4
ASA F							F5
ASA G							G5
ASA H							H2
ACM 1				1.3	1.3		
ACM 2				2.5	2.5, 2.6		
ACM 3							3.6, 3.7

Table 2.5.5. ASA and ACM support for addressing DEFW questions: *Which DEFW (2018) Principles are relevant for interpretation, and which ASA and ACM elements support addressing these DEFW questions?*

As noted earlier, "interpretation" is not a particular aspect of ethical computing practice – not to the same extent that it is for statistical and data science practice. Thus, we only consider what else the ASA GLs, and not the ACM CEs, can contribute to ethical interpretation.

ASA

All of the ASA Principles except H (Responsibilities relating to alleged misconduct) offer relevant guidance on interpretation, but not all of the DSEC areas, nor all of the DEFW (2018) Principles, do.

Chapter 2.6
Documenting your Work

The documentation of work – outlining what and how the system or analysis works, and including methodological features and assumptions – is a critical part of our solo work, to remind us of exactly what we did, when, why, and how; and is also essential to team science and teamwork. In either context, documentation can ensure that mistakes that might occur are not repeated; provide context for peer review of our work; and in the best cases, ensure that the most efficient approach to any problem is identified. Documentation should be done throughout any project. Essential in the beginning and in planning and design, the documentation of decisions can be revisited later to ensure that justifications for early decisions are still sound. Supportive throughout the execution of designs, documentation supports ongoing engagement with the project and ensures that assumptions and tests of those assumptions are recorded. At the end of a project, the documentation enables transparency and accountability while also ensuring that innovations have a full and correct report for the wider community. Because we also consider "reporting and communicating" (next chapter), the documentation of work is here considered to be everything short of reporting – the documentation of our work is what the ultimate report will utilize/be based on.

DSEC: The documentation of a data science project should begin with the plan; the first point where it can be seen to be relevant in the DSEC is C5, auditability. Clearly, a system or program (or analysis) cannot be audited if it hasn't been fully documented. Similarly, the explainability of any decision made by a system (DSEC D4) requires careful documentation, as does the communication of limitations and biases of any model that a system implements (DSEC D5). DSEC area E (deployment) requires full documentation, otherwise, none of the aspects mentioned (redress; roll back; concept drift; unintended use) can be addressed.

DEFW: Understanding the limitations of the data (Principle 4), ensuring robust practices (Principle 5), and making work transparent (Principle 6) clearly require full documentation of the system and the justifications behind decisions in all aspects of its design and implementation. Principle 7 (embed data use responsibly) also requires documentation, so that potential tradeoffs between data use and harms can be considered.

	C.1 Missing perspectives: Have we sought to address blind spots in the analysis through engagement with relevant stakeholders (e.g., checking assumptions and discussing implications with affected communities and subject matter experts)?	C.2 Dataset bias: Have we examined the data for possible sources of bias and taken steps to mitigate or address these biases (e.g., stereotype perpetuation, confirmation bias, imbalanced classes, or omitted confounding variables)?	C.3 Honest representation: Are our visualizations, summary statistics, and reports designed to honestly represent the underlying data?	C.4 Privacy in analysis: Have we ensured that data with PII are not used or displayed unless necessary for the analysis?	C.5 Auditability: Is the process of generating the analysis well documented and reproducible if we discover issues in the future?
ASA A					A5, A11
ASA B					B4, B5
ASA D					D6, D8
ASA F					F2
ACM 2					(2.4)

Table 2.6.1. ASA and ACM support for addressing DSEC C. Analysis questions. *Which DSEC Area C questions are relevant for documenting your work, and which ASA and ACM elements support addressing these DSEC questions?*

	D.1 Proxy discrimination: Have we ensured that the model does not rely on variables or proxies for variables that are unfairly discrimenatory?	D.2 Fairness across groups: Have we tested model results for fairness with respect to different affected groups (e.g., tested for disparate error rates)?	D.3 Metric selection: Have we considered the effects of optimizing for our defined metrics and considered additional metrics?	D.4 Explainability: Can we explain in understandable terms a decision the model made in cases where a justify-cation is needed?	D.5 Communicate bias: Have we communicated the short-comings, limitations, and biases of the model to relevant stakeholders in ways that can be generally understood?
ASA A				A1, A4	A2
ASA B				B2, B3, B4	B1, B2, B3
ASA C				C3, C4, C6	C2
ASA D				D10	D6, D10
ASA E					E4
ACM 2				2.5, 2.9	2.5

Table 2.6.2. ASA and ACM support for addressing DSEC D. Modeling questions. *Which DSEC Area D questions are relevant for documenting your work, and which ASA and ACM elements support addressing these DSEC questions?*

	E.1 Redress: Have we discussed with our organization a plan for response if users are harmed by the results (e.g., how does the data science team evaluate these cases and update analysis and models to prevent future harm)?	E.2 Roll back: Is there a way to turn off or roll back the model in production if necessary?	E.3 Concept drift: Do we test and monitor for concept drift to ensure the model remains fair over time?	E.4 Unintended use: Have we taken steps to identify and prevent unintended uses and abuse of the model and do we have a plan to monitor these once the model is deployed?
ASA A	A2, A3, A4, A9	A3, A4, A9	A2, A3	A2, A3, A4, A9
ASA B	B5	B5, B6	B1, B2, B5, B6	B3, B5, B6
ASA C	C2, C4	C2		C7
ASA D	D1, D4, D5, D6, D7, D8			D; D4, D5, D6, D7, D10, D11
ASA E		E4		E4
ASA F		F2		
ASA G		G5	G5	G5
ASA H		H2		H2
ACM 2	2.5	2.5, 2.9	2.5	2.5, 2.9

Table 2.6.3. ASA and ACM support for addressing DSEC E. Deployment questions. *Which DSEC Area E questions are relevant for documenting your work, and which ASA and ACM elements support addressing these DSEC questions?*

	Principle 1. Start with clear user need and public benefit	Principle 2. Be aware of relevant legislation and codes of practice	Principle 3. Use data that is proportionate to the user need	Principle 4. Understand the limitations of the data	Principle 5. Ensure robust practices and work within your skillset	Principle 6. Make your work transparent and be accountable	Principle 7. Embed data use responsibly
ASA PRINCIPLES HAVING ELEMENTS IN ALIGNMENT WITH DEFW							
ASA A				A2	A1, A2	A2, A5, A6	A2, A4
ASA B				B1-B4, B6	B1, B2, B4, B6	B1-B4, B6	B1-B4, B6
ASA C				C3	C1, C3	C1, C2, C4, C5	C1-C4, C7, C8
ASA D				D4, D5, D10		D6	D1, D2, D3, D11
ASA E					E4	E3, E4	E4
ASA F							F5
ASA G						G5	G5
ASA H							H2
ACM 1				1.3	1.3	1.2	
ACM 2				2.5	2.5, 2.6	2.1, 2.4	
ACM 3							3.6, 3.7

Table 2.6.4. ASA and ACM support for addressing DEFW questions: *Which DEFW questions are relevant for documenting your work, and which ASA and ACM elements support addressing these DEFW questions?*

Transparent and complete documentation is a foundational principle in all scientific endeavors, so it is not surprising that for statistics and data science - which are essential in many modern scientific disciplines, the DSEC, DEFW, ASA GLs, and ACM CE all support documentation as essential in ethical practice. What is essential to note in this chapter is that, although it is so critical to professional practice, the reliance of the DSEC questions *on careful documentation* is neither identified nor implied. DEFW Principle 6 does specify that work must be accountable and transparent – i.e., documentation must be complete and correct. Documentation is the substance of much of ASA GLs Principle B (why it features extensively in every table here), and is a consistent, if implicit, theme in the ACM CE – documentation of systems and their capabilities and limitations is what 2.4 and 2.5 (among other CE elements) require.

Chapter 2.7
Reporting your Results/Communication

As noted in the previous chapter, all of our work must be carefully and transparently documented, but this documentation would then be used to report or communicate what was done/how our systems or analyses work. Maintaining careful documentation can only facilitate reporting and communication. The documentation of methods, assumptions, checks and tests of assumptions, and evaluation of the function of the system by which data were collected would be combined transparently with results and the interpretations of those results, in order to yield the final report (or even an interim report). Thus, communication of function, findings, or both would be incomplete without full documentation of methods, assumptions, and caveats that describe what was done and why.

DSEC: Specific attention must be given to privacy (DSEC C4) and the auditability (DSEC C5) of the report of an analysis. Consideration of the potential for bias and discrimination that a model or system creates (DSEC D1) and its fairness (DSEC D2) and the explainability of a model's decisions (DSEC D4) are critical features of how these results are reported and communicated to stakeholders (DSEC D5). Whether or not a model can be rolled back (DSEC E2) is also a feature of how it is reported, and how its results would be communicated to stakeholders.

DEFW: Reporting and communication of the functioning and outputs of any system permit the creators and maintainers – as well as all stakeholders – to understand the limitations of the data (Principle 4), ensure that robust practices were used (Principle 5), that all work (system functioning and results) are transparent (Principle 6) and how/that the use of the data was necessary, and that users know how the system works (Principle 7).

	C.4 **Privacy in analysis**: Have we ensured that data with PII are not used or displayed unless necessary for the analysis?	C.5 **Auditability**: Is the process of generating the analysis well documented and reproducible if we discover issues in the future?
ASA A		A2
ASA B	B10	ALL B
ASA C	C5	
ASA F	F1	
ACM 1	1.6	
ACM 2		(2.4)

Table 2.7.1. *ASA and ACM support for addressing DSEC C. Analysis questions.* *Which DSEC Area C questions are relevant for reporting your results/communication, and which ASA and ACM elements support addressing these DSEC questions?*

	D.1 Proxy discrimination: Have we ensured that the model does not rely on variables or proxies for variables that are unfairly discriminatory?	D.2 Fairness across groups: Have we tested model results for fairness with respect to different affected groups (e.g., tested for disparate error rates)?	D.3 Metric selection: Have we considered the effects of optimizing for our defined metrics and considered additional metrics?	D.4 Explainability: Can we explain in understand-able terms a decision the model made in cases where a justification is needed?	D.5 Communicate bias: Have we communicated the shortcomings, limitations, and biases of the model to relevant stakeholders in ways that can be generally understood?
ASA A	A2, A3, A4	A2, A3, A4		A1, A4	A2
ASA B	B1, B2, B5, B6	B1, B2, B3, B7		B2, B3, B4	B1, B2, B3
ASA C				C3, C4, C6	C2
ASA D	D6	D2, D5, D6, D10, D11		D10	D6, D10
ASA E	E4	E4			E4
ACM 1	1.4	1.2, 1.4			
ACM 2	2.4, 2.5			2.5, 2.9	2.5

Table 2.7.2. ASA and ACM support for addressing DSEC D. Modeling. *Which DSEC Area D questions are relevant for reporting your results/communication, and which ASA and ACM elements support addressing these DSEC questions?*

	E.1 Redress: Have we discussed with our organization a plan for response if users are harmed by the results (e.g., how does the data science team evaluate these cases and update analysis and models to prevent future harm)?	E.2 Roll back: Is there a way to turn off or roll back the model in production if necessary?	E.3 Concept drift: Do we test and monitor for concept drift to ensure the model remains fair over time?	E.4 Unintended use: Have we taken steps to identify and prevent unintended uses and abuse of the model and do we have a plan to monitor these once the model is deployed?
ASA A		A3, A4, A9		
ASA B		B5, B6		
ASA C		C2		
ASA E		E4		
ASA F		F2		
ASA G		G5		
ASA H		H2		
ACM 2		2.5, 2.9		

Table 2.7.3. ASA and ACM support for addressing DSEC E. Deployment questions. *Which DSEC Area E questions are relevant for reporting your results/communication, and which ASA and ACM elements support addressing these DSEC questions?*

	Principle 1. Start with clear user need and public benefit	Principle 2. Be aware of relevant legislation and codes of practice	Principle 3. Use data that is proportionate to the user need	Principle 4. Understand the limitations of the data	Principle 5. Ensure robust practices and work within your skillset	Principle 6. Make your work transparent and be account-table	Principle 7. Embed data use responsibly
ASA PRINCIPLES HAVING ELEMENTS IN ALIGNMENT WITH DEFW							
ASA A				A2	A1, A2	A2, A5, A6	A2, A4
ASA B				B1-B4, B6	B1, B2, B4, B6	B1-B4, B6	B1-B4, B6
ASA C				C3	C1, C3	C1, C2, C4, C5	C1-C4, C7, C8
ASA D				D4, D5, D10		D6	D1, D2, D3, D11
ASA E					E4	E3, E4	E4
ASA F							F5
ASA G						G5	G5
ASA H							H2
ACM 1				1.3	1.3	1.2	
ACM 2				2.5	2.5, 2.6	2.1, 2.4	
ACM 3							3.6, 3.7

Table 2.7.4. ASA and ACM support for addressing DEFW questions: *Which DEFW (2018) Principles are relevant for reporting your results/communication, and which ASA and ACM elements support addressing these DEFW questions?*

Comment:

Reporting results and communication are fundamental and crucial parts of statistics and data science practice, therefore the extensive representation in these tables from all ASA GL Principles and many from the ACM CE is not surprising. Without the expert's communication, non-experts are unlikely to transmit and report the results correctly and completely. In some cases, even with the expert's communication, non-experts may seek to transmit their own report of results without the critical documentation that the ethical expert provides.

For this reason, ASA Principle E (Responsibilities to members of multidisciplinary teams) seeks to ensure that the ASA GLs – and full, transparent reporting to promote valid and reproducible interpretations. The GLs support the ethical practitioner's efforts to ensure that all reports and communications are correct and transparent (avoiding detrimental research practices per NASEM 2017 p. 206). Computing professionals' responsibilities relating to reporting and communication depend on the stakeholders to/with whom they are reporting, with principal emphasis placed on the communication of capabilities and limitations of systems to stakeholders (2.1 and 2.5). This fundamental role - communication - for the ethical practitioner is not included in DSEC, but is alluded to in DEFW Principle 6 (echoed in ACM Principle 2.1).

Chapter 2.8
Engaging in Team Science/Teamwork

The most significant gap between the professional practice standards (ACM CE, ASA GLs) and the DEFW and DSEC is that the practice standards do focus on working with "others", while DEFW (2018) and DSEC do not. DEFW (2018) does include considerations of "users" and it also includes peer review/access as part of transparency. DSEC considers stakeholders for their perspectives (to limit bias) or for communication. However, the ASA and ACM standards recognize far more explicitly that, while there may be only one statistician or data scientist (or computing professional) involved on any project, statistics and data science are rarely done by individuals outside of *teams*. Whether the overall objective is to contribute new knowledge to the scientific literature (scholarship or team science) or to complete projects in the workplace (teamwork), ethical practice by the statistician/data scientist is essential to both cooperative and individual efforts. Similar to communication, consideration of the expertise of others on the team and in the wider scientific community - particularly when it is not in statistics and data science, is an essential feature of ethical teamwork.

DSEC: Similar to communication, the statistician/data scientist must consider how to describe how the system works to team members who bring diverse stakeholder perspectives to the system's design so that missing perspectives can be identified (DSEC C1). The system and its results must be communicated clearly and honestly to members of the team (DSEC C3) – as well as other stakeholders, so that its effects can be fully appreciated and evaluated. Considerations of privacy (DSEC C4) are relevant even for communication within the team, and the auditability of the system (DSEC C5) and explainability of decisions (DSEC D4) must be accessible to team members who are, as well as those who are not, experts in the field. The work should also be describable to other stakeholders (including those on the team, DSEC D5). The team or organization will also need to consider a plan for responding if users or other stakeholders are harmed by a program, algorithm, or system (DSEC E1).

DEFW: Data Work should always be transparent, and workers must consider themselves responsible and accountable (Principle 6).

	C.1 Missing perspectives: Have we sought to address blind spots in the analysis through engagement with relevant stakeholders (e.g., checking assumptions and discussing implications with affected communities and subject matter experts)?	**C.2 Dataset bias**: Have we examined the data for possible sources of bias and taken steps to mitigate or address these biases (e.g., stereotype perpetuation, confirmation bias, imbalanced classes, or omitted confounding variables)?	**C.3 Honest representation**: Are our visualizations, summary statistics, and reports designed to honestly represent the underlying data?	**C.4 Privacy in analysis**: Have we ensured that data with PII are not used or displayed unless necessary for the analysis?	**C.5 Auditability**: Is the process of generating the analysis well documented and reproducible if we discover issues in the future?
ASA A	A2, A4		A3, A4, A5, A7, A8, 11		A5, A11
ASA B	B1, B2		B1, B2, B3, B5, B6	B4	B4, B5
ASA C	C1, C2		C2, C3, C5	C4, C7	
ASA D	D, D1, D6		D5, D6, D7, D10, D11	D4, D5, D9	D6, D8
ASA E	E3		E4		
ASA F	F4		F4	F4	F2
ACM 1	1.4			1.6	
ACM 2			2.1		(2.4)

Table 2.8.1. ASA and ACM support for addressing DSEC C. Analysis questions. *Which DSEC Area C questions are relevant for engaging in team science/teamwork, and which ASA and ACM elements support addressing these DSEC questions?*

	D.1 Proxy discrimination: Have we ensured that the model does not rely on variables or proxies for variables that are unfairly discriminatory?	D.2 Fairness across groups: Have we tested model results for fairness with respect to different affected groups (e.g., tested for disparate error rates)?	D.3 Metric selection: Have we considered the effects of optimizing for our defined metrics and considered additional metrics?	D.4 Explainability: Can we explain in understandable terms a decision the model made in cases where a justification is needed?	D.5 Communicate bias: Have we communi-cated the shortcomings, limitations, and biases of the model to relevant stakeholders in ways that can be generally understood?
ASA A				A1, A4	A2
ASA B				B2, B3, B4	B1, B2, B3
ASA C				C3, C4, C6	C2
ASA D				D10	D6, D10
ASA E					E4
ACM 1					
ACM 2				2.5, 2.9	2.5

Table 2.8.2. ASA and ACM support for addressing DSEC D. Modeling questions. *Which DSEC Area D questions are relevant for engaging in team science/teamwork, and which ASA and ACM elements support addressing these DSEC questions?*

	E.1 Redress: Have we discussed with our organization a plan for response if users are harmed by the results (e.g., how does the data science team evaluate these cases and update analysis and models to prevent future harm)?	E.2 Roll back: Is there a way to turn off or roll back the model in production if necessary?	E.3 Concept drift: Do we test and monitor for concept drift to ensure the model remains fair over time?	E.4 Unintended use: Have we taken steps to identify and prevent unintended uses and abuse of the model and do we have a plan to monitor these once the model is deployed?
ASA A	A2, A3, A4, A9			
ASA B	B5			
ASA C	C2, C4			
ASA D	D1, D4, D5, D6, D7, D8			
ACM 2	2.5			

Table 2.8.3. ASA and ACM support for addressing DSEC E. Deployment questions. Which DSEC Area E questions are relevant for engaging in team science/teamwork, and which ASA and ACM elements support addressing these DSEC questions?

	Principle 1. Start with clear user need and public benefit	Principle 2. Be aware of relevant legislation and codes of practice	Principle 3. Use data that is proportionate to the user need	Principle 4. Understand the limitations of the data	Principle 5. Ensure robust practices and work within your skillset	Principle 6. Make your work transparent and be account-table	Principle 7. Embed data use responsibly
ASA A						A2, A5, A6	
ASA B						B1-B4, B6	
ASA C						C1, C2, C4, C5	
ASA D						D6	
ASA E						E3, E4	
ASA G						G5	
ACM 1						1.2	
ACM 2						2.1, 2.4	

Table 2.8.1. ASA and ACM support for addressing DEFW questions: *Which DEFW (2018) Principles are relevant for engaging in team science/teamwork, and which ASA and ACM elements support addressing these DEFW questions?*

ASA

The ASA Ethical Guidelines include many elements that are specific to working on teams, whether these are made up of non-statisticians (Principle E) or other practitioners (Principle F). There is also a Principle specific for leadership (Principle G). Even if a statistical or data science practitioner is actually filling different roles on different teams or projects, the Guidelines still offer multiple options for practicing ethically in all contexts. These complexities in ethical practice are absent from the DSEC Areas and questions, although Principle 6 of the DEFW (2018) offers opportunities for the ethical practitioner to leverage six of the eight ASA Principles.

ACM

The ACM CE specify (ACM 2.2) that "Professional competence starts with technical knowledge and with awareness of the social context in which their work may be deployed. Professional competence also requires skill in communication, in reflective analysis, and in recognizing and navigating ethical challenges. Upgrading skills should be an ongoing process and might include independent study, attending conferences or seminars, and other informal or formal education. Professional organizations and employers should encourage and facilitate these activities." The CE also specifies (ACM 2.3) that computing professionals should "Know and respect existing rules pertaining to professional work. "Rules" here include local, regional, national, and international laws and regulations, as well as any policies and procedures of the organizations to which the professional belongs." ACM 2.3 further contextualizes the computing professional's work to include both the wider "social context" (ACM 2.2) and the professional context (ACM 2.3). Both of these are essential dimensions to consider when working on teams.

Comment:

As noted earlier, DSEC and DEFW include considerations of users and some of the stakeholders in projects. The first group of stakeholders that should be considered is the team completing the project for which the DSEC and DEFW questions are being answered; this is not explicit in either set of questions. This implies that all teams will function equally well; however, this is not plausible. Thus, both the ASA and ACM practice standards inclusion of specific elements addressing teamwork.

Chapter 2.9
Summary of Section 2

Chapters 2.1 and 2.2-2.8 all show that the ASA GLs and ACM CE can be used to answer the questions posed in DSEC and DEFW. However, discussions of how the DSEC and DEFW questions may serve to promote ethical practice on each of seven common tasks in statistics and data science suggest a few conclusions:

DSEC and DEFW questions can only be answered when there is a specific project in mind. They cannot be used to guide decisions made outside of projects. By contrast, the ASA and ACM professional practice standards are designed to be used throughout practice, whether or not a specific project is in question/in progress.

ASA Principle D is involved in every DSEC Area and DEFW Principle, but personal (ACM 1.1) and professional integrity (ASA Principle A) are not involved in every DSEC Area. The ASA and ACM include specific focus on the effects of ethical practice on the public and the wider scientific community; while directing attention to answering key project-specific questions relating to ethical practice, DSEC and DEFW do not contextualize the importance of the answers to these questions. They therefore have somewhat limited potential to support the development of ethical practitioners.

Further, considerations of other practitioners are missing from DSEC and DEFW. The role of the statistician and/or data scientist in the wider scientific community is lacking from DSEC and DEFW, although the DEFW (2018) Principle 1 focuses on ensuring that the user need for data is sufficiently well articulated so that all other decisions (Principles 2-7) can be balanced against that need. This attribute of the DEFW (2018) Principles means that it treats broader considerations beyond just the project in mind.

By contrast, the ASA GLs include a Principle relating to other research team members (E) and other practitioners (F). In particular, ASA E3 and E4 specify that, when working on teams, the ethical practitioner works to promote excellence in science as well as all collaborators' understanding of the role of statistics in that science (emphasis added):

E3 Ensures that all communications regarding statistical practices are consistent with these Guidelines. *Promotes* transparency in all statistical practices.

E4 Avoids *compromising* validity for expediency. Regardless of pressure on or within the team, does not use inappropriate statistical practices

Similarly, ASA GL F focuses attention on the contributions that ethical practice, and the ethical practitioner, make to the profession and others who use statistical practices (emphasis added):

F2 Helps *strengthen*, and does not undermine, the work of others through appropriate peer review or consultation. Provides feedback or advice that is impartial, constructive, and objective.

F5 *Serves as an ambassador* for statistical practice by promoting thoughtful choices about data acquisition, analytic procedures, and data structures among non-practitioners and students. *Instills appreciation* for the concepts and methods of statistical practice.

Relating ethical practice to supporting the profession, and working respectfully with others (particularly for leadership roles) is included in the ACM CE and ASA GLs, but is missing from DSEC and DEFW (2018). The ACM CE and ASA GLs were created to capture and describe how ethical practitioners – "professionals" in their respective disciplines – can be recognized, including the 'orientation' of the practice standards. This makes sense for those reference documents, and perhaps explains why they are missing from the DSEC and DEFW (2018), which were created to provide structure and a set of project-specific questions, and so are oriented to projects rather than practitioners.

The ACM CE specify the importance of engaging with peers and other experts in the field to consider and evaluate one's work (2.4, 2.5). Moreover, the ACM CE Preamble firmly contextualizes both the computing professional and the importance of ethical work in this domain:

> Computing professionals' actions change the world. To act responsibly, they should reflect upon the wider impacts of their work, consistently supporting the public good. The ACM Code of Ethics and Professional Conduct ("the Code") expresses the conscience of the profession.

The ASA GLs focus attention on stakeholders, the practice/professional, and other stakeholders. This serves to embed the practitioner and their considerations of ethical practice in these wider contexts. The ASA GLs promote stewardship of the profession of statistics and data science, as well as of science itself. The ACM CE promote stewardship of the computing profession by articulating specific roles for leaders/those in leadership positions to preserve and promote ethical practice (ACM Principle 3) and by highlighting the importance of following, and striving to improve, their CE (ACM Principle 4).

In Section 2 we saw that and how the DSEC, DEFW questions, and ASA and ACM practice standards, can support everyday ethical practice in statistics and data science. These documents make up the majority of our prerequisite knowledge (Ethical Reasoning KSA 1). Although several of the tasks and questions mentioned stakeholders, we will return to the stakeholder analysis (Chapter 1.8) in Section 3, enriching our prerequisite knowledge further, as we utilize all six of the ER KSAs to work through 47 vignettes – mini "cases" requiring ethical decision making – sorted according to task (one task with 5-9 vignettes per chapter). Given that the ASA GLs and ACM CE constitute the current consensus on what characterizes an ethical practitioner -i.e., are essential and foundational prerequisite knowledge, readers should consider how the DEFW and DSEC can/should be used in any given project – either to organize, or to augment, the prerequisite knowledge needed to practice ethically. In the next section, the roles of the GLs, CE, DEFW, and DSEC will be explored further – not only to provide additional engagement with these materials, but also to practice utilizing all of this knowledge to identify ethical problems and then come to a decision about the most appropriate response.

Section 3. Using all Six Ethical Reasoning KSAs with the GLs/CE in Practice (case analyses with discussion)

Chapter 3.1 Introduction to Section 3.

Chapter 3.2 Planning/Designing

Chapter 3.3 Data collection/munging/wrangling

Chapter 3.4 Analysis (perform or program to perform)

Chapter 3.5 Interpretation

Chapter 3.6 Documenting your work

Chapter 3.7 Reporting your results/communication

Chapter 3.8 Engaging in team science/ teamwork

Chapter 3.9 Summary chapter: Section 3

Chapter 3.10 Summary of Book

Chapter 3.1
Introduction to Section 3[16]

"In addition to teaching the special obligations of scientists and engineers, ethics instruction can prepare students for challenges that might arise in their professional lives. Ethics education programs typically help students develop the skills to recognize ethical problems, to reason about conflicts between important values, and to evaluate possible actions to address those problems." (National Academy of Engineering, NAE, 2013: p2)

The proliferation of checklists like the DSEC and DEFW can direct the practitioners' or teams' attention to some project-specific issues that need to be discussed, but there is a myriad of other ways that ethically challenging situations can arise in the workplace for the statistician and/or data scientist. Moreover, Loukides et al. (2018) argue that "hard work", i.e., thinking carefully about ethical issues, as is required to utilize the DEFW, **ensures** that "ethical thought doesn't happen". However, to accomplish NAS Recommendation 2.4, would require a working understanding of the professional ethical practice standards in order to *select and effectively* implement the best checklist; otherwise, it might be more likely to take shortcuts and simply go on record as saying that the work has been done (per Loukides et al. 2018). To be ethical practitioners in statistics and data science, one needs to know a) how to utilize the practice standards to figure out what both the "yes" and the "no" answers to any checklist item imply for the practitioner's next steps; and b) what *else* constitutes "ethical practice"; namely, the ethical practice standards themselves.

Both ASA and ACM preambles state that their practice standards offer guidance, not rules; and that each practitioner – while responsible for following their disciplinary guidelines and following local rules/laws – must also utilize their judgement in order to practice ethically. While the DSEC and DEFW are compilations of important considerations, both are project-specific and do not offer any guidance on how to practice ethically or how to recognize an ethical practitioner. The practice standards that the ASA and ACM offer are useful in more situations than just specific projects; the rather limited alignment in DSEC

---

[16] This material is adapted from Tractenberg RE. (2020, February 19). Concordance of professional ethical practice standards for the domain of Data Science: A white paper. Published in the *Open Archive of the Social Sciences* (SocArXiv), 10.31235/osf.io/p7rj2

and DEFW with ASA Principle A and ACM Principle 2 are telling in that A and 2 both discuss professional practice and integrity at all times, and not with respect to a specific project. Ethical reasoning (Chapter 1.1) provides a method by which you can make decisions about how to respond in such situations. As discussed in Chapter 1.1, the first step in this process is to compile and characterize your prerequisite knowledge. We have seen that the ethical practice guidelines from the ASA and the ACM are comprehensive and complex. Knowing that this guidance exists is clearly insufficient for ethical practice, and memorizing their content is also incompatible with the level of engagement with these standards required by the statistical, data science, or computing professional. However, in Section 2, we have seen that the DSEC and DEFW questions can be useful to support ethical practice. In Section 3 we will explore their utility in responding when ethical challenges arise.

Case analysis is a formal method for learning complex qualitative reasoning (see Dow et al. 2015). Simply memorizing the ASA and ACM professional practice guidelines cannot really contribute to "the skills to recognize ethical problems, to reason about conflicts between important values, and to evaluate possible actions to address those problems." (NAE 2013, p2). That is, the abilities to utilize the ethical practice standards in case analyses will require a degree of *ethical reasoning* (Briggle & Mitcham, 2012; see also Tractenberg & FitzGerald, 2012): "Ethics is the effort to guide one's conduct with careful reasoning. One cannot simply claim "X is wrong." Rather, one needs to claim, "X is wrong because (fill in the blank)"." (Briggle & Mitcham, 2012, p. 38) While you may "instinctively" or intuitively know "what to do" or "what is best" in a given situation, the steps or KSAs of ethical reasoning (ER) enable you to do two critical things: 1. Walk through the same process each time you encounter an ethical challenge and decide what to do about it; and 2. Demonstrate or discuss for any observer (yourself; your colleagues; your boss; a potential employer, etc.) that and how you arrive at the decisions you make. This book is intended to allow you to document your thinking, so that once decisions are made, they can be shared and even revisited – possibly when new information becomes available. Sharing your ethical reasoning is an important part of earning the trust of others. The summary chapter of Section 3 explores how documenting your earning of this trust makes your profession, as well as your own professional identity, stronger. For those who do not (yet) feel they can instinctually or intuitively "know what is right" in the practice of statistics and data science, the KSAs of ER can help you become more comfortable with, and confident in, your abilities to reason in the face of uncertainty, and make decisions that you can justify.

The fact that there is rarely "one right answer" in any situation where there is an ethical problem makes it very difficult to teach – and learn - "ethical practice". Therefore, the case analyses featured in Section 3 emphasize the use of the ER KSAs as a dynamic *process*. The process allows you to consider how ethical challenges can be identified by using the ASA Ethical Guidelines and ACM Code of Ethics, and especially your summary of the benefits and harms accruing to stakeholders for each case. The decisions that a "virtue" (using the Guidelines/Code) and a "utilitarian" (considering effects on stakeholders) perspective suggest for each vignette will be considered.

Each of the 47 case analyses included in Section 3 present the ER KSAs as they can be applied to simple vignettes to:

1. **Identify and 'quantify' your prerequisite knowledge:** ASA GLs, ACM EC, whether DSEC or DEFW contribute (and how), and stakeholder analysis.
2. **Identify decision-making frameworks.** The ASA GLs and ACM EC represent a "virtue ethics" decision making framework, which can generally be summarized as, "what would the ethical practitioner do in this case?" Direct appeal to the GL/CE facilitates making decisions that are consistent with "the ethical practitioner", but only when there is an obvious match between the GL/CE and features of the case. When the guidance of the GL/CE is less clear, another framework can be used: The utilitarian framework. The utilitarian perspective can help sort out the positive and negative effects that a decision can have on each of the stakeholders. The utilitarian perspective can generally be summarized as, "how can benefits be maximized while harms are minimized?" While the virtue ethics framework is a more direct appeal to the GL/CE practice standards, the utilitarian framework instead requires both the GL/CE and the stakeholder analysis.
3. **Identify or recognize the ethical issue.** What, in the vignette, is inconsistent with the GLs or EC? What seems "questionable"? Does something about the vignette represent an undue imbalance in the harms and benefits identified in the stakeholder analysis?
4. **Identify and evaluate alternative actions** (on the ethical issue). Just like real life, every ethics case analysis requires that a decision be made. To ensure that viable options are considered, the ER KSAs specify that at least two plausible alternative options (decisions, actions) are both identified *and evaluated*. Each analysis in the next Section considers three decisions that can be made in any circumstance:
   a) do nothing.

    b)   consult or confer with a peer or a supervisor – using the professional guidelines or other resources (e.g., the DSEC) about what to do.

    c)   report violations of policy, procedure, ethical guidelines, or law (and do not do what is asked).

There are always two options in every situation: "do nothing" and "do something". In each case we consider whether "do nothing" is consistent with the ethical practice standards, but we also explore these two variations on "do something" – either consult with others as to how to "do something" or, in some cases, "what <something> to do". In some cases, the ethical issue is clear – and there is a reporting mechanism available, so this is a plausible option ("report"). It is probably only a first step, though, even in those situations where it is in fact available. Simply notifying someone (in authority) that someone has done something unethical will not necessarily address the harms that the unethical behavior has created, so there may be other actions that are also needed. So, for every vignette we consider these three options as a starting point, but the case analyses discuss how to modify these options so that they are relevant responses to the vignette as well as being plausible responses to unethical behaviors.

5.   **Make and justify a decision.** Articulating the decision –what to do in the face of the ethical challenge that was identified – must include at least some discussion of how stakeholder effects were considered.

6.   **Reflect on the decision.** What makes this case "hard"? What additional information would be/would have been helpful? How can you get better at these challenging features of ethical reasoning? How does a case/analysis like this one help create the culture that promotes fluency in ethical reasoning and/or a more ethical workplace?

A critical aspect of KSA 4 (identify and evaluate alternative actions) is that every situation automatically creates two alternative actions: "do *nothing*" and "do *something*". As you can see from both sets of three possible decisions listed above, "do *something*" might be "consult with someone knowledgeable".

The simple vignettes have been abstracted from actual experiences. As we have discussed throughout the book so far, there are seven different types of general tasks in which any statistician or data scientist will engage, and a variety of simple situations ("vignettes") are presented for each task. These vignettes are analyzed following the steps of ethical reasoning to help readers develop an understanding of how they can identify their own options in any given situation. This set of vignettes is not intended to be comprehensive, nor are the analyses of these cases intended to direct you to the "correct" response or decision, even in cases similar to these. Instead, the purpose is to provide more practice and opportunities for you to see how the prerequisite knowledge that was outlined throughout Sections 1 and 2 can be useful to you in the event you ever need to move past KSAs 1-2 and into the decision making KSAs (3-6).

The first case in each set per task has a reflection on the decision. At a minimum, the reader should construct a reflection on each of the subsequent decisions (prompts are provided) in that task's section. Revisit the end of Chapter 1.1 for more information on reflections/how to construct a reflection on a decision and its justification. Readers are invited to work through the six KSAs using different decision-making frameworks (we chose utilitarian and virtue because they are very closely aligned with the ACM and ASA practice standards, respectively). Discussion questions at the end of each section invite reviewers to consider different alternatives than the ones considered. There might be stronger justifications using different dimensions of the same prerequisite knowledge (e.g., a stronger focus on the SHA results; emphasizing a different principle or element of the GLs or CE, etc.), or a justifiable selection of a different one of the alternatives presented in any given case. Readers can re-do any of the analyses or reflect on these possibilities in their own self-constructed reflections. Readers are invited to construct their own reflections of the case as it is analyzed to document their perceptions of how the decision and its justification relate to their practice or experience.

We focused on KSA 1, prerequisite knowledge, throughout Sections 1 and 2 because your knowledge of ethical practice standards is essential for everyday work, and because 90% of the time you will simply be following these practice standards and will not need to move beyond KSA 2! However, in rare cases, you will actually need the other KSAs, starting with needing to identify or recognize an ethical issue or challenge – to which you must respond. Therefore, although the chapters in Section 3 are organized according to task, just like Section 2, in Section 3, each case describes an actual problem arising while you do that task. You will need to apply all the KSAs to contemplate, and then formulate, a justifiable response in each case. All of the KSAs require that you can identify your prerequisite knowledge, so much of the work done in Section 2 will be utilized again.

You will see highlighting in the SHA that was not included when we first discussed stakeholders in Section 1. Dark grey highlights feature potentially-substantial harms – which happen to accrue to the public/public trust; the profession; and unknown individuals in the majority of cases. Light grey highlights draw your attention to two aspects of the harms they identify: firstly, the harms highlighted in light grey do *not* have the potential to be as damaging that the dark grey-highlighted harms do. Thus, the harms in the SHA are not exchangeable (as we mentioned back in Section 1, Chapter 1.7). Secondly, not only are the harms not exchangeable, but also, the harms to you, your boss & employer, and your colleagues may be considered *minor* –particularly as compared to the other harms/potential harms. Different highlighting of the harms points out these differences/their non-exchangeability. Moreover, you may perceive in the light grey highlights that avoiding these minor harms to you, your boss/employer, and your colleagues could be conceptualized as *incentives* to prioritize these stakeholders in behaviors or decisions that are made. Clearly, avoiding inconvenience (to yourself/your colleagues) is a straightforward incentive; our analyses of *all stakeholders* are included to underscore the importance of recognizing that a relatively minor inconvenience is not an adequate tradeoff against a potentially major harm to other stakeholders.

You will also see throughout the cases that the stakeholders, or more correctly, potential harms and benefits to stakeholders, play a different role in the analysis of each case. In some of these cases, there is a fairly clear Guideline (ASA) or Code (ACM) principle that the situation contradicts or with which the situation is inconsistent. In those cases, benefits and harms to stakeholders are less of a concern than ensuring that you respond in a manner that is consistent with professional practice standards or their overarching intentions. In other cases, the potential harms to stakeholders are most clear, and to avoid those harms, guidance from the standards can help. Engaging with all of the cases is intended to help you both solidify your familiarity with professional practice standards and practice ethical reasoning – so that, just in case anything like these cases happens to you in the future, you will be prepared to reason your way through it to a justifiable and defensible decision.

Importantly, you may be familiar with, or have already completed, some federally mandated training in "the responsible conduct of research" or "responsible conduct *in* research". This training is required by US Federal law (America Creating Opportunities to Meaningfully Promote Excellence in Technology, Education, and Science (COMPETES) Act (42 U.S.C. 18620–1)) to be completed by all students whose training or research experience is funded or supported by federal grant funds. A list of topics, deemed to be essential to "responsible conduct of research" by the National Institutes of Health in the United States (https://grants.nih.gov/grants/guide/notice-files/NOT-OD-10-019.html), is shown below. These topics are discussed in Steneck (2009), and used to organize the entire text in Macrina (2014), but are not included in the *Fostering Research Integrity* report from the National Academies of Sciences,

Engineering, and Medicine (2017). These topics are only listed here, and the interested reader should consult the Macrina (2014) book or Steneck (2009). Additionally, online resources have been marshalled by the Resources for Research Ethics Education (RREE) program that was funded at the University of California, San Diego (http://research-ethics.org/topics/overview/). The reason why these topics are not considered fully in this book is that they are primarily motivated by NIH mandate or institutional habit/policies to meet the NIH mandate rather than ethical quantitative practice considerations. Furthermore, a great deal of training in "responsible conduct of research" are tightly consistent with the four main principles of biomedical ethics (Beauchamp & Childress 2001 (B&C):

- **Respect for autonomy:** respecting the decision-making capacities of autonomous persons; enabling individuals to make reasoned informed choices. (Informed consent and privacy/confidentiality fall here)

- **Beneficence:** balancing benefits of treatment against risks/costs.

- **Non maleficence:** avoiding the causation of harm…(do) not harm the patient. All treatment involves some/risk of harm, even if minimal, but the harm should not be disproportionate to the benefits of treatment.

- **Justice:** distributing benefits, risks and costs fairly; patients in similar positions should be treated in a similar manner. (Privacy/ confidentiality falls here)

The cases in this book are, instead, based on actual experiences with statistics and data science. Although they are analyzed with respect to the ER KSAs and intended to reinforce how to reason ethically using the comprehensive practice standards of the ASA and ACM, augmented as relevant by the DSEC and DEFW checklists, the cases also provide opportunities to explore the topics list that are favored/required to be covered by federally funded training in the responsible conduct of research in the US (as of February 2022):

- †conflict of interest – personal, professional, and financial **– and conflict of commitment, in allocating time, effort, or other research resources**
- ‡policies regarding human subjects, live vertebrate animal subjects in research, and safe laboratory practices
- †mentor/mentee responsibilities and relationships
- *collaborative research, including collaborations with industry **and investigators and institutions in other countries**
- *†peer review, including the **responsibility for maintaining confidentiality and security in peer review**

- †data acquisition **and analysis;** laboratory tools (**e.g., tools for analyzing data and creating or working with digital images); recordkeeping practices, including methods such as electronic laboratory notebooks**
- **†secure and ethical data use; data confidentiality,** management, sharing, and ownership
- research† misconduct and policies for handling misconduct
- *responsible authorship and publication
- the scientist as a †responsible member of society

I have included markers in the list above that are not considered in the NIH training lists/materials. Every topic on the list is considered required for US federally funded training **in research,** but not everyone whose training is supported by federal funds will go on to do research. The three topics on the list that would primarily be of interest to academics and researchers are indicated with a *. The principles that are (also) of importance to everyone – irrespective of whether they conduct research – are indicated with a †. You can see that for some topics, the † appears at the initiation of the topic, while in others, it is placed after the word "research" or "scientist" – that is, while this topic list is promulgated for "training in responsible conduct of research", most of these topics are also relevant for those *not doing research.* The four bioethics principles are contained within the single topic indicated with a ‡. You may recongize that the ASA and ACM guidelines include practice-specific ethical guidance for all topics except *conflict of interest.* Since identifying and managing conflicts of interest are typically addressed in institutional policies, the elements in the ASA and ACM practice standards that specify that the ethical practitioner is aware of, and follows, applicable policies and laws/rules is ideal for addressing this topic in a professionally relevant way, rather than in a general, abstract way.

In short, this topics list is simply insufficient to support ethical statistics and data science: the list includes only a few topics in common with the practice standards (the responsibility to know rules and policies; the importance of peer review; misconduct (ASA); features of authorship; and the scientist/ practitioner as a responsible member of society). For example, a focus on the topic, "policies regarding human subjects, live vertebrate animal subjects in research, and safe laboratory practices" suggests that other policies that **are** relevant to the statistician and data scientist – including the ACM membership principle (4) stating that violating the ACM CE is inconsistent with membership - is actually *not relevant according to NIH training* in responsible conduct of research.

Table 3.1.1 shows alignment between the NIH topics list and the case analyses featuring ASA and ACM (and DSEC and DEFW) guidance for each task, but as you will see throughout this Section, the ASA and ACM practice standards include far more – and more directly relevant – material than this topics list suggests is essential for ethical quantitative practice. The ASA and ACM practice standards state that all relevant policies are to be followed – not only those identified for human subjects and animal subjects. That is why the practice standards, and not this topic list, have guided the organization of this book. However, to ensure that these relevant topics can be included within discussions of the 47 cases and the ethical practice standards, questions follow the case analyses in each chapter of Section 3 to prompt the reader to consider these specific topics (should this be necessary) in the development of the functional set of KSAs that can support learning and demonstrating how to respond ethically to challenges in the workplace.

	Planning/ Designing Ch 3.2	Data collection/ munging/ wrangling Ch 3.3	Analysis (perform or program to perform) Ch 3.4	Interpretation Ch 3.5	Documenting your work Ch 3.6	Reporting your results/ communication Ch 3.7	Engaging in team science/ team work Ch 3.8
conflict of interest – personal, professional, and financial – **and conflict of commitment, in allocating time, effort, or other research resources**			x			x	
policies regarding human subjects, live vertebrate animal subjects in research, and safe laboratory practices	x	x					
mentor/mentee responsibilities and relationships					x	x	

	Planning/ Designing Ch 3.2	Data collection/ munging/ wrangling Ch 3.3	Analysis (perform or program to perform) Ch 3.4	Interpretation Ch 3.5	Documenting your work Ch 3.6	Reporting your results/ communication Ch 3.7	Engaging in team science/ team work Ch 3.8
safe research environments (e.g., those that promote inclusion and are free of sexual, racial, ethnic, disability and other forms of discriminatory harassment)							x
collaborative research, including collaborations with industry and investigators and institutions in other countries							
peer review, including the responsibility for maintaining confidentiality and security in peer review					x	x	

	Planning/ Designing Ch 3.2	Data collection/ munging/ wrangling Ch 3.3	Analysis (perform or program to perform) Ch 3.4	Interpretation Ch 3.5	Documenting your work Ch 3.6	Reporting your results/ communication Ch 3.7	Engaging in team science/ team work Ch 3.8
data acquisition and analysis; laboratory tools (e.g., tools for analyzing data and creating or working with digital images); recordkeeping practices, including methods such as electronic laboratory notebooks	x	x					
secure and ethical data use; data confidentiality, management, sharing, and ownership	x	x					

	Planning/ Designing Ch 3.2	Data collection/ munging/ wrangling Ch 3.3	Analysis (perform or program to perform) Ch 3.4	Interpretation Ch 3.5	Documenting your work Ch 3.6	Reporting your results/ communication Ch 3.7	Engaging in team science/ team work Ch 3.8
research misconduct and policies for handling misconduct				x			x
responsible authorship and publication				x		x	
the scientist as a responsible member of society			x				x

Table 3.1.1 *NIH RCR training topics and discussion questions in Section 3 by chapter/task*

The table is simply a suggestion for matching cases within each chapter of Section 3 to the NIH RCR training topics list. For readers who are receiving federal (US) funds for their training in statistics, data science, or research involving these; and for individuals who seek to describe how self-directed learning with this book will satisfy federal requirements for training in the responsible conduct of research (in the United States, see https://grants. nih.gov/grants/guide/notice-files/not-od-10-019.html and https://grants.nih. gov/grants/guide/notice-files/NOT-OD-22-055.html for the updated topics list), the table above can be used with any relevant ancillary materials to round out discussions or considerations of either the US federal policies governing any of those topics (e.g., human subjects and consent). Note that there are no cases directly related to "collaborative research"; this is because the ASA and ACM practice standards do not differ depending on collaborative partners. Any of the cases can be reframed for discussions about collaboration specifically. Resources that are not specific to statistics and data science, including the freely available book by the National Academy of Sciences, National Academy of Engineering, and Institute of Medicine, *"On being a scientist, 3E"* (https://www.nap.edu/download/12192), can be utilized. Another resource (not free!) is the book, *"Scientific Integrity, 4E"* by Macrina (2014) (https://www.amazon.com/Scientific-Integrity-Responsible-Conduct-Research/dp/1555816614). These resources which have not been updated to reflect the 2022 topics list modifications, can augment a Google or Wikipedia search on any of the topics listed above – and can be utilized by an instructor or a self-directed learner to amplify the prerequisite knowledge brought to bear on the analysis of any of the 47 vignettes in this Section.

Chapter 3.2
Planning/Designing

Case 1. You recognize during the planning stage that there is/you have/the team has an incomplete understanding of the problem to be addressed.

Case 2. You are asked to create one computational step in a multi-step process, and *no one will tell you* what will happen with your results.

Case 3. You seek to incorporate sensitivity checks along the planning/development process but meet with resistance.

Case 4. You recognize a better way to achieve a computational result than the proprietary way you were told to follow. Your way takes longer, so there is resistance to trying your method; but you can show it uses less data and results are less biased.

Case 5. You are asked to use a specific analysis or system design that is methodologically inappropriate given the research question or objective.

Case 6. You are asked to design a study or system that will collect either implausible/unreasonably low amounts of data (small sample size) or unnecessarily high amounts of data.

Case 1. You recognize during the planning stage that there is/you have/the team has an incomplete understanding of the problem to be addressed.

1. Identify and 'quantify' prerequisite knowledge:

Which DEFW and DSEC Principles or Areas seem most relevant to this vignette?

This situation is inconsistent with DEFW Principle 1 (and therefore, none of the other DEFW questions can be answered).

Not addressed in DSEC.

Which ASA and ACM Principles and elements seem most relevant to this vignette?

ASA:

Principle A. Professional integrity and accountability: A1, A2, A4, A9, A11
Principle B. Integrity of data and methods: B1, B3, B6
Principle C. Responsibilities to stakeholders: C1, C2, C3, C4, C5
Principle D. Responsibilities to research subjects, data subjects, or those directly affected by statistical practices: D2, D3, D7, D10

ACM:

1. General Ethical Principles
2. Professional Responsibilities
3. Professional Leadership Principles

Potential result:	HARM	BENEFIT	UNKNOWN	UNKNOWABLE
Stakeholder:				
YOU	Figuring out the problem takes time.	No benefits to starting a project without a good understanding of its purpose.		
Your boss/client	Takes time; may alienate client if you suggest they don't know what they want/are doing.	Without clear purpose, we are free to do what we want/what's fastest.		
Unknown individuals	Ineffective design/ poor planning can create bias or permit privacy breaches		Without knowing the problem, a "solution" could be misused and go undetected.	Not all risks/harms can be foreseen and mitigated, but NONE of them can be if you don't know what problem your system/analysis is intended to solve.
Employer	Takes time to make sure problem is concretely defined.			Could limit business/profit

Potential result:	HARM	BENEFIT	UNKNOWN	UNKNOWABLE
Stakeholder:				
Colleagues	Your "need" for a specific problem to solve may complicate colleagues' work, particularly if they are not interested in following, or not required to follow, the GL/CE			Commitment to GL/CE strengthens colleagues' trust in ethical practice
Profession	Failure of project due to lack of focus can reflect badly on, and undermine, the profession			Commitment to GL/CE strengthens trust in profession and enables solutions/conversations
Public/public trust	Ineffective design/poor planning –especially if they lead to misuse/abuse can undermine public trust in systems and the profession		Transparency and focus not be possible if project isn't targeted; misuse could occur and be undetected.	Engagement with GL/CE promotes trust even if misuse does occur

Table 3.2.1. *Stakeholder Analysis: Planning/designing with incomplete understanding of problem*

2. Identify decision-making frameworks. The virtue and utilitarianism frameworks are used.

ASA GLs tend to support a virtue approach: the ethical statistical practitioner cannot fulfill Principles A, B, and D without a focused project objective. Most of Principle C is also impossible without a stated purpose. Thus, planning requires a pre-specified project objective.

ACM CE takes the utilitarian perspective, such that decisions made in practice reduce harms. Since a lack of focus to the project limits the ability to identify and anticipate harms and risks of harms, they cannot be minimized and/or mitigated as required by the CE. Clients or colleagues who do not recognize this need to better understand the ACM CE (Principle 4).

3. Identify or recognize the ethical issue.

Both the ASA and ACM standards require that projects have articulated purposes. The SHA suggests that harms accrue for all stakeholders when there is an incomplete understanding of a project, and while the harms for you and your employer/colleagues may be minor (highlighted in light grey), other harms (highlighted in dark grey) have more serious implications.

4. Identify and evaluate alternative actions (on the ethical issue).

Three decisions that can be made in any circumstance are: a) do nothing; b) consult or confer with a peer (ASA) or a supervisor (ACM); and c) report violations of policy, procedure, ethical guidelines, or law (ASA) or refuse to implement the system (ACM). The practice standards in Chapters 1.4 (ASA) and 1.5 (ACM) are quite specific: ethical practitioners should follow the standards; so, "do nothing" is never consistent with the practice standards. In this case we could change option a) from "do nothing" to "make and document your own decisions", because it is not possible to design or evaluate the designs of any system or analysis that has no firm objective. Although a failure to specify a clear goal makes it impossible for the ethical practitioner to follow their guidelines, this failure itself is not a violation of either standard, so option c "report violation" or "refuse to implement" are also not feasible. Thus, our options in this case are:

a) make and document your own decisions about the objectives;
b) confer with a peer or supervisor to make (and document) the group's decisions about objectives;
c) refuse to engage in the project (at least until the client/project has a firm goal).

5. Make and justify a decision.

Decision: *depending on your role in the project, options a and b are most consistent with the practice standards*. Without a clear understanding of the project goals, no one on the team can follow ASA Principles A and B or ACM Principle 2. The virtue (ASA) and utilitarian (ACM) perspectives both support the decision (of either a) or b), depending on your role) because it results in appropriate ethical practice so you (the practitioner) can continue to follow the relevant practice standards. Note that:

> 1. In case you select option b) and do confer with others and they are unwilling to make decisions leading to a focus/goal, you will either need to choose option a) or c). Again, the choice here will depend on your seniority and experience.

> 2. All three of the options result in either the specification of the problem or in not engaging until the problem is more fully specified. All options honor the ASA and ACM requirements that there be some purpose to which work is directed but that can be assessed for its harms/potential harms and risks.

6. Reflect on the decision.

This case represents a challenge that arises – threatening to prevent you from following the ethical guidelines – from sources that are *not you*. However, clearly you must be prepared to react to challenges like these. In this case, there is no single individual creating the ethical problem; it simply is the case that this situation (no clear project/objective) makes ethical practice impossible. Therefore, the decision about what to do in response to the situation must be a variant of, "make a decision before you begin to practice/do anything". As noted, the exact option will include this, either alone (option a) or in consultation with others (option b). Option c) is to let the client (or wait for the client) to articulate their objective – but still, that must occur before you begin to do anything.

Importantly, refusing to engage until an objective is articulated (or articulating one yourself) passes a *test of justice* because – as the GLs and CE state – ethical work can only proceed with a good understanding of what the objective is. The *test of publicity* might be somewhat embarrassing because it seems like resources are being dedicated to a project with no stated purpose, and so there is no way to determine if the resources were appropriately allocated (thus, whomever allocated the resources may look a bit foolish or unprepared). This is not a failure on the test of publicity! It only underscores how important objectives are! Finally, a *test of universality* is easily passed by all three of these alternative actions, because they all embody the GL and CE perspectives that ethical work can only- ever - proceed with a good understanding of what the objective is.

The ER KSAs, particularly KSA 4, suggest that there might be separate alternative decisions to make, and yet in this case, we could not really choose between them without more information. However, we did see that "do nothing" is never an option that is consistent with the practice standards, but some form of that (refuse to engage in the project) – *can be*, because it represents you taking action to ensure that you are allowed/able to follow the practice standards.

Reflection on the decision is important so that the next time someone encounters a similar situation, they will have a better idea of what to do. In this case, recognizing that sometimes the client's objectives (or the project) are insufficiently well-specified could become a feature of your "intake" process: for example, instituting a policy that "work should not begin on the plan or design until a concrete description of the project/problem to be solved has been given." Creating that kind of policy is consistent with ACM Principle 3, and ASA Principle G, but is also consistent with ASA Principles that articulate that the ethical statistical practitioner needs to follow the ASA GLs, and not other groups' practice standards (e.g., where project definition might not be as foundational) if they conflict with the GLs. Note also that DEFW Principle 1 cannot be addressed in this situation, but DSEC questions can actually be answered even if you don't know the real reason for implementing your project, or the problem it is intended to solve. That represents an opportunity for reflection on the utility of the DSEC when there is no, or an unknown, purpose for the statistical and data science practice that you might engage in.

Case 2. You are asked to create one computational step in a multi-step process, and *no one will tell you* what will happen with your results.

1. Identify and 'quantify' prerequisite knowledge:

Which DEFW and DSEC Principles or Areas seem most relevant to this vignette?

This situation is inconsistent with DEFW Principle 1 (and therefore, none of the other DEFW questions can be answered).
Not specifically addressed in DSEC, but this situation makes it impossible for this individual to be confident about answering any DSEC items, especially in DSEC C (analysis), DSEC D (modeling), and DSEC E (deployment).

Which ASA and ACM Principles seem most relevant to this vignette?

ASA:

Principle A. Professional integrity and accountability
Principle B. Integrity of data and methods
Principle C. Responsibilities to stakeholders
Principle E. Responsibilities to members of multidisciplinary teams

With considerations also described in:
Principle F. Responsibilities to fellow statistical practitioners and the Profession
Principle H. Responsibilities regarding potential misconduct

ACM:

1. General Ethical Principles
2. Professional Responsibilities
3. Professional Leadership Principles

Potential result: Stakeholder:	HARM	BENEFIT	UNKNOWN	UNKNOWABLE
YOU	You may be unqualified (security wise) to know where your work will lead/end up. You may appear less knowledgeable if you design a "generic" system or analysis.	Without clear purpose, I am free to do what I want/what's fastest.	If you innovate "and release", your work could lead to unpredicted harms (or benefits).	
Your boss/client	You may not design the most efficient/effective system or analysis without knowing the ultimate use/goal.	We can re-use any materials the team creates, any way we want, if they don't know why they created it.		
Unknown individuals	Ineffective design/ poor planning can create bias or permit privacy breaches. Overall system design may be weakened by its generality.		Without knowing the problem, a "solution" could be misused and go undetected.	Not all risks/harms can be foreseen and mitigated, but NONE of them can be if you don't know how your system/analysis is intended to be used.

Potential result: Stakeholder:	HARM	BENEFIT	UNKNOWN	UNKNOWABLE
Employer	Takes time to ensure that every one of the computational steps is as efficient as they can be/all team members know where their work will end up.	We can re-use any materials the team creates, any way we want, if they don't know why they created it.	If everyone on the team creates a generic "part", the end result may be too general for utility.	
Colleagues	If everyone on the team creates a generic "part", the end result may be too general for utility. It could take longer to create something totally generic, particularly if colleagues also want to design to be "misuse-proof".	Any collaborator can re-use any materials that anyone on the team creates, any way they want, if the creator doesn't know what is "supposed" to happen next.		

Potential result: Stakeholder:	HARM	BENEFIT	UNKNOWN	UNKNOWABLE
Profession	Your creation of a generic "part", especially if it ends up inefficient or less effective than it could have been, will reflect badly on the profession.			
Public/public trust	Ineffective design can lead to misuse/abuse. Even if it is only one small part of a system, a generic (less efficient) component can undermine public trust in systems and the profession		Transparency and accountability are not truly possible for what your component does in a larger system; misuse of just your part could occur and be undetected.	Full documentation of your plans & the context may help offset any unanticipated abuses, even without knowing what happens before or after your component.

Table 3.2.2. *Stakeholder Analysis: Planning/designing with no knowledge of what happens after your work is done.*

2. Identify decision-making frameworks. The virtue and utilitarianism frameworks are used.

ASA/virtue approach: the ethical statistical practitioner *can* follow Principles A and B without knowing the ultimate purpose of their work, but they cannot ensure that their ethical performance will accompany that work to the ultimate use/user – specifically, responsibilities to stakeholders (ASA C); members of multidisciplinary teams (ASA E); and fellow statistical practitioners and the Profession (ASA F) cannot be met. However, in order to fulfill Responsibilities to research subjects, data subjects, or those directly affected by statistical practices (ASA D), and to ensure confidentiality and privacy obligations are met (ASA B), the strongest protections should be utilized – even if this results in lower efficiency. In this situation, it is possible that there may be some misconduct (statistical, scientific, or professional, ASA H2) that you are being kept unaware of.

ACM/utilitarian perspective: To reduce harms, the strongest protections should be utilized – even if this results in lower efficiency and effectiveness of the work. Full documentation, to ensure any future user can identify and anticipate harms and risks of harms arising from the work, is essential. However, without knowledge of the ultimate use of this work, harms and potential harms/risks cannot be effectively minimized and/or mitigated as required by the CE. Because of this limitation, the ACM CE suggests that the system should not be deployed.

3. Identify or recognize the ethical issue.

SHA suggests that harms accrue for all stakeholders when it is not known how one's work will ultimately be utilized. Just as when there is an incomplete understanding of a project, if the ultimate use of a project/system or component is unknown, there may be risks of harms or actual harms that cannot be predicted – or prevented. In this case, however, benefits for some stakeholders constitute additional, potentially substantial, harms to other stakeholders.

4. Identify and evaluate alternative actions (on the ethical issue).

Three decisions that can be made in any circumstance are: a) do nothing; b) consult or confer with a peer (ASA) or a supervisor (ACM); and c) report violations of policy, procedure, ethical guidelines, or law (ASA) or refuse to implement the system (ACM). As we have seen, "do nothing" is never consistent with the practice standards. In this case we could change option a) from "do nothing" to "do not participate in this project"; which is consistent with the ACM CE and also, enables full compliance with the ASA GLs. Option c "report violation" might be feasible for those following the ACM, since failure to inform an innovator or worker of the ultimate purpose or disposition of their work literally makes it both impossible to predict and mitigate harms, and also could be evidence of someone deliberately trying to create the harms you might avoid if you knew what the ultimate use of your work was. Thus, our options in this case are:

a) refuse to engage in the project (or at least until you are better (fully) informed about the ultimate objective).

b) consult or confer with a peer (ASA) or a supervisor (ACM) on how to obtain, or justify a demand for, full disclosure on what your work will be used for (and do not engage until you obtain that).

c) report the violation of ethical guidelines (ACM) and formally object to the project on the grounds that harms cannot be predicted (and so, do not engage).

5. Make and justify a decision.

Decision: All three options include "do not engage", the context of your withdrawal or refusal depend on your role in the project. Options a and b are similar – you will not engage, and you will ask for more information in both options. Option c is more consistent with ACM – because CE Principles 1 and 2 specify that withdrawing or refusal should be considered whenever full evaluation is impossible or when harms are predictable. The virtue (ASA) and utilitarian (ACM) perspectives both support the decision (of either a) or b), depending on your role), although they do so for slightly different reasons. The ethical statistical practitioner cannot meet their responsibilities under ASA GL Principles C, E, and F, nor can they be sure that their participation will not violate H2, "Avoids condoning or appearing to condone statistical, scientific, or professional misconduct". The computing professional cannot effectively evaluate a system or its real/potential harms.

Note that:

1. ACM 2.3 stipulates that "A computing professional who decides to violate a rule because it is unethical, or for any other reason, must consider potential consequences and accept responsibility for that action." In case you select option b) and do confer with others and they are unwilling to support your requirement of full disclosure, you will either need to choose option a) or c). Again, the choice here will depend on your seniority and experience.

2. All options honor the ASA and ACM requirements that work can be assessed for its harms/potential harms and risks.

6. Reflect on the decision.

Construct your own reflection on this decision.

Case 3. You seek to incorporate sensitivity checks along the planning/ development process but meet with resistance.

1. Identify and 'quantify' your prerequisite knowledge:

Which DEFW and DSEC Principles or Areas seem most relevant to this vignette?

This situation is inconsistent with DEFW Principles 5 (ensure robust practices) and 6 (make your work transparent and accountable). Without sensitivity analyses as the project develops, Principle 4 (understand limitations in the data) will also not be addressable.

The resistance makes it impossible to answer questions in DSEC areas C (analysis) and D (modeling). DSEC items that involve fairness and bias will also not be answerable without sensitivity checks.

Which ASA and ACM Principles seem most relevant to this vignette?

ASA:

Principle A. Professional Integrity and Accountability
Principle B. Integrity of data and methods
Principle E. Responsibilities to members of multidisciplinary teams

With considerations also described in:
Principle D. Responsibilities to research subjects, data subjects, or those directly affected by statistical practices
Principle H. Responsibilities regarding potential misconduct

ACM:

1. General Ethical Principles
2. Professional Responsibilities

Potential result: / Stakeholder:	HARM	BENEFIT	UNKNOWN	UNKNOWABLE
YOU	You may appear less knowledgeable if you design a system or analysis with no checks – and it doesn't work as intended.	I am free to do what I want/what's fastest, there won't be any way to show it wasn't right/fair/effective.	Without sensitivity analyses or checks, your work could lead to unpredicted harms, bias, or unfair results.	Unpredicted harms, biases, or unfair results may be taken up into policy or other systems/analyses and yield additional, unpredictable/ unpredicted harms/bias/ unfair results.
Your boss/client	Without sensitivity checks, the system or analysis may be biased or unfair.	If there is no documentation of bias or unfairness, then no one will know these features were built into the system.	Untested bias can be propagated through any system that relies on this one – even if later systems do include sensitivity checks.	
Unknown individuals	Ineffective design/ poor planning can create bias or permit privacy breaches – if untested, a system could be unfair, insecure, or both. Untested assumptions can lead to incorrect decisions and policy.		If there is no documentation of bias or unfairness, then no one will know these features were built into the system.	Not all risks/harms can be foreseen and mitigated, but NONE of them can be if you don't check for bias, unfairness, and sensitivity to assumptions in your system/ analysis.

Potential result: Stakeholder:	HARM	BENEFIT	UNKNOWN	UNKNOWABLE
Employer	Without sensitivity checks, the system or analysis may be biased or unfair. Decisions based on such a system could be undermined.	If there is no documentation of bias or unfairness, then no one will know these features were built into the system	Untested bias can be propagated through any system that relies on this one – even if later systems do include sensitivity checks.	Undiscovered bias could limit business/profit and ultimately undermine the company/ group.
Colleagues	Colleagues may assume sensitivity checks were done and incorrectly build on biased or unfair foundations.	Work will be completed faster if sensitivity checks are not requested. Liabilities will be (remain) unknown if un-tested – simplifying work.		
Profession	Your creation of a generic "part", especially if it ends up inefficient or less effective than it could have been, will reflect badly on the profession.			Commitment to GL/CE strengthens trust in profession and enables solutions/ conversations

Potential result:	HARM	BENEFIT	UNKNOWN	UNKNOWABLE
Stakeholder:				
Public/public trust	Ineffective design can lead to misuse/abuse. Even if it is only one small part of a system, a generic (less efficient) component can undermine public trust in systems and the profession.		Transparency and accountability are not truly possible if the system has not been checked for bias and sensitivity to assumptions; misuse of a biased system could be undetected or undetectable.	Full documentation of your plans & the context may help offset some unanticipated abuses.

Table 3.2.3. *Stakeholder Analysis: Planning/designing without sensitivity checks*

2. Identify decision-making frameworks.

ASA GLs/virtue approach: This resistance to the inclusion of sensitivity checks may prevent you from following ASA Principles A, B, C, and D and may arise in the context of teamwork – so that you are prevented from following Principle E. The ethical statistical practitioner "avoids condoning or appearing to condone statistical, scientific, or professional misconduct" (ASA H2) – and follows the ASA GLs, even when others on their team seek to follow other guidelines or practice standards (ASA E1). The resistance encountered in this vignette undermines many key aspects of ethical statistical practice.

ACM CE takes the utilitarian perspective, and the SHA shows that many potential harms accrue, without many real benefits, if the resistance to sensitivity checks in your system planning is not overcome. This is the only/best way to anticipate harms and risks of harms, so that they can be minimized and/or mitigated, so this resistance creates and does not address any harms (contrary to most of the ACM CE). Clients or colleagues who do not recognize this – who are resistant to sensitivity checks - need to better understand the ACM CE.

3. Identify or recognize the ethical issue.

Both the ASA and ACM practice standards support the incorporation of sensitivity checks in systems for data analysis. While this is the limit of the ethical issue according to the ASA GL, the ACM also suggests that the computing professional also has an obligation to get the resistant person(s) to comply with the CE, as well as ensuring that the system they design or contribute to is consistent with the CE. The ethical issue is that a person preventing sensitivity checks being built into a system is behaving contrary to the GLs and CE. Moreover, that person is preventing others from following the GLs and CE.

4. Identify and evaluate alternative actions (on the ethical issue).

The three decisions that can be made in any circumstance are: a) do nothing. b) consult or confer with a peer (ASA) or a supervisor (ACM); and c) report violations of policy, procedure, ethical guidelines, or law (ASA) or refuse to implement the system (ACM). Naturally, "do nothing" is never consistent with the practice standards, but we can change option a) from "do nothing" to "ignore the resistance and build sensitivity checks into the system – but do not consult colleagues/report this behavior". Then option b) would become, "consult or confer with a peer (ASA) or a supervisor (ACM) as to how best to respond to a resister, while ignoring that resistance and building sensitivity checks into the system". Option c) must also be modified, because simply reporting the resister (ASA) or refusing to implement the system (without sensitivity checks) (ACM) would likely mean that your time on the project would end, and another person who has no problem designing a system that has no sensitivity checks built in will be brought on. This is clearly inconsistent with the virtue and utilitarian perspectives, not to mention the CE and GLs. So, option c) needs to change to "report violations of policy, procedure, ethical guidelines, or law to a peer (ASA) or a supervisor (ACM), and ignore the resistance and build sensitivity checks into the system."

5. Make and justify a decision.

Decision: *Note that all options include* "build sensitivity checks into the system". That is because the core of response to this situation is to follow the practice standards, and ignore the resistance to what must be considered "best practices". What is less clear is what *else* to do. Option a) says "follow best practices, and beyond that, do nothing (but ignore the resistance)". This would be the *least desirable option* because the person/people who are encouraging you to ignore your professional practice standards need to understand that this is inappropriate. However, if such behavior is common in your work context, it might be impossible for you to do anything except seek another team, other colleagues, or a different job. Option b) would be the *most desirable option* in that consulting with a colleague/supervisor will ensure that this kind of resistance – inappropriate interference with your professional practice - is documented. There may not be policies in your workplace against such interference/violation of the ASA GLs or ACM CE, so reporting (option c) may not be an option. However, in contexts where such policies do exist, it can be difficult to report such behaviors; however, you (the practitioner) can and should continue to follow the other relevant practice standards (i.e., do your job competently!). While it is also important to make sure that you don't get into trouble for ignoring "input" like resistance to incorporating sensitivity checks – even if it is contrary to ethical practice standards – it is also important *to the profession* to support the next practitioner who ends up in a similar situation (or better, to prevent such interference in the future). When you present the plan for your system and it features the sensitivity checks, it might be an opportunity to point out to everyone involved that this is the only way to meet ASA GLs/ACM CE and also enable the team to answer DEFW and DSEC questions.

6. Reflect on the decision.

Construct your own reflection on this decision.

Case 4. You recognize a better way to achieve a computational result than the proprietary way you were told to follow. Your way takes longer, so there is resistance to trying your method; but you can show it uses less data and results are less biased.

1. Identify and 'quantify' prerequisite knowledge:

> Which DEFW and DSEC Principles or Areas seem most relevant to this vignette?

Your solution is consistent with DEFW Principle 1 (and therefore, Principles 3 and 4).

Not specifically addressed in DSEC, but this situation is especially relevant for DSEC C2 and DSEC D3.

> Which ASA and ACM Principles seem most relevant to this vignette?

ASA:

Principle A. Professional integrity and accountability
Principle B. Integrity of data and methods
Principle C. Responsibilities to stakeholders
Principle D. Responsibilities to research subjects, data subjects, or those directly affected by statistical practices

ACM:

1. General Ethical Principles
2. Professional Responsibilities

Potential result: Stakeholder:	HARM	BENEFIT	UNKNOWN	UNKNOWABLE
YOU	Your method takes longer. You may "get in trouble" if you design a system or analysis that doesn't use proprietary methods. If you embed your own methods in a system, you may lose rights to your innovation.	If you use your method, results will be less biased and there will be lower risk of data security issues.		
Your boss/client	Taking longer to create the system could translate to missed deadlines or milestones. New methods may be risky even if they produce less biased results.	Using less data may limit liability and exposure to risk (and may cost less for data access/storage).		
Unknown individuals	Continuing to use existing proprietary methods means risks will not change (will not decrease) relative to "now".	Results from the system with your innovation will be less biased and there will be lower risk of data security issues.		
Employer	Taking longer to create the system could translate to missed deadlines or milestones. New methods may be risky even if they produce less biased results and/or are less expensive data-wise.	Employer may look "innovative" and sensitivity to data security concerns of users if they adopt your method instead of the "old way".		

Potential result: Stakeholder:	HARM	BENEFIT	UNKNOWN	UNKNOWABLE
Colleagues	Your new method may create a requirement for adaptations (requiring extra time and effort from everyone).	Your new method may create opportunities for innovations from everyone.		
Profession	Continuing to use existing proprietary methods means risks of harms will not change (will not decrease) relative to "now".	Demonstrating that this discipline "innovates" in ways that protect the data contributor would show sensitivity to data security concerns of users.		
Public/public trust	Public sentiment about proprietary methods will continue to decline if concerns about bias and data security remain unaddressed.	Demonstrating that this discipline "innovates" in ways that protect the data contributor would show sensitivity to data security concerns of users and might improve public trust.		

Table 3.2.4. *Stakeholder Analysis: Planning/designing with a "new" approach*

2. Identify decision-making frameworks.

The ethical statistical practitioner has responsibilities to "Use methodology and data that are valid, relevant, and appropriate, without favoritism or prejudice, and in a manner intended to produce valid, interpretable, and reproducible results." (ASA A2) and "Make informed recommendations for sample size and statistical practice methodology in order to avoid the use of excessive or inadequate numbers of subjects and excessive risk to subjects." (ASA D2)/ "For animal studies, seek to leverage statistical practice to reduce the number of animals used, refine experiments to increase the humane treatment of animals, and replace animal use where possible" (ASA D3). Moreover, the ethical statistical practitioner "Refrains from collecting or using more data than is necessary. Uses confidential information only when permitted and only to the extent necessary. Seeks to minimize the risk of re-identification when sharing de-identified data or results where there is an expectation of confidentiality. Explains any impact of de-identification on accuracy of results." (ASA D7) Thus, pursuing the new method is consistent with the ASA GLs.

There are more, and more 'significant', benefits to utilizing a new method that conserves data and reduces bias than there are harms that might accrue if the old method continues. The ACM CE, with its utilitarian perspective, typically focuses on limiting harms but the tradeoff of harms and benefits must also be considered (as it is, implicitly, in DEFW Principle 1). Assuming that the proprietary method has "acceptable" data use and bias – which is *unknown* in this case - the SHA shows that many potential real benefits accrue, without many real harms, if a new method that takes longer, but "is better", is employed. The innovation that leads to this method may or may not be part of your "job description", but the prevention of harms and particularly, the avoidance of risk/risks of harms is part of the ACM CE.

3. Identify or recognize the ethical issue.

Both the ASA and ACM practice standards state explicitly and repeatedly that professionals perform only those tasks for which they are competent, and that they do not yield to pressures to behave contrary to their respective practice standards (**ASA**: A; A1, A2, A4, A9, A12; C1, C2; E4; G1; H2; **ACM**: 1.2; 1.3; 2.1; 2.2; 2.3; 3.4; 4). Moreover, both the ASA (ASA Principle G) and ACM CE (Principle 3) also suggest that the practitioner and those in leadership positions have an obligation to ignore the person(s) who want you to violate the ethical practice standards. Both practice standards also articulate a responsibility to encourage others to comply with the standards, if possible. These are specified together with responsibilities to ensure that the system you design or contribute to is correct and appropriate. The primary ethical issue is that a person preventing a professional from doing their job competently is behaving contrary to both the GLs and CE. Moreover, that person is preventing you *and* *others* from following the GLs and CE, but both the GLs and CE specify your responsibility to both follow and encourage others to follow the GLs and CE.

4. Identify and evaluate alternative actions (on the ethical issue).

The three decisions that can be made in any circumstance are: a) do nothing. b) consult or confer with a peer (ASA) or a supervisor (ACM); and c) report violations of policy, procedure, ethical guidelines, or law (ASA) or refuse to implement the system (ACM). Clearly, "do nothing" is both totally inappropriate in this case, and also a violation of the practice standard. We can change option a) from "do nothing" to "ignore the "suggestion" and design the system/analysis appropriately – but do not consult colleagues/report this behavior". Then option b) would become, "consult or confer with a peer (ASA) or a supervisor (ACM) as to how best to respond to a person who wants you to do your job incompetently, while ignoring that "suggestion" and designing the system/analysis appropriately". Option c) must also be modified, because simply reporting the resister (ASA) or refusing to implement the system (using inappropriate methodology) (ACM) could mean that your time on the project would end, and another person who has no problem designing a system or analysis using whatever methods the client requests will be brought on. This is clearly an outcome that is inconsistent with both the virtue and utilitarian perspectives (particularly in light of all the harms in the SHA), not to mention violating the CE and GLs. So, option c) needs to change to "report violations of policy, procedure, ethical guidelines, or law to a peer (ASA) or a supervisor (ACM), and ignore that "suggestion" and design the system/analysis appropriately."

5. Make and justify a decision.

Decision: *Note that all options include* "ignore that "suggestion" and design the system/analysis appropriately". That is because the core of response to this situation is to follow the practice standards – do your job competently - and ignore the resistance to what must be considered "best practices". What is less clear is what *else* to do. Option a) says "follow best practices, and beyond that, do nothing (besides ignore the encouragement to violate practice standards)". This would be the *least desirable option* because the person/people who are encouraging you to ignore your professional practice standards need to understand that this is inappropriate and stop that behavior. However, if such behavior is common in your work context, it might be impossible for you to do anything except seek another team, other colleagues, or a different job. Option

b) would be the *most desirable option* in that consulting with a colleague/ supervisor will ensure that this kind of frankly unethical behavior by a collaborator (or boss) – inappropriate interference with your professional practice - is documented. There may not be policies in your workplace against such interference/violation of the ASA GLs or ACM CE, so reporting (option c) may not be an option. However, in contexts where such policies do exist, it may still be difficult to report such behaviors; but you (the practitioner) can demonstrate your professional competence and follow the relevant practice standards. It is important *to the profession* to support the next practitioner who ends up in a similar situation (or better, to prevent such unethical interference in the future). When you present the plan for your system and it features the appropriate methodology, it might be an opportunity to point out to everyone involved that this is the only way to meet ASA GLs/ACM CE, limit harms/risks of harms, and also enable the team to answer DEFW and DSEC questions.

6. Reflect on the decision.

Construct your own reflection on this decision.

Case 5. You are asked to use a specific analysis or system design that is methodologically inappropriate given the research question or objective.

1. Identify and 'quantify' prerequisite knowledge:

   Which DEFW and DSEC Principles or Areas seem most relevant to this vignette?

This situation is not aligned with DEFW Principle 1, and is inconsistent with Principle 2 (since using *appropriate* **methodology is the hallmark of competent professional practice) making it impossible to address any of the other DEFW questions.**

Not specifically addressed in DSEC, but this situation will specifically make DSEC C and D impossible to address. This situation technically constitutes DSEC E4 ("Have we taken steps to identify and prevent unintended uses and abuse of the model.") **– and the answer here is "no".**

   Which ASA and ACM Principles seem most relevant to this vignette?

ASA:

Principle A. Professional integrity and accountability
Principle B. Integrity of data and methods
Principle C. Responsibilities to stakeholders
Principle D. Responsibilities to research subjects, data subjects, or those directly affected by statistical practices

With considerations also described in:
Principle E. Responsibilities to members of multidisciplinary teams
Principle F. Responsibilities to fellow statistical practitioners and the Profession
Principle H. Responsibilities regarding potential misconduct

ACM:

1. General Ethical Principles
2. Professional Responsibilities

Potential result: / Stakeholder:	HARM	BENEFIT	UNKNOWN	UNKNOWABLE
YOU	You will appear incompetent if you design a system or analysis using inappropriate methodology – *especially* if it appears to work as intended.	There is **no benefit** to following directions that direct you to violate professional practice standards of competence.	Your work could lead to unpredicted harms, bias, or unfair results, and although you were told to design the system/analysis incorrectly, *only you* will appear at fault.	Incorrect/ inappropriate results may be taken up into other applications, or policy and yield additional inappropriate and incompetent work as well as unpredictable/ unpredicted harms/bias/unfair results.
Your boss/client	Inappropriate methodology may lead to "desired" results, but they will not be reproducible or reliable, or correct	Inappropriate methodology may lead to "desired" results (but they will not be reproducible or reliable, or correct.)		Incorrect/inappropriate results may be taken up into other applications, or policy and yield additional inappropriate and incompetent work as well as unpredictable/ unpredicted harms/bias/unfair results.

Potential result: Stakeholder:	HARM	BENEFIT	UNKNOWN	UNKNOWABLE
Unknown individuals	Ineffective design/ poor planning can create bias or permit privacy breaches. Results may be unfair, insecure, or both – and decisions made by incorrectly-designed systems may be unauditable. Untested – known to be incorrect - assumptions can lead to incorrect decisions and policy.			Incorrect/inappropriate results may be taken up into other applications, or policy and yield additional inappropriate and incompetent work as well as unpredictable/ unpredicted harms/bias/ unfair results.
Employer	Ineffective design/ poor planning can create bias or permit privacy breaches. The company will appear to employ incompetent workers if it is discovered that inappropriate methods were used. Results may be unfair, insecure, or both – and decisions made by incorrectly-designed systems may be unauditable. Untested – known to be incorrect - assumptions can lead to incorrect decisions and policy.			The company will appear to employ incompetent workers- and supervisors - if it is discovered that inappropriate methods were directed to be used.

Potential result: Stakeholder:	HARM	BENEFIT	UNKNOWN	UNKNOWABLE
Colleagues	Colleagues may assume the work was done competently and incorrectly build on inappropriate, biased, or otherwise unfair foundations.	Colleagues who are *not* instructed to use inappropriate methods may carry out tests that demonstrate the system or analysis cannot move forward with this design.		
Profession	The profession may appear incompetent when it is discovered that practitioners can or do design systems or analyses using inappropriate methodology.	There is **no benefit** to the profession that accrues from following directions that violate professional practice standards of competence.	Inappropriate work could lead to unpredicted harms, bias, or unfair results, and although you were told to design the system/analysis incorrectly, *only the practitioner* will appear at fault.	Incorrect/inappropriate results may be taken up into other applications, or policy and yield additional inappropriate and incompetent work as well as unpredictable/unpredicted harms/bias/unfair results. Trust in the profession will be undermined (why not use amateurs – they're cheaper and would be expected to make the same mistakes – for less!)

Potential result:	HARM	BENEFIT	UNKNOWN	UNKNOWABLE
Stakeholder:				
Public/public trust	Inappropriate methodology can lead to conflicting results (best case) or misuse/abuse (worst case). Even if it is only one small part of a system, in inappropriate component can lead to many harms and undermine public trust in systems and the profession.			Trust in the profession will be undermined (it doesn't matter if you have a degree in that field, "those people" cannot be trusted to perform competently).

Table 3.2.5. *Stakeholder Analysis: Planning/designing with inappropriate methodology*

2. Identify decision-making frameworks.

The ethical statistical practitioner has responsibilities to "Use methodology and data that are valid, relevant, and appropriate, without favoritism or prejudice, and in a manner intended to produce valid, interpretable, and reproducible results." (ASA A2) and "Avoid condoning or appearing to condone statistical, scientific, or professional misconduct. Encourage other practitioners to avoid misconduct or the appearance of misconduct." (ASA H2) Thus, following the directions to use an inappropriate method is fully inconsistent with the ASA GLs.

The SHA shows that there are no legitimate benefits to following directions to use inappropriate methodology, and the harms are significant – to the practitioner as well as other stakeholders. ACM CE, with its utilitarian perspective, typically focuses on limiting harms but the tradeoff of harms and benefits must also be considered (as it is, implicitly, in DEFW Principle 1).

3. Identify or recognize the ethical issue.

We could *assume* that the method or system has "acceptable" properties, although in most cases, what is "accept**ed**" is not generally what users or other stakeholders - including data contributors or those put at risk for harm due to biased results – would call "accept**able**". However, in this case, "what has always been done", i.e., using the specific analysis method or system that is known to be inappropriate, is inconsistent with the responsibilities to use best practices that are described in the GLs and CE. That is, knowingly using inappropriate methodology is unethical. Using an appropriate method or system is the *ethical* choice.

4. Identify and evaluate alternative actions (on the ethical issue).

The three decisions that can be made in any circumstance are: a) do nothing. b) consult or confer with a peer (ASA) or a supervisor (ACM); and c) report violations of policy, procedure, ethical guidelines, or law (ASA) or refuse to implement the system (ACM). In this case, "do nothing" means "use the specified –inappropriate- method", and we have seen above that this is frankly unethical –so obviously, not a legitimate option. We can change option a) to "use a different, appropriate, method or refuse to do the analysis/design the system". Option b) would become, "consult or confer with a peer (ASA) or a supervisor (ACM) as to how best to document that the recommended method is inappropriate, and use a different, appropriate, method". Option c) can be modified to, "report the use (and recommendation) of inappropriate methodology –which may require documentation that the recommended method is inappropriate, and use a different, appropriate, method." Note that there have been well-publicized cases where experts using appropriate methodology generate results contrary to what was expected using the inappropriate methodology, and are harassed or otherwise punished for not violating the law and the professional practice standards (Note also that intimidation to coerce violations of practice standards are also violations of law and the GLs).

5. Make and justify a decision.

Decision: *Note that all options include* "use an *appropriate* method". The level of pressure that is brought to bear on the request to use inappropriate methodology must be kept in mind as one of the options is selected. Option a) is supported if you are able to make such decisions without justification or approval; so, its viability depends on your seniority or role on the team (in case you need approval to make any changes to the analysis or design). Option b) might be the *most desirable option* because consulting with a colleague/supervisor will raise awareness of the fact that the specified methodology is inappropriate, but will also generate some documentation that this is the case. Option b) is the most desirable because it results in both a conversation about what should be done (that is appropriate) and some documentation about what should *not* be done, and why. This will be particularly useful if you need to convince others on the team to both utilize a different method and also engage in the opportunities for them to also innovate (changing this from a harm to a benefit from the SHA). Option c) is an important one to consider if you encounter pressure, intimidation, or harassment from those who insist on obtaining specific results (which are guaranteed by the inappropriate method), or who insist that inappropriate method be used because of some expedience rather than rigor, reproducibility, or the public good. Thus, option b) may be the best choice generally, your situation may make option c) a better choice. Certainly, the documentation of why the specified methodology is inappropriate will help others make decisions about method choice in the future.

6. Reflect on the decision.

Construct your own reflection on this decision.

Case 6. You are asked to design a study or system that will collect either implausible/unreasonably low amounts of data (small sample size) or unnecessarily high amounts of data.

1. Identify and 'quantify' prerequisite knowledge:

> Which DEFW and DSEC Principles or Areas seem most relevant to this vignette?

This situation is inconsistent with DEFW Principle 1, but is also specifically contrary to Principle 3. Because of these issues, none of the other DEFW questions can be answered.

Not specifically addressed in DSEC, but this situation creates challenges for data security (DSEC B) and DSEC E4 – unintended uses of too little or too much data are difficult to predict.

> Which ASA and ACM Principles seem most relevant to this vignette?

ASA:

Principle A. Professional Integrity and Accountability
Principle B. Integrity of data and methods
Principle C. Responsibilities to stakeholders
Principle D. Responsibilities to research subjects, data subjects, or those directly affected by statistical practices
Principle H. Responsibilities regarding potential misconduct

ACM:

1. General Ethical Principles
2. Professional Responsibilities

Potential result: Stakeholder:	HARM	BENEFIT	UNKNOWN	UNKNOWABLE
YOU	Too little data suggests you are not competent. Too much data suggests you did not understand the objective of the analysis or system (but designed it anyway – so, you appear incompetent).	There is **no benefit** to following directions that direct you to violate professional practice standards of competence.	Your work could lead to unpredicted harms, bias, or unfair results, and although you were told to design the system/analysis incorrectly, *only you will appear at fault*	Incorrect/ inappropriate results may be taken up into other applications, or policy and yield additional inappropriate and incompetent work as well as unpredictable/ unpredicted harms/bias/ unfair results.
Your boss/client	Too little data will yield irreproducible results. Too much data can create risks for the data contributors that are unwarranted.	Instructions to collect an unreasonable sample size (too big or too small) may arise from "good intentions" or be consistent with prior work.		

Potential result:	HARM	BENEFIT	UNKNOWN	UNKNOWABLE
Stakeholder:				
Unknown individuals	Inappropriate sample sizes create risks of irreproducible (and incorrect) results (too small) or, increase risks associated with data storage and security (too big).	If the justification for the incorrect sample size is "historical", the use of previously used methods *may mean* risks will not change (will not decrease) relative to "now".		Incorrect/ inappropriate results may be derived from the inappropriate sample size; the wrong sample size or results may then be taken up into other applications, or policy and yield additional inappropriate and incompetent work as well as unpredictable/ unpredicted harms/bias/ unfair results.
Employer	Too little data will yield irreproducible results. Too much data can create risks for the data contributors that are unwarranted.	There is **no benefit** to insisting that professionals follow directions that to violate professional practice standards of competence.		

Potential result:	HARM	BENEFIT	UNKNOWN	UNKNOWABLE
Stakeholder:				
Colleagues	Too little data will yield irreproducible results. Too much data can create risks for the data contributors that are unwarranted. If others on the team are not aware that the directions to collect inappropriate sample sizes came from supervisors and not the statistician/data scientist, colleagues may mistakenly believe the sample size is correct – and erroneously base their work on that.	There is **no benefit** to other colleagues when the data to be collected is too little or too much.		

Potential result: Stakeholder:	HARM	BENEFIT	UNKNOWN	UNKNOWABLE
Profession	The profession may appear untrustworthy when it is discovered that practitioners can or do design systems or analyses using inappropriate methodology –particularly when the sample size is too large, and people's data are put at risk (for security/ privacy breaches).	There is **no benefit** to the profession that accrues from following directions to violate professional practice standards of competence.	Inappropriate sample sizes could lead to unpredicted harms, bias, or unfair results, and although you were told to design the system/ analysis incorrectly, *only the practitioner* will appear at fault. Harms will and do accrue to the profession when those entrusted to act professionally fail to do so, even when they are "just following orders".	Incorrect/ inappropriate results may be taken up into other applications, or policy and yield additional inappropriate and incompetent work as well as unpredictable/ unpredicted harms/bias/unfair results. Trust in the profession will be undermined – if competent professionals cannot be trusted to sample appropriately, then why bother with them at all?
Public/ public trust	Public sentiment about contributing or sharing data will continue to decline if concerns about bias, reproducibility of results, and data security remain unaddressed.	There are only harms, *and no benefits,* to the public and public trust when too much or too little data are purposefully sought		

Table 3.2.6. *Stakeholder Analysis: Unreasonable sample size*

2. Identify decision-making frameworks.

The ethical statistical practitioner has responsibilities to "Use methodology and data that are valid, relevant, and appropriate, without favoritism or prejudice, and in a manner intended to produce valid, interpretable, and reproducible results." (ASA A2) and "Strives to avoid the use of excessive or inadequate numbers of research subjects—and excessive risk to research subjects" (ASA D2). Moreover, the ethical statistical practitioner must follow ASA C4 "Informs stakeholders of the potential limitations on use and re-use of statistical practices in different contexts and offers guidance and alternatives, where appropriate, about scope, cost, and precision considerations that affect the utility of the statistical practice." Thus, designing a system with an inappropriate sample, whether too big (ASA D7) or too small, is *inconsistent* with the ASA GLs.

There are no real benefits whatsoever to collecting too little data – even cost savings cannot be counted a benefit when the result is too little data for generalizable or reproducible results. If there is too little data to support sufficient rigor and reproducibility, the analyses will only need to be done again with a larger sample. Similarly, too much data may be highly attractive to data thieves; or those who were entrusted with data based on informed consent may succumb to desires to utilize the data for other – not-approved, not-consented-to uses of that data (ASA D5).

The ACM CE, with its utilitarian perspective, typically focuses on limiting harms but following the directions to design a system that collects an inappropriate sample size has no benefits, with many harms. The prevention of harms and particularly, the avoidance of risk/risks of harms is part of the ACM CE, making it totally inconsistent with the CE to follow these directions.

3. Identify or recognize the ethical issue.

Both the ASA and ACM practice standards state explicitly and repeatedly that professionals perform only those tasks for which they are competent, and that they do not yield to pressures to behave contrary to their respective practice standards (**ASA**: A; A1, A2, A4, A9, A12; C1, C2; E4; G1; H2; **ACM**: 1.2; 1.3; 2.1; 2.2; 2.3; 3.4; 4). Moreover, both the ASA (ASA Principle G) and ACM CE (Principle 3) also suggest that the practitioner and those in leadership positions have an obligation to ignore the person(s) who want you to violate the ethical practice standards. Both practice standards also articulate a responsibility to encourage others to comply with the standards, if possible. Leaders should also ensure that the system, analysis, or modeling they and other statistics and data science practitioners design or contribute to is correct and appropriate (ASA Principle G, ACM Principle 3). The primary ethical issue is that a person is trying to prevent a professional from doing their job competently; the individual directing you to collect too much or too little data is behaving contrary to the GLs and CE. Moreover, that person is preventing others (or trying to) from following the GLs and CE.

4. Identify and evaluate alternative actions (on the ethical issue).

As usual, the three decisions that can be made in any circumstance are: a) do nothing. b) consult or confer with a peer (ASA) or a supervisor (ACM); and c) report violations of policy, procedure, ethical guidelines, or law (ASA) or refuse to implement the system (ACM). Clearly, "do nothing" is both totally inappropriate in this case, and also a violation of the practice standard. We can change option a) from "do nothing" to "ignore the request and design the system/analysis to collect an appropriate sample size – but do not consult colleagues/report this behavior". Then option b) would become, "consult or confer with a peer (ASA) or a supervisor (ACM) as to how best to respond to a person who wants you to do your job incompetently, while ignoring that request and designing the system/analysis to collect the appropriate amount of data". Option c) must also be modified, because simply reporting the individual who directs you to violate the GLs (ASA) or refusing to implement the system (to collect an inappropriate sample size) (ACM) could mean that your time on the project would end, and another person who has no problem designing a system or analysis to do whatever is requested irrespective of the violation of the ASA and ACM practice standards will be brought on. This is clearly an outcome that is inconsistent with both the virtue and utilitarian perspectives (particularly in light of all the harms in the SHA), not to mention violating the CE and GLs. So, option c) needs to change to "report violations of policy, procedure, ethical guidelines, or law to a peer (ASA) or a supervisor (ACM), and ignore that request and design the system/analysis to collect an appropriate sample size."

5. Make and justify a decision.

Decision: *Note that all options include* "ignore that request and design the system/analysis to collect an appropriate sample size". That is because the core of response to this situation is to **follow the practice standards** – do your job competently - and ignore the resistance to, or direction to violate, both the guidelines and what must be considered "best practices". What is less clear is what *else* to do. Option a) says "follow best practices, and beyond that, do nothing (besides ignore the encouragement to violate practice standards)". This would be the *least desirable option* because the person/people who are encouraging you to ignore your professional practice standards need to understand that this is inappropriate and should stop that behavior. However, if such behavior is common in your work context, it might be impossible for you to do anything except seek another team, other colleagues, or a different job. Option b) would be the *most desirable option* in that consulting with a colleague/supervisor will ensure that this kind of frankly unethical behavior by a collaborator (or boss) – inappropriate interference with your professional practice - is documented. There may not be policies in your workplace against such interference/violation of the ASA GLs or ACM CE, so reporting (option c) may not be an option, making it a slightly less desirable option than b. However, in contexts where such policies do exist, it may still be difficult to report such behaviors; but you (the practitioner) can – and should - still demonstrate your professional competence and follow the relevant practice standards. One thing worth considering is that the person directing you to violate your professional practice standards may not be aware of them/this fact. So, option c may not be as much about "reporting" that person as it is about "educating" them. It is important *to the profession* to support the next practitioner who ends up in a similar situation (or better, to prevent such unethical interference in the future), so it is definitely worth considering politely notifying the individual that their direction is a violation of the ACM CE, and you being directed to violate the CE and GLs is impossible. When you present the plan for your system and it features the appropriate sample size, it would be an opportunity to point out to everyone involved –and everyone in attendance - that this is the only way to meet ASA GLs/ACM CE, limit harms/risks of harms, and also enable the team to answer DEFW and DSEC questions.

6. Reflect on the decision.

Construct your own reflection on this decision.

Summary of six case analyses for Planning/designing

We have seen in the six case analyses of Chapter 3.2 that, although the 'theme' of these cases is "planning and design", and we followed the same 6 KSAs of ethical reasoning, and used all the same prerequisite knowledge, each analysis was slightly different. Some similarities in these analyses should be pointed out, specifically, harms *always* accrue, from unethical behavior (or direction/encouragement to disregard best practices/professional practice standards), but benefits *do not always* accrue. We also saw that harms and benefits are definitely not exchangeable – although the reader is encouraged to review each case and think of other harms and benefits for each stakeholder in each case. Whether you are more inclined to follow the ASA GLs or the ACM CE, consideration of harms is important (although explicitly to be avoided/minimized by ACM CE).

We also saw that, when discussing how to perform "planning and design" ethically in Section 2, general principles from the GLs or CE, and particular questions from the DSEC and DEFW, could be used to ensure that our general behavior is ethical. However, in response to specific challenges and ethical violations, DSEC and DEFW were *not* helpful.

Finally, each of the cases in this chapter were the result of *others* trying to impede professional and ethical practice. Exploring how to identify what those impediments actually are requires a good working understanding of the GLs and CE. Responding also features the GLs and CE, and it might be useful (as we discussed in case 6 in this chapter) to share the GLs and CE as widely as possible in your working context – to ensure that people are not surprised when you seek to adhere to professional practice standards. The emphasis in ACM CE Principles 3 and 4 (and ASA Principle G and Appendix) on ensuring that all professionals are allowed to work in ethical contexts suggests that publicising the GLs and CE in the workplace is an important aspect of ethical statistics and data science practice.

Questions for Discussion:

1. Review at least two cases in this chapter.

> A. Do you agree with the decisions in these case analyses? Why or why not? Which parts of the ER process are most and least acceptable for each case? Do these parts differ for different cases in this chapter? Were there different ASA or ACM Principles or elements that you identified, and if so, how do these affect your case analyses? Are the elements and/or Principles you identify represented in your reflections?

B. Discuss your results for the tests of universality, justice, and publicity for any (or all) case(s) in this chapter.

C. Discuss the relevance to each case – including the analysis and your reflection – of federal, state, or local/organizational policies regarding human subjects, live vertebrate animal subjects in research, and safe laboratory practices. How can you incorporate what you know about these policies into any part of the case analysis (prerequisite knowledge; identifying the ethical issue; determining alternative actions; making or justifying your decision; or reflection)? Note that, if there is no alignment between federal, state, or your local/institutional policies and a given case, you can comment on that (e.g., do policies in your organization focus on *misconduct* – and ignore detrimental practices?). Do you feel that these policies could have helped in the reasoning, decision or justification?

D. Discuss the relevance to each case - including the analysis and your reflection – of federal, state, or local/organizational policies relating to data acquisition and laboratory tools; management, sharing and ownership. How can you incorporate what you know about these policies into any part of the case analysis (prerequisite knowledge; identifying the ethical issue; determining alternative actions; making or justifying your decision; or reflection)? If there are no relevant policies, you should comment on that, particularly when a) data are obtained using federal monies (grants); or b) data are collected (acquired) from humans, such that Federal Regulations about the treatment of human subjects and the acquisition of tissues and data from humans are relevant.

2. Review at least two (different) cases in this chapter.

Keeping in mind that doing nothing/not responding are *not plausible responses,* are there other plausible alternatives (KSA 4) that you can think of for any case? If not, discuss that; if so, list them and discuss their evaluation (are they equally consistent with GL/CE, do they lead to similar decisions, etc.).

3. Choose any case in this chapter. Redo the analysis, but feature a different GL or CE principle in your reasoning process. Make sure the *justification* is different from what is given. Is the *decision* also different? Discuss your decision with its justification and the tests of universality, justice, and publicity.

4. Consider all six cases in this chapter.

A. Comment on the role of the stakeholder analysis in these decisions. Were stakeholder impacts featured in justifications? Were they features in your own reflections?

B. Comment on the importance of your reflections on the decisions you were asked for: for others doing a similar task at work; for those using results from this task; and for those who might join the profession (and be learning about planning and design).

C. Consider your professional identity and the profession itself: planning/designing projects and data collection/analysis projects is obviously foundational in statistics and data science. How can these six cases and the decisions reached via the analyses presented earn you (the decider) the trust of the practicing community or the public for the profession and its future?

Chapter 3.3 Data Collection/Munging/Wrangling

Case 7. A plan is created to collect data that cannot possibly be housed securely.

Case 8. Data collection is carried out by scraping the Internet; you notice that at least some of the time, the results of confidentiality and privacy breaches get swept up in the scraping.

Case 9. Your supervisor directs you to assume that if *any* of the data in your collection was obtained with any level of consent (whether none *or* opt-out), then treat *all* of the data as if it was obtained "with consent".

Case 10. Standard Operating Procedures (SOP) manuals direct you to ignore data provenance.

Case 11. You discover that there has been no consent obtained for any of the data you are asked to collect/wrangle/munge.

Case 12. You have collected/wrangled data from multiple sources and provenance information about the data is inconsistent – different people at work describe it differently and there's no real evidence about the provenance of *any* of the data.

Case 7. A plan is created to collect data that cannot possibly be housed securely.

1. Identify and 'quantify' prerequisite knowledge:

Which DEFW and DSEC Principles or Areas seem most relevant to this vignette?

This situation is inconsistent with DEFW Principle 2, but may also be specifically contrary to Principle 1, in case the benefit to the public is unknown, unclear, or limited. Without a clear benefit to offset the obvious risk to the security of the data in this case, none of the other DEFW questions can be answered satisfactorily.

Note that this situation means that, even if you answer "yes" to DSEC B1 ("do we have a *plan* to protect and secure data"), that plan <u>cannot possibly work</u> in this case. The point in this case is that even if you have a plan, *it will not work*. Thus, DSEC E2 cannot possibly work, and the answer to DSEC E4 must be "no".

Which ASA and ACM Principles and elements seem most relevant to this vignette?

ASA:

Principle A. Professional integrity and accountability (A3; A9; A11; A12)
Principle B. Integrity of data and methods (B1; B3)
Principle C. Responsibilities to stakeholders (C1; C4; C7)
Principle D. Responsibilities to research subjects, data subjects, or those directly affected by statistical practices (D1; D4; D5; D9; D10; D11)
Principle H. Responsibilities regarding potential misconduct: (H1, H2)

ACM:

1. General Ethical Principles (1.2; 1.7)
2. Professional Responsibilities (2.1; 2.3; 2.5; 2.9)

Potential result: Stakeholder:	HARM	BENEFIT	UNKNOWN	UNKNOWABLE
YOU	Planning to collect data that cannot be secured violates the data contributor rights as well as ethical practice standards.	While a violation of multiple specific elements of the professional practice standards, unsecured data are less expensive to maintain, and may be accessed by multiple users (potentially increasing risks to data contributors). IF **all contributors acknowledge that their data will be unsecured and consent anyway,** lack of security can simplify systems.	Failing to secure data could lead to unpredicted harms, bias, or unfair results, with the statistician or data scientist bearing responsibility for misuse, unauthorized access, or losses of that data.	Failure to secure data could lead to unpredictable/ unpredicted harms/bias/ unfair results, and may also suggest to others/other system developers that collecting data that cannot be secured is OK, even though this directly violates practice standards.

Potential result: Stakeholder:	HARM	BENEFIT	UNKNOWN	UNKNOWABLE
Your boss/client	Failing to secure data could lead to unpredicted harms, bias, or unfair results, and can create risks for the data contributors that were foreseeable.	IF all contributors acknowledge that their data will be unsecured and consent anyway, lack of security can simplify systems and enable extra or additional data collection.		
Unknown individuals	Failure to secure data could lead to unpredicted harms, bias, or unfair results, as well as to misuse by others.	IF all contributors acknowledge that their data will be unsecured and consent anyway, then the data collection can be made more seamless for the data contributor.		Incorrect/inappropriate results may be derived from the inappropriate sample size; the wrong sample size or results may then be taken up into other applications, or policy and yield additional inappropriate and incompetent work as well as unpredictable/ unpredicted harms/bias/unfair results.

Potential result: Stakeholder:	HARM	BENEFIT	UNKNOWN	UNKNOWABLE
Employer	Failing to secure data could lead to unpredicted harms, bias, or unfair results, and can create risks for the data contributors that were foreseeable, thus incurring liability to the employer.	IF all contributors acknowledge that their data will be unsecured and consent anyway, lack of security can simplify systems, make data storage more affordable, and enable extra or additional data collection.		
Colleagues	If others on the team are not aware that the data are not secure, colleagues may mistakenly refuse to share data, or may take the usual –but in this case, unnecessary- precautions to secure it on their end.	IF all contributors acknowledge that their data will be unsecured and consent anyway, all colleagues can freely access all the data, and it can also be shared as widely as is desired.		

Potential result: Stakeholder:	HARM	BENEFIT	UNKNOWN	UNKNOWABLE
Profession	The profession may appear untrustworthy when it is discovered that practitioners can or do design systems or analyses that collect, but do not secure, data – particularly if consent is obtained but the lack of data security is not part of the informedness. Collecting data that cannot be secured, without fully informing data contributors of this fact, is a violation of multiple specific elements of the professional practice standards.	Unsecured data may be accessed by multiple users, and so would be available for evaluation or use by others in the profession. **IF all contributors acknowledge that their data will be unsecured and consent anyway**, lack of security can simplify sharing and access of the data.	Failing to secure data could lead to unpredicted harms, bias, or unfair results, with the statistician or data scientist bearing responsibility for misuse, unauthorized access, or losses of that data. This would harm the profession.	Failure to secure data could lead to unpredictable/ unpredicted harms/bias/ unfair results, and may also suggest to others/other system developers that collecting data that cannot be secured is OK, even though this directly violates practice standards.
Public/ public trust	Public sentiment about contributing or sharing data will continue to decline if public concerns about data security, or the lack of honesty in data collection systems, continue.	**IF all contributors acknowledge that their data will be unsecured and consent anyway,** each individual's contributed data may go further, promote more innovation or discovery.		

Table 3.3.1. *Stakeholder Analysis: Collecting unsecurable data*

2. Identify decision-making frameworks.

The ethical statistical practitioner has responsibilities to "Understand and conform to confidentiality requirements for data collection, release, and dissemination and any restrictions on its use established by the data provider (to the extent legally required). Protects the use and disclosure of data accordingly. Safeguards privileged information of the employer, client, or funder." (ASA C7). Moreover, the ethical statistical practitioner "Understands the provenance of the data, including origins, revisions, and any restrictions on usage, and fitness for use prior to conducting statistical practices." (ASA D10). Thus, designing a system to collect data that cannot possibly be secured, violates the ASA GLs – unless all contributors acknowledge that their data will be unsecured *and consent anyway*. Obviously, promising security of data– or failing to fully inform data contributors of the lack of data security, violates both ASA C7 and D10, but also violates ASA GL Principle H2 ("Avoid condoning or appearing to condone statistical, scientific, or professional misconduct. Encourage other practitioners to avoid misconduct or the appearance of misconduct."). In fact, failing to inform data donors that their data cannot be secured may be considered misconduct, but telling data donors their data *will be secure* when it is known to be impossible is not only inconsistent with the ASA GLs, but it can also be considered scientific, statistical, and professional misconduct.

There are some benefits to the system designers and those responsible for data security that accrue from collecting data with no plan to secure it, including both cost savings and the potential to share the data for other – not-approved, not-consented-to - uses of that data. However, none of these benefits is realized if there *are plans to secure the data* but those plans predictably **cannot succeed**. That is the situation in this case: it is known that the data cannot possibly be secured.

The ACM CE, with its utilitarian perspective, focuses on limiting harms and in this case, there are many harms, and only benefits if all data contributors acknowledge the lack of data security and consent to contribute data anyway. In any case where that does not happen, then *all harms would accrue and no benefits*. The ACM CE specify that, "In cases where misuse or harm are predictable or unavoidable, the best option may be to not implement the system." (ACM 2.9).

3. Identify or recognize the ethical issue.

Both the ASA and ACM practice standards state explicitly that professionals must secure data and honor agreements to maintain that data securely – as long as this is a condition of data contribution. The primary ethical issue is that both GLs and CE require that data be obtained with consent and the contributors' knowledge of what the data will be used for – but the expectation of privacy and security is a feature of securing that consent, more often than not. In this case we do not know what type of data is being collected– nor what types of expectations of privacy contributors have. If all contributors acknowledge that their data will be unsecured *and consent anyway*, or if contributors have no reasonable expectation of privacy of their data (e.g., they are counted entering a public place by a turnstile, or their image is captured using public space cameras), then a failure to secure the data is not a concern for either the data collector or contributor, and **there is no ethical issue**.

4. Identify and evaluate alternative actions (on the ethical issue).

As usual, the three decisions that can be made in any circumstance are: a) do nothing. b) consult or confer with a peer (ASA) or a supervisor (ACM); and c) report violations of policy, procedure, ethical guidelines, or law (ASA) or refuse to implement the system (ACM). Clearly, "do nothing" is both totally inappropriate in this case, and also a violation of the practice standards relating to respecting data contributors – unless you can be certain that all contributors acknowledge that their data will be unsecured *and consent anyway*, or the data are contributed with no expectation of privacy or security. Thus, we can change option a) from "do nothing" to "do nothing if all contributors acknowledge that their data will be unsecured *and consent anyway* or there is no expectation of privacy or security"; clearly option a) is only a reasonable option in that specific situation. Then, **only in the situation where option a) is NOT viable**, option b) would become, "consult or confer with a peer (ASA) or a supervisor (ACM) as to how best to ensure there is no expectation of data privacy or security; AND if this cannot be ensured, then consult/confer as to how to prevent this data collection effort". Option c) must also be modified, because simply reporting the project that collects data in violation of the practice standards will not necessarily result in preventing that data from being collected in violation of the data contributors' expectations and the standards. So, option c) needs to change to "report violations of policy, procedure, ethical guidelines, or law to a peer (ASA) or a supervisor (ACM), and prevent this data collection effort."

5. Make and justify a decision.

There are really two viable decisions in this case, and the selection depends on whether or not the contributors of data have an expectation of privacy and data security. If they do not, **and that is consistent with institutional and local/federal law** (rather than simply being how organizations like yours treat data contributors), then there is no ethical issue in collecting data that cannot be secured – making option a) your choice: as long as all contributors acknowledge that their data will be unsecured *and consent anyway* or there is no expectation of privacy or security. As noted, if the conditions that make option a) reasonable are not met, then it there **is** an ethical problem and option a) cannot be considered. This response began with "there are *two* viable decisions" because options b) and c) both include the same action: "prevent this data collection effort". In the case that data contributors do not know their data cannot be secured, or are misled into believing that it will be when it can't be, then it is unethical to collect that data, so your decision must be to prevent the data from being collected. This can be accomplished either by notifying contributors that their data will not be secured (so there is no expectation of privacy/security, and they are able to consent anyway or to withhold consent because of this lack of security) or, by ensuring that such data are not collected. Note that options b) and c) are not the same as "withdrawing from the project", but are consistent with "refusing to deploy the system", which is one option repeatedly referred to in the ACM CE when predictable risks cannot be avoided.

6. Reflect on the decision.

Note that there are cases where data might be collected and where people who contribute that data would have no problems with it not being secured; one example might be the being-counted-by-a-turnstile; or cameras on roads in the US that capture people exceeding the speed limit. Driving down a public road is a public act, and while people might not want it publicized that they got a speeding ticket, individuals can have no claim or expectation to privacy when breaking the law in such a public way. Note also, however, that if some individuals do design data collection systems that cannot be secured, but there *is* a promise, and an expectation of privacy and security of the data, but lawless individuals steal ("collect") that data and fail to secure it, the data contributors still do have that expectation of privacy. No ethical practitioner would then analyze or even consider working with data with such provenance; this is the specific subject of ASA GLs D10 and D11. ACM CE 1.6 also addresses the considerations of provenance that computing professionals should give before utilizing data.

This case is important because of the fact that in some situations, the exact same case may lead to an ethical dilemma (where the only options are "prevent the data from being collected" or "ensure that data contributors are aware that their data are not secured, and hope they consent anyway") or to *no ethical dilemma* at all (in the case where there can be no reasonable expectation of privacy or security of their data). This is one example of what both the GLs and

CE note in their preambles: there is no specific rule or algorithm that can apply in every case, and the ethical practitioner is familiar with the ethical practice standards and also knows how and when to apply them.

A final note: In some contexts, data are collected without the knowledge of the data contributor. This is contrary to both the ASA GLs and the ACM CE: both state that human data contributors must consent to the collection and use of their data when this is specifically collected. There are federal and state/local regulations about informing data contributors, and ensuring they consent to that data contribution; there are also institutional policies –particularly in academia and the pharmaceutical industry – relating to considerations of the rights of all humans to determine what data is collected from them. Even if a group, organization, or company "usually" collects data without informed consent from the contributors (e.g., online sellers collecting data on purchases), it does not relieve them of the burden of securing that data if the contributor has a reasonable expectation that the data should be secured; just as their usual practice of not honouring basic human rights to determine what of their data is collected does not obviate the responsibilities of any and all data scientists and statisticians – and those who use computing and statistics, per the CE and GLs – to respect these fundamental human rights. This is one very important reason why the first item on the DSEC is **A.1 Informed consent**: If there are human subjects, have they given informed consent, where subjects affirmatively opt-in and have a clear understanding of the data uses to which they consent? And the first question after ensuring that collecting any data is justifiable (principle 1) in DEFW is **Principle 2.** Be aware of relevant legislation and codes of practice. Note that neither DSEC nor DEFW articulates any priority for "your organizational practice" – both DSEC and DEFW recognize the fundamental human right to direct and determine whether and how their personal/ contributed data are used.

Case 8. Data collection is carried out by scraping the Internet; you notice that at least some of the time, the results of confidentiality and privacy breaches get swept up in the scraping.

1. Identify and 'quantify' prerequisite knowledge:

> Which DEFW and DSEC Principles or Areas seem most relevant to this vignette?

This situation is inconsistent with DEFW Principle 2, since "relevant legislation" as well as "codes of practice" require that human subjects give consent and are notified about the uses to which their data will be put. If the extent of data from breaches –i.e., data for which you do not have consent – is unknown, then it is impossible to fulfill Principle 4 (understand the limitations of the data).

In this case, the answer to the first question in the DSEC set, A1 (have human subjects given informed consent) is "no" – because it cannot be answered confidently in the affirmative.

> Which ASA and ACM Principles seem most relevant to this vignette?

ASA:

Principle C. Responsibilities to stakeholders
Principle D. Responsibilities to research subjects, data subjects, or those directly affected by statistical practices
Principle H. Responsibilities regarding potential misconduct

ACM:

1. General Ethical Principles
2. Professional Responsibilities

Potential result: Stakeholder:	HARM	BENEFIT	UNKNOWN	UNKNOWABLE
YOU	Using data with unknown provenance, particularly if some of it is known not to have been contributed with informed consent, violates the data contributor rights as well as ethical practice standards.	While a violation of multiple specific elements of the professional practice standards and some laws, scraping data is a relatively inexpensive way to collect it.	Failing to obtain consent to use data and using stolen data (the results of breaches) could lead to unpredicted harms, bias, or unfair results, as well as risks to data contributors - with the statistician or data scientist bearing responsibility for misuse, unauthorized access, or losses of that data.	Using stolen data and data that was not contributed with informed consent may suggest to others/other system developers that collecting stolen/breached data is OK, even though this directly violates practice standards.
Your boss/client	Data with unknown provenance can lead to unpredicted harms, bias, or unfair results, and can create risks for the data contributors that were foreseeable.	Ignoring the provenance of data, while a violation of multiple specific elements of the professional practice standards, is simpler and cheaper than ensuring data are obtained with proper consent.		

Potential result: Stakeholder:	HARM	BENEFIT	UNKNOWN	UNKNOWABLE
Unknown individuals	Data that is stolen or accessed without authorization and consent (i.e., from breaches) may expose data contributors to risks, as well as to (further) misuse by others.	There are no benefits that accrue when stolen data are utilized, and no one stops the use of that kind of data.		Using stolen data and data that was not contributed with informed consent may suggest to others/other system developers that collecting stolen/breached data is OK, even though this directly violates practice standards.
Employer	Using data with unknown provenance could lead to unpredicted harms, bias, or unfair results, and can create risks for the data contributors that were foreseeable, thus incurring liability to the employer.	Ignoring the provenance of data, while a violation of multiple specific elements of the professional practice standards, is simpler and cheaper than ensuring data are obtained with proper consent.		

Potential result: Stakeholder:	HARM	BENEFIT	UNKNOWN	UNKNOWABLE
Colleagues	If others on the team are not aware that the data provenance is mixed (and some is stolen), colleagues may mistakenly share –i.e., further the misuse of- the data.	While a violation of multiple specific elements of the professional practice standards and some laws, scraping data is a relatively inexpensive way to collect it.		
Profession	The profession may appear untrustworthy when it is discovered that practitioners use whatever data they collect, even if no consent was given or even if the collected data was "available" because of a data breach. Using data with unknown provenance is a violation of multiple specific elements of the professional practice standards.	There are no benefits that accrue when stolen data are utilized, and no one stops the use of that kind of data.	Failing to obtain consent to use data and using stolen data (the results of breaches) could lead to unpredicted harms, bias, or unfair results, as well as risks to data contributors - with the statistician or data scientist bearing responsibility for misuse, unauthorized access, or losses of that data.	Using stolen data and data that was not contributed with informed consent may suggest to others/other system developers that collecting stolen/breached data is OK, even though this directly violates practice standards. This decrements professional integrity in a concrete way.

Potential result:	HARM	BENEFIT	UNKNOWN	UNKNOWABLE
Stakeholder:				
Public/public trust	Public sentiment about the security of their data will continue to worsen, and people will become less inclined to contribute or sharing data if public concerns about data security, or the lack of honesty in data collection systems, continue.	There are no benefits that accrue when stolen data are utilized, and no one stops the use of that kind of data.		

Table 3.3.2. *Stakeholder Analysis: Collecting data that includes un-consented and/or stolen data*

2. Identify decision-making frameworks.

Data that are "scraped from the Internet" cannot be considered to have been contributed to the planned analyses or system with consent. However, some data (e.g., counts of persons walking through a turnstile) simply cannot be deemed "data you must consent to contribute". The problem in this case is that some of the data being scraped is the result of confidentiality and security breaches, meaning that it is unlikely that innocuous data like turnstile counts are among what else is being scraped/collected in this vignette.

The ethical statistical practitioner "understands the provenance of the data, including origins, revisions, and any restrictions on usage, and fitness for use prior to conducting statistical practices." (ASA D10), and "does not conduct statistical practice that could reasonably be interpreted by subjects as sanctioning a violation of their rights. Seeks to use statistical practices to promote the just and impartial treatment of all individuals." (ASA D11) Thus, utilizing data with unknown provenance violates the ASA GLs. Obviously, failing to protect basic human rights by using data you do not have consent to use violates ASA D10 and D11, but also violates ASA H2 ("Avoid condoning or appearing to condone statistical, scientific, or professional misconduct. Encourage other practitioners to avoid misconduct or the appearance of misconduct.")

The ACM CE, with its utilitarian perspective, focuses on limiting harms and in this case, there are many harms, and no real benefits, to using data that was contributed to your analysis without consent, or data with unknown provenance.

3. Identify or recognize the ethical issue.

Both the ASA and ACM practice standards state explicitly that professionals must respect laws and practice standards to ensure data are collected with the knowledge and consent of contributors. The primary ethical issue in this case is that both GLs and CE require that data be obtained with consent and the contributors' knowledge of what the data will be used for, but when data are scraped, this is unlikely. That is compounded by the fact that the scraping may incorporate stolen data into the final data set. In this case we do not know what type of data is being collected– nor what types of expectations of privacy contributors have – although the fact that data from breaches is in the mix means there must be some expectation of privacy – at least for that stolen/breached data – and possibly, for the rest of the data as well.

4. Identify and evaluate alternative actions (on the ethical issue).

As usual, the three decisions that can be made in any circumstance are: a) do nothing. b) consult or confer with a peer (ASA) or a supervisor (ACM); and c) report violations of policy, procedure, ethical guidelines, or law (ASA) or refuse to implement the system (ACM). Clearly, "do nothing" is both totally inappropriate in this case, and also a violation of the practice standards relating to respecting data contributors. We can change option a) from "do nothing" to "do not proceed with any use of the data until provenance can be established or data from breaches can be removed". Then option b) would become, "consult or confer with a peer (ASA) or a supervisor (ACM) as to how best to stop the collection and any use of the data until provenance can be established or data from breaches can be removed from the scraping". Option c) must also be modified, because simply reporting the fact that the data collection violates the GLs (ASA), or refusing to implement the system (that scrapes up stolen/breached data) (ACM) might not ensure that the data with unknown provenance is both not used and also, its collection ceases. So, option c) needs to change to "report violations of policy, procedure, ethical guidelines, or law to a peer (ASA) or a supervisor (ACM), and stop the collection and any use of the data until provenance can be established or data from breaches can be removed from the scraping."

5. Make and justify a decision.

Decision: *Note that all options include* "do not proceed with any use of the data". That is because the core of response to this situation is to **follow the practice standards** and do not utilize data without the consent of the contributors. What is less clear is what *else* to do. Option a) says "follow the ethical practice standards, and do not use inappropriately-obtained data". This would be the *least desirable option* because the data collection – as well as its use – also needs to stop, because both are inappropriate. However, if data with unknown provenance is commonly collected in your work context, it might be impossible for you to do anything ethical except seek another team, other colleagues, or a different job. Option b) would be the *most desirable option* in that consulting with a colleague/supervisor will ensure at least some kind of notification that, or publicity for the fact that, data collection without consent is frankly unethical, as is the use of such inappropriately obtained data. There may not be policies in your workplace against such interference/violation of the ASA GLs or ACM CE, so reporting (option c) may not be a viable option, making it a slightly less desirable option than b. However, in contexts where such policies do exist, it may still be difficult to report such behaviors; but you (the practitioner) can – and should - still demonstrate your professional competence and follow the relevant practice standards. Option c may not be as much about "reporting" a person or practice as it is about "educating" people or your company/

organization about the practice standards. It is important *to the profession* to support the next practitioner who ends up in a similar situation (or better, to prevent such unethical data collection in the future), so it is definitely worth considering politely notifying relevant individuals that the data collection system violates the ACM CE, and you cannot be directed to violate the CE and GLs (see ASA GL Principle H). When you present the system and it features modifications that may limit the amount of data but also limit the likelihood of including stolen/breached data, it would be an opportunity to point out to everyone involved –and everyone in attendance - that this is the only way to meet ASA GLs/ACM CE, limit harms/risks of harms, and also enable the team to answer DEFW and DSEC questions.

6. Reflect on the decision.

Construct your own reflection on this decision.

Case 9. Your supervisor directs you to assume that if *any* of the data in your collection was obtained with any level of consent (whether none *or* opt-out), then treat *all* of the data as if it was obtained "with consent".

1. Identify and 'quantify' prerequisite knowledge:

> Which DEFW and DSEC Principles or Areas seem most relevant to this vignette?

This situation is inconsistent with DEFW Principle 2, since "relevant legislation" as well as "codes of practice" require that human subjects give consent and are notified about the uses to which their data will be put. If the extent of data from breaches –i.e., data for which you do not have consent – is unknown, then it is impossible to fulfill Principle 4 (understand the limitations of the data).

In this case, the answer to the first question in the DSEC set, A1 (have human subjects given informed consent) is "no" – because it cannot be answered confidently in the affirmative.

> Which ASA and ACM Principles seem most relevant to this vignette?

ASA:

Principle C. Responsibilities to stakeholders
Principle D. Responsibilities to research subjects, data subjects, or those directly affected by statistical practices
Principle H. Responsibilities regarding potential misconduct

ACM:

1. General Ethical Principles
2. Professional Responsibilities

Potential result: Stakeholder:	HARM	BENEFIT	UNKNOWN	UNKNOWABLE
YOU	Using data with unknown provenance, particularly if some of it is known not to have been contributed with informed consent, violates the data contributor rights as well as ethical practice standards.	There are no benefits that accrue when data obtained without appropriate consent. This supervisor is directing you to violate law as well as practice standards.	Failing to obtain consent to use data could lead to unpredicted harms, bias, or unfair results, as well as risks to data contributors - with the statistician or data scientist bearing responsibility for misuse, unauthorized access, or losses of that data.	Following these directions, and using data that was not contributed with informed consent, may suggest to others that ignoring data provenance is OK, even though this directly violates practice standards.
Your boss/client	Data with unknown provenance can lead to unpredicted harms, bias, or unfair results, and can create risks for the data contributors that were foreseeable.	Ignoring the provenance of data, while a violation of multiple specific elements of the professional practice standards, is simpler and cheaper than ensuring data are obtained with proper consent.		

Potential result: Stakeholder:	HARM	BENEFIT	UNKNOWN	UNKNOWABLE
Unknown individuals	Data that is accessed without authorization and consent may expose data contributors to risks, as well as to (further) misuse by others.	There are **no benefits** that accrue when data are utilized in unauthorized ways, and no one stops this type of inappropriate behavior.		Using data that was not contributed with informed consent may suggest to others that data provenance is not important, even though this directly violates practice standards as well as basic human rights.
Employer	Data with unknown provenance can lead to unpredicted harms, bias, or unfair results, and can create risks for the data contributors that were foreseeable, thus incurring liability to the employer.	Ignoring the provenance of data, while a violation of multiple specific elements of the professional practice standards, is simpler and cheaper than ensuring data are obtained with proper consent.		
Colleagues	If others on the team are not aware that the data provenance is mixed, colleagues may mistakenly share –i.e., further the misuse of- the data.	There are **no benefits** that accrue when data are utilized in unauthorized ways, and no one stops this type of inappropriate behavior.		

Potential result: Stakeholder:	HARM	BENEFIT	UNKNOWN	UNKNOWABLE
Profession	The profession may appear untrustworthy when it is discovered that practitioners use whatever data they collect, even if no consent was given or even if the collected data was "available" because of a data breach. Using data with unknown provenance is a violation of multiple specific elements of the profess-sional practice standards.	There are no benefits that accrue when stolen data are utilized, and no one stops the use of that kind of data.	Failing to obtain consent to use data and using stolen data (the results of breaches) could lead to unpredicted harms, bias, or unfair results, as well as risks to data contributors - with the statistician or data scientist bearing responsibility for misuse, unauthorized access, or losses of that data.	Using stolen data and data that was not contributed with informed consent may suggest to others/other system developers that collecting stolen/breached data is OK, even though this directly violates practice standards. This decrements professional integrity in a concrete way.
Public/public trust	Public sentiment about the procedures by which they do and do not give their consent will continue to worsen, and people will become less inclined to contribute or share their data if public concerns about the lack of honesty in data collection systems, continue.	There are **no benefits** that accrue when consents are not obtained lawfully, and data are inappropriately utilized, and no one stops the use of that kind of data.		

Table 3.3.3. *Stakeholder Analysis: Assuming any observed level of consent pertains to all data*

2. Identify decision-making frameworks.

The ethical statistical practitioner "informs stakeholders of the potential limitations on use and re-use of statistical practices in different contexts and offers guidance and alternatives, where appropriate, about scope, cost, and precision considerations that affect the utility of the statistical practice." (ASA C4). Moreover, the ethical statistical practitioner "understands the provenance of the data, including origins, revisions, and any restrictions on usage, and fitness for use prior to conducting statistical practices." (ASA D10). Thus, utilizing data with unknown provenance violates the ASA GLs. Obviously, failing to protect basic human rights by using data you do not have consent to use violates both ASA C4 and D10, but also violates ASA H2 ("Avoid condoning or appearing to condone statistical, scientific, or professional misconduct. Encourage other practitioners to avoid misconduct or the appearance of misconduct.").

The ACM CE, with its utilitarian perspective, focuses on limiting harms (ACM 1.2) and in this case, there are many harms, and no benefits, to using data that was contributed to your analysis without consent, or data with unknown provenance. In this case, a person in a position of leadership is directing you to violate the ACM CE, which is *itself* a violation of the ACM CE.

3. Identify or recognize the ethical issue.

Both the ASA and ACM practice standards state explicitly that professionals must respect laws and practice standards to ensure data are collected with the knowledge and consent of contributors. The primary ethical issue in this case is that both GLs and CE require that data be obtained with consent and the contributors' knowledge of what the data will be used for, but your supervisor is directing you to ignore whatever level of consent – or the lack thereof – was obtained and use some "other" level of provenance for all data. In this case we do not know what type of data is being collected– nor what types of expectations of privacy contributors have – but it is clear that at least some people do have expectations (i.e., the ones that chose not to opt-out), while at least some others were never informed, and their data were simply taken without consent. Utilizing data that was obtained without appropriate consent – or data with unknown or questionable provenance is unethical. In this case, a person in a position of leadership is directing you to violate the practice standards as well as possibly other applicable laws about informed consent – also unethical.

4. Identify and evaluate alternative actions (on the ethical issue).

As usual, the three decisions that can be made in any circumstance are: a) do nothing. b) consult or confer with a peer (ASA) or a supervisor (ACM); and c) report violations of policy, procedure, ethical guidelines, or law (ASA) or refuse to implement the system (ACM). Clearly, "do nothing" is both totally inappropriate in this case, and also a violation of the practice standards relating to respecting data contributors and data provenance. We can change option a) from "do nothing" to "do not proceed with any use of the data until provenance can be established". Then option b) would become, "consult or confer with a peer (ASA) or a supervisor (ACM) as to how best to stop the collection and any use of the data until provenance can be established or data obtained without consent can be removed." Option c) must also be modified, because simply reporting the fact that the data collection violates the GLs (ASA), or refusing to implement the system (that scrapes up stolen/breached data) (ACM) might not ensure that the data with unknown provenance is both not used and also, its collection ceases. So, option c) needs to change to "report violations of policy, procedure, ethical guidelines, or law to a peer (ASA) or a supervisor (ACM), and stop the collection and any use of the data until provenance can be established or data obtained without consent can be removed."

5. Make and justify a decision.

Decision: *Note that all options include* "do not proceed with any use of the data". That is because, like the previous case, the core of response to this situation is to **follow the practice standards** and to utilize data according to the consent of the data contributors. Option a) says "follow the ethical practice standards, and do not use inappropriately-obtained data". This would be the *least desirable option* because the data collection – as well as its use – also needs to stop, because both are inappropriate. Moreover, the person instructing you to violate practice guidelines is also behaving inappropriately. This direction may be given because data with unknown or mixed provenance is commonly collected in your work context; it might be impossible for you to do anything ethical in this case/context except seek another team, other colleagues, or a different job. Option b) would be the *most desirable option* in that consulting with a colleague/supervisor will ensure some notification of leadership or your colleagues (preferably, both) that data collection without consent is frankly unethical, as is the use of such inappropriately obtained data. There may not be policies in your workplace against such interference/violation of the ASA GLs or ACM CE, so *reporting* (option c) may not be a viable option, making it a slightly less desirable option than b. However, in contexts where such policies do exist, it may still be difficult to report such behaviors; but you (the practitioner) can – and should - still demonstrate your professional competence and follow the relevant practice standards. Option c may not be as much about "reporting" a person or practice as it is about "educating" people or your company/organization about the practice standards. It is important *to the profession* to support the next practitioner who ends up in a similar situation (or better, to prevent such unethical data collection in the future), so it is definitely worth considering politely notifying relevant individuals that the data collection system violates the ACM CE, and you cannot be directed to violate the CE and GLs. Engaging in discussion of alternative data collection methodologies –that *at a minimum* ensure that all data are in fact collected with the same actual – and not assumed (as the supervisor is directing you to do) – level of consent by data contributors would be an opportunity to point out to everyone involved that this is the only way to meet ASA GLs/ACM CE, limit harms/risks of harms, and also enable the team to answer DEFW and DSEC questions.

6. Reflect on the decision.

Construct your own reflection on this decision.

Case 10. Standard Operating Procedures (SOP) manuals direct you to ignore data provenance.

1. Identify and 'quantify' prerequisite knowledge:

Which DEFW and DSEC Principles or Areas seem most relevant to this vignette?

This situation is inconsistent with DEFW Principle 2, since "relevant legislation" as well as "codes of practice" require that human subjects give consent and are notified about the uses to which their data will be put. If the extent of data from breaches –i.e., data for which you do not have consent – is unknown, then it is impossible to fulfill Principle 4 (understand the limitations of the data).

In this case, the answer to the first question in the DSEC set, A1 (have human subjects given informed consent) is "no" – because it cannot be answered confidently in the affirmative.

Which ASA and ACM Principles seem most relevant to this vignette?

ASA:

Principle C. Responsibilities to stakeholders
Principle D. Responsibilities to research subjects, data subjects, or those directly affected by statistical practices
Principle E. Responsibilities to members of multidisciplinary teams
Principle H. Responsibilities regarding potential misconduct

ACM:

1. General Ethical Principles
2. Professional Responsibilities

Potential result: Stakeholder:	HARM	BENEFIT	UNKNOWN	UNKNOWABLE
YOU	Using data with unknown provenance, particularly if some of it is known not to have been contributed with informed consent, violates the data contributor rights as well as ethical practice standards.	There are **no benefits** that accrue when data obtained without appropriate consent. This supervisor is directing you to violate law as well as practice standards.	**Failing to obtain consent to use data could lead to unpredicted harms, bias, or unfair results, as well as risks to data contributors** – with the statistician or data scientist bearing responsibility for misuse, unauthorized access, or losses of that data.	Following these directions, and using data that was not contributed with informed consent, may suggest to others that ignoring data provenance is OK, even though this directly violates practice standards.
Your boss/client	Data with unknown provenance can lead to unpredicted harms, bias, or unfair results, and can create risks for the data contributors that **were** foreseeable.	Ignoring the provenance of data, while a violation of multiple specific elements of the professional practice standards, is simpler and cheaper than ensuring data are obtained with proper consent.		

Potential result: Stakeholder:	HARM	BENEFIT	UNKNOWN	UNKNOWABLE
Unknown individuals	Data that is accessed without authorization and consent may expose data contributors to risks, as well as to (further) misuse by others.	There are **no benefits** that accrue when data are utilized in unauthorized ways, and no one stops this type of inappropriate behavior.		Using data that was not contributed with informed consent may suggest to others that data provenance is not important, even though this directly violates practice standards as well as basic human rights.
Employer	Data with unknown provenance can lead to unpredicted harms, bias, or unfair results, and can create risks for the data contributors that were foreseeable, thus incurring liability to the employer.	Ignoring the provenance of data, while a violation of multiple specific elements of the professional practice standards, is simpler and cheaper than ensuring data are obtained with proper consent.		

Potential result: Stakeholder:	HARM	BENEFIT	UNKNOWN	UNKNOWABLE
Colleagues	If others on the team are not aware that the data provenance is mixed, colleagues may mistakenly share –i.e., further the misuse of- the data.	There are **no benefits** that accrue when data are utilized in unauthorized ways, and no one stops this type of inappropriate behavior.		
Profession	The profession may appear untrustworthy when it is discovered that practitioners use whatever data they collect, even if no consent was given or even if the collected data was "available" because of a data breach. Using data with unknown provenance is a violation of multiple specific elements of the professional practice standards.	There are no benefits that accrue when stolen data are utilized, and no one stops the use of that kind of data.	Failing to obtain consent to use data and using stolen data (the results of breaches) could lead to unpredicted harms, bias, or unfair results, as well as risks to data contributors – with the statistician or data scientist bearing responsibility for misuse, unauthorized access, or losses of that data.	Using stolen data and data that was not contributed with informed consent may suggest to others/other system developers that collecting stolen/breached data is OK, even though this directly violates practice standards. This decrements professional integrity in a concrete way.

Potential result:	HARM	BENEFIT	UNKNOWN	UNKNOWABLE
Stakeholder:				
Public/public trust	Public sentiment about the procedures by which they do and do not give their consent will continue to worsen, and people will become less inclined to contribute or share their data if public concerns about the lack of honesty in data collection systems, continue.	There are **no benefits** that accrue when consents are not obtained lawfully, and data are inappropriately utilized, and no one stops the use of that kind of data.		

Table 3.3.4. *Stakeholder Analysis: Policy directs you to ignore data provenance*

2. Identify decision-making frameworks.

The ethical statistical practitioner "Informs stakeholders of the potential limitations on use and re-use of statistical practices in different contexts and offers guidance and alternatives, where appropriate, about scope, cost, and precision considerations that affect the utility of the statistical practice." (ASA C4). Moreover, the ethical statistical practitioner "Understands the provenance of the data, including origins, revisions, and any restrictions on usage, and fitness for use prior to conducting statistical practices." (ASA D10). Thus, utilizing data with unknown provenance violates the ASA GLs. Obviously, failing to protect basic human rights by using data you do not have consent to use violates both ASA C4 and D10, but also violates ASA GL Principle H2 ("Avoid condoning or appearing to condone statistical, scientific, or professional misconduct. Encourage other practitioners to avoid misconduct or the appearance of misconduct.").

The ACM CE, with its utilitarian perspective, focuses on limiting harms and in this case, there are many harms, and no benefits, to using data that was contributed to your analysis without consent. Moreover, a system that collects data without obtaining consent may include methodologies or other features that are insecure or otherwise create risks (in addition to risks of confidentiality and privacy breaches) that are not addressed – because the manner in which data are collected may not be legitimate or sufficiently specified.

3. Identify or recognize the ethical issue.

The primary ethical issue in this case is that both GLs and CE require that data be obtained with consent and the contributors' knowledge of what the data will be used for, and in this case, the policy is directing you to violate these guidelines. In this case we do not know what type of data is being collected– nor what types of expectations of privacy contributors have – but it is clear that at least some people do have expectations – otherwise there would be no cause to consider provenance. Utilizing data that was obtained *without appropriate consent* is unethical. The case suggests that there *is cause for concern*, because you are instructed (by policy) to ignore data provenance.

4. Identify and evaluate alternative actions (on the ethical issue).

As usual, the three decisions that can be made in any circumstance are: a) do nothing. b) consult or confer with a peer (ASA) or a supervisor (ACM); and c) report violations of policy, procedure, ethical guidelines, or law (ASA) or refuse to implement the system (ACM). Clearly, "do nothing" is both totally inappropriate in this case, and also a violation of the practice standards relating to respecting data contributors and data provenance. We can change option a) from "do nothing" to "do not proceed with any use of the data until all data provenance can be established". Then option b) would become, "consult or confer with a peer (ASA) or a supervisor (ACM) as to how best to stop the collection and any use of the data until provenance can be established or data obtained without consent can be removed." Option c) must also be modified, because simply reporting the fact that the data collection violates the GLs (ASA), or refusing to implement the system (that scrapes up stolen/breached data) (ACM) might not ensure that the data with unknown provenance is both not used and also, its collection ceases. So, option c) needs to change to "report violations of policy, procedure, ethical guidelines, or law to a peer (ASA) or a supervisor (ACM), and stop the collection and any use of the data until provenance can be established or data obtained without consent can be removed."

5. Make and justify a decision.

Decision: *Note that all options include* "do not proceed with any use of the data". Like the previous cases, the core of response to this situation is to **follow the practice standards** and to utilize data according to the consent of the data contributors. Option a) says "follow the ethical practice standards, and do not use inappropriately-obtained data". This would be the *least desirable option* because the data collection – as well as its use – **also** needs to stop, because both are inappropriate. Moreover, the policy (SOP) instructing you to violate practice guidelines is also inappropriate and possibly unlawful. This policy may exist because data with unknown or mixed provenance is commonly collected and utilized in your work context; it might be impossible for you to do anything ethical in this case/context except seek another team, other colleagues, or a different job. However, that leaves the SOP in place, instructing others to violate practice standards. Thus, Option a) is least desirable.

Option b) would be the next *most desirable option* - not the most desirable one in this case - in that consulting with a colleague/supervisor will ensure some notification of leadership or your colleagues (preferably, both) that the SOP is directing employees to collect and use data without consent – which is frankly unethical, and may also be illegal. There may not be policies in your workplace against such interference/violation of the ASA GLs or ACM CE, but *reporting* (option c) is actually *the most desirable option* in this case. The key in this case that differentiates these options from others in this chapter is that the organization has a policy in place – the SOP – that reflects unethical and possibly illegal behavior. This creates both a professional and an individual/ civic obligation to report. You (the practitioner) can – and should - still demonstrate your professional competence and follow the relevant practice standards. With its "report" feature, option c) has the key benefit of permitting "educating" people or your company/organization about the practice standards but also notifying others that the SOP and policy are illegal (if this is the case). It is important *to the profession* to support the next practitioner who ends up in a similar situation (or better, to prevent such unethical data collection in the future), which is why option c) with the reporting feature is so important in this case: notifying relevant individuals in your organization that the policy around data collection is in direct violation the ACM CE and the ASA GLs (as well as possible laws). Furthermore, reporting may permit you to point out that you cannot ethically be directed by a policy to violate the CE and GLs. Engaging in discussion of alternative data collection methodologies –that *at a minimum* ensure a known provenance for all data would also be an opportunity to point out to everyone involved that this is the only way to meet ASA GLs/ACM CE, limit harms/risks of harms, and also enable the team to answer DEFW and DSEC questions.

6. Reflect on the decision.

Construct your own reflection on this decision.

Case 11. You discover that there has been no consent obtained for any of the data you are asked to collect/wrangle/munge.

1. Identify and 'quantify' prerequisite knowledge:

Which DEFW and DSEC Principles or Areas seem most relevant to this vignette?

This situation may be inconsistent with DEFW Principle 2, since "relevant legislation" as well as "codes of practice" require that human subjects give consent and are notified about the uses to which their data will be put. Data for which you do not have consent may make it difficult or impossible to fulfill Principle 4 (understand the limitations of the data).

In this case, the answer to the first question in the DSEC set, A1 (have human subjects given informed consent) is "no".

Which ASA and ACM Principles seem most relevant to this vignette?

ASA:

Principle A. Professional integrity and accountability
Principle C. Responsibilities to stakeholders
Principle D. Responsibilities to research subjects, data subjects, or those directly affected by statistical practices
Principle H. Responsibilities regarding potential misconduct

ACM:

1. General Ethical Principles
2. Professional Responsibilities

Potential result: Stakeholder:	HARM	BENEFIT	UNKNOWN	UNKNOWABLE
YOU	Using data obtained without consent violates the data contributor rights as well as ethical practice standards.	While a violation of multiple specific elements of the professional practice standards and applicable laws, data collected without con-sent is simpler to collect and creates no obligations for security (since users, not knowing their data were collected, cannot expect it to be protected).	Failing to obtain consent to use data could lead to unpredicted harms, bias, or unfair results, as well as risks to data contributors - with the statistician or data scientist bearing responsibility for misuse, unauthorized access, or losses of that data.	Following these directions, and using data that was not contributed with informed consent, may suggest to others that ignoring data provenance is OK, even though this directly violates practice standards.
Your boss/client	Data with unknown provenance can lead to unpredicted harms, bias, or unfair results, and can create risks for the data contributors that were foreseeable.	Ignoring the provenance of data, while a violation of multiple specific elements of the professional practice standards, is simpler and cheaper than ensuring data are obtained with proper consent.		

Potential result: Stakeholder:	HARM	BENEFIT	UNKNOWN	UNKNOWABLE
Unknown individuals	Data that is accessed without authorization and consent may expose data contributors to risks, as well as to (further) misuse by others.	There are **no benefits** that accrue when data are obtained in unauthorized ways, and no one stops this type of inappropriate behavior.		Using data that was not contributed with informed consent may suggest to others that data provenance is not important, even though this directly violates practice standards as well as basic human rights.
Employer	Data with unknown provenance can lead to unpredicted harms, bias, or unfair results, and can create risks for the data contributors that were foreseeable, thus incurring liability to the employer.	Collecting data without consent, while a violation of multiple specific elements of the professional practice standards and laws, is simpler and cheaper than ensuring data are obtained with proper consent.		
Colleagues	If others on the team are not aware that the data provenance is mixed, colleagues may mistakenly share –i.e., further the misuse of– the data.	There are **no benefits** that accrue when data are utilized in unauthorized ways, and no one stops this type of inappropriate behavior.		

Potential result: Stakeholder:	HARM	BENEFIT	UNKNOWN	UNKNOWABLE
Profession	The profession may appear untrustworthy when it is discovered that practitioners use whatever data they collect, even if no consent was given or even if the collected data was "available" because of a data breach. Using data with unknown provenance is a violation of multiple specific elements of the professional practice standards.	There are **no benefits** that accrue when stolen data are utilized, and no one stops the use of that kind of data.	Failing to obtain consent to use data could lead to unpredicted harms, bias, or unfair results, as well as risks to data contributors – with the statistician or data scientist bearing responsibility for misuse, unauthorized access, or losses of that data.	Using data that was not contributed with informed consent may suggest to others/other system developers that collecting stolen/breached data is OK, even though this directly violates practice standards. This decrements professional integrity in a concrete way.
Public/public trust	Public sentiment about the procedures by which they do and do not give their consent will continue to worsen, and people will become less inclined to contribute or share their data if public concerns about the lack of honesty in data collection systems, continue.	There are **no benefits** that accrue when consents are not obtained lawfully, and data are inappropriately utilized, and no one stops the use of that kind of data.		

Table 3.3.5. *Stakeholder Analysis: Using data obtained without consent*

2. Identify decision-making frameworks.

The ethical statistical practitioner has responsibilities to "Understands and conforms to confidentiality requirements for data collection, release, and dissemination and any restrictions on its use established by the data provider (to the extent legally required). Protects the use and disclosure of data accordingly. Safeguards privileged information of the employer, client, or funder." (ASA C7). Moreover, the ethical statistical practitioner "Understands the provenance of the data, including origins, revisions, and any restrictions on usage, and fitness for use prior to conducting statistical practices." (ASA D10). Obviously, failing to obtain consent to collect data– or failing to fully inform data contributors of the planned uses for their data, violate both ASA C7 and D10, but also violates ASA H2 ("Avoid condoning or appearing to condone statistical, scientific, or professional misconduct. Encourage other practitioners to avoid misconduct or the appearance of misconduct.").

The ACM CE, with its utilitarian perspective, focuses on limiting harms (ACM 1.2) and in this case, there are many harms, and no real benefits when data are collected without consent.

However, in the event that there was no consent to be obtained (e.g., the turnstile count data we have mentioned in other cases), then there is no violation of ASA GL principles, although the harms that accrue according to the SHA above may still be present and outweigh any benefits.

3. Identify or recognize the ethical issue.

Both the ASA and ACM practice standards (and some laws) state explicitly that professionals must inform data contributors about the data being collected and the uses to which that data will be put; this has clearly not been done in this case, creating an ethical problem. In this case we do not know what type of data is being collected– nor what types of expectations contributors have about the data being collected. If contributors have no reasonable expectation of privacy of their data (e.g., they are counted entering a public place by a turnstile), then a failure to secure the data is not a concern for either the data collector or contributor, and **there is no ethical issue**. However, given the structure of the vignette, the fact that "none of the data was obtained with consent" is a worry suggests that at least some of it *should have been* collected with consent –or else the data should not have been collected at all.

Note also that the task in this chapter includes data collection *and* wrangling *and* munging – while data should be collected with appropriate consent, as noted, wrangling and munging are only done ethically with data of known and recognized provenance. Thus, the identification of "no consent" – i.e., unknown provenance – for all the data to be wrangled and/or munged creates an ethical challenge, even though the data were already collected and most of the practice standards relate to ensuring that the data are collected with consent, but are possibly less specific about what to do with already-collected data. ASA GLs D10 and D11 specifically relate to analyzing or working with data that were collected without appropriate consent, and so are specifically informative in this case, i.e., they strongly suggest that working on data with unknown provenance is unethical, whether the work is in the form of an analysis, or of data munging or wrangling.

4. Identify and evaluate alternative actions (on the ethical issue).

As usual, the three decisions that can be made in any circumstance are: a) do nothing. b) consult or confer with a peer (ASA) or a supervisor (ACM); and c) report violations of policy, procedure, ethical guidelines, or law (ASA) or refuse to implement the system (ACM). Clearly, "do nothing" is totally inappropriate in this case – as with every case- and also a violation of the practice standards relating to respecting data contributors – unless you can be certain that the data are contributed with no expectation of privacy, security, or consent. Thus, we can change option a) from "do nothing" to "do nothing if all data were contributed with no expectation of privacy, security, or consent"; clearly option a) is only viable in that specific situation. Then, **in the situation where option a) is NOT viable**, option b) would become, "consult or confer with a peer (ASA) or a supervisor (ACM) as to how best to ensure data were contributed with no expectation of privacy, security, or consent; AND if this cannot be ensured, then consult/confer as to how to prevent use of this data". Option c) must also be modified, because simply reporting that the project has collected data in violation of the practice standards (and some laws) will not necessarily result in preventing that data from being used or shared in violation of the data contributors' expectations and the standards. So, option c) needs to change to "report violations of policy, procedure, ethical guidelines, and/or law to a peer (ASA) or a supervisor (ACM), and prevent the use of this data."

5. Make and justify a decision.

There are really two viable decisions in this case, and the selection depends on whether or not consent should have been obtained. If no consent was needed **and that is consistent with institutional and local/federal law** (rather than simply being how organizations like yours treat data contributors), then there is no ethical issue associated with utilizing the data – making option a) your choice: only as long as there is basic human right, law, policy, or expectation on the part of data contributors that their consent should have been obtained. Note that data contributors may belong to a culture where their consent for any data collection is essential to their individual rights, and this will have to be respected even if there are no laws or policies – or expectations on the part of the data collectors. *The basic human rights of the individual always take priority over the rights of those seeking to use an individual's data.*

As noted, if the conditions that make option a) reasonable are not met, then there **is** an ethical problem and option a) cannot be considered. This response began with "there are *two* viable decisions" because options b) and c) both include the same action: "prevent this data from being used".

Even though the data have already been collected, the data collected without appropriate consent cannot be used ethically. The use of such data is unethical, no matter what the policy or norm is utilized in the workplace.

In the case that data contributors were not informed that their data was being collected, were misled into believing that it would not be, or were not given the choice to contribute or not, then it is unethical to use that data, so your decision must be to prevent the data from being used. This can be either by conferring with colleagues/supervisors (option b) or by reporting the unethical and possibly illegal collection of that data (option c) and possibly also notifying contributors that their data was collected without their consent – making this fact public, and blowing the whistle on the practice and policies that led to the collection of this data in this unethical way. Note that options b) and c) are consistent with "refusing to deploy the system", which is one option repeatedly referred to in the ACM CE when predictable risks cannot be avoided. However, and more to the point, the data was collected in an unethical manner – possibly also illegally – and use of that data will always be unethical, even if the organizational policy is less strict or does not recognize the violations of the ASA GLs and ACM CE.

6. Reflect on the decision.

Construct your own reflection on this decision.

Case 12. You have collected/wrangled data from multiple sources and provenance information about the data is inconsistent – different people at work describe it differently and there's no real evidence about the provenance of *any* of the data.

1. Identify and 'quantify' prerequisite knowledge:

Which DEFW and DSEC Principles or Areas seem most relevant to this vignette?

This situation is inconsistent with DEFW Principle 4 (understand the limitations of the data) but if some of the data were illegally or unethically derived (leading to missing or incorrect provenance), then DEFW Principle 2 will also not be addressed.

In this case, the answer to the first question in the DSEC set, A1 (have human subjects given informed consent) is "no" – because it cannot be answered confidently in the affirmative. Moreover, because there is no evidence about the data provenance, there might also be inconsistencies in evidence for answering other DSEC questions, including whether a "right to be forgotten" (DSEC B2) can even be honoured. Depending on the extent of poor documentation, the data may also be missing key perspectives (DSEC C1), be biased (DSEC A2, C2), be unauditable (DSEC C5), and be both unfair across groups (DSEC D2) and un-explainable (DSEC D4).

Which ASA and ACM Principles seem most relevant to this vignette?

ASA:

Principle C. Responsibilities to stakeholders
Principle D. Responsibilities to research subjects, data subjects, or those directly affected by statistical practices
Principle H. Responsibilities regarding potential misconduct

ACM:

1. General Ethical Principles
2. Professional Responsibilities

Potential result: Stakeholder:	HARM	BENEFIT	UNKNOWN	UNKNOWABLE
YOU	Using data with unknown provenance, particularly if some of it is known not to have been contributed with informed consent, violates the data contributor rights as well as ethical practice standards.	There are **no benefits** that accrue when data are obtained without appropriate consent. Without understanding the provenance of the data, you cannot be sure you had appropriate consent/the data were collected ethically.	Failing to obtain consent to use data could lead to unpredicted harms, bias, or unfair results, as well as risks to data contributors – with the statistician or data scientist bearing responsibility for misuse, unauthorized access, or losses of that data.	Using data with unknown provenance may mean that data that was not contributed with informed consent could have been included. The use of this type of data may suggest to others that ignoring data provenance is OK, even though this directly violates practice standards.
Your boss/client	Data with unknown provenance can lead to unpredicted harms, bias, or unfair results, and can create risks for the data contributors that **were** foreseeable.	Ignoring the provenance of data, while a violation of multiple specific elements of the professional practice standards, is simpler and cheaper than ensuring data are obtained with proper consent.		

Potential result: Stakeholder:	HARM	BENEFIT	UNKNOWN	UNKNOWABLE
Unknown individuals	Data that is accessed without authorization and consent may expose data contributors to risks, as well as to (further) misuse by others.	There are **no benefits** that accrue when data are utilized in unauthorized ways, and no one stops this type of inappropriate behavior.		Using data that was not contributed with informed consent may suggest to others that data provenance is not important, even though this directly violates practice standards as well as basic human rights.
Employer	Data with unknown provenance can lead to unpredicted harms, bias, or unfair results, and can create risks for the data contributors that were foreseeable, thus incurring liability to the employer.	Ignoring the provenance of data, while a violation of multiple specific elements of the professional practice standards, is simpler and cheaper than ensuring data are obtained with proper consent.		

Potential result: Stakeholder:	HARM	BENEFIT	UNKNOWN	UNKNOWABLE
Colleagues	If others on the team are not aware that the data provenance is mixed, colleagues may mistakenly share –i.e., further the misuse of– the data.	There are **no benefits** that accrue when data are utilized in unauthorized ways, and no one stops this type of inappropriate behavior.		
Profession	The profession may appear untrustworthy when it is discovered that practitioners use whatever data they collect, even if no consent was given or even if the collected data was "available" because of a data breach. Using data with unknown provenance is a violation of multiple specific elements of the professional practice standards.	There are no benefits that accrue when stolen data are utilized, and no one stops the use of that kind of data.	Failing to obtain consent to use data and using stolen data (the results of breaches) could lead to unpredicted harms, bias, or unfair results, as well as risks to data contributors - with the statistician or data scientist bearing responsibility for misuse, unauthorized access, or losses of that data.	Using stolen data and data that was not contributed with informed consent may suggest to others/other system developers that collecting stolen/breached data is OK, even though this directly violates practice standards. This decrements professional integrity in a concrete way.

Potential result: Stakeholder:	HARM	BENEFIT	UNKNOWN	UNKNOWABLE
Public/public trust	Public sentiment about the procedures by which they do and do not give their consent will continue to worsen, and people will become less inclined to contribute or share their data if public concerns about the lack of honesty in data collection systems, continue.	There are **no benefits** that accrue when consents are not obtained lawfully, and data are inappropriately utilized, and no one stops the use of that kind of data.		

Table 3.3.6. *Stakeholder Analysis: Unknown data provenance*

2. Identify decision-making frameworks.

The ethical statistical practitioner "informs stakeholders of the potential limitations on use and re-use of statistical practices in different contexts and offers guidance and alternatives, where appropriate, about scope, cost, and precision considerations that affect the utility of the statistical practice." (ASA C4). Moreover, the ethical statistical practitioner "Understands the provenance of the data, including origins, revisions, and any restrictions on usage, and fitness for use prior to conducting statistical practices." (ASA D10). Thus, utilizing data with unknown provenance violates these ASA GLs. Obviously, failing to protect basic human rights by using data you do not have consent to use violates both ASA C4 and D10, but also violates ASA GL Principle H2 ("Avoid condoning or appearing to condone statistical, scientific, or professional misconduct. Encourage other practitioners to avoid misconduct or the appearance of misconduct.") as well as federal laws in some cases.

The ACM CE, with its utilitarian perspective, focuses on limiting harms and in this case, there are many harms, and no benefits, to using data that was contributed to your analysis without consent. Moreover, a system that collects data without obtaining consent may include methodologies or other features that are insecure or otherwise create risks (in addition to risks of confidentiality and privacy breaches) that are not addressed – because the manner in which data are collected may not be legitimate or sufficiently specified. In this case, you simply do not know what the provenance is of the data. Note also that, because of the poor documentation of the data and its provenance, there may be other key features of the data collection or other aspects of the system/pipeline or workflow that are similarly poorly documented. These weaknesses impede your ability to evaluate a system effectively to identify strengths and liabilities (i.e., following ACM CE Principles).

3. Identify or recognize the ethical issue.

The primary ethical issue in this case is that both GLs and CE require that data be obtained with consent and the contributors' knowledge of what the data will be used for, and in this case, you simply cannot tell if this did happen. In this case we do not know what type of data is being collected– nor what types of expectations of privacy contributors have – but it is clear that at least some people do have expectations – otherwise there would be no cause to consider provenance. Utilizing data that was obtained *without appropriate consent* is unethical. The case suggests that there *is cause for concern*, because you cannot obtain a correct report of data provenance. Although this case does not include other documentation also being missing or inconsistent, that would also be a serious challenge to implementing key features of the ASA and ACM practice standards.

4. Identify and evaluate alternative actions (on the ethical issue).

As usual, the three decisions that can be made in any circumstance are: a) do nothing. b) consult or confer with a peer (ASA) or a supervisor (ACM); and c) report violations of policy, procedure, ethical guidelines, or law (ASA) or refuse to implement the system (ACM). Clearly, "do nothing" is totally inappropriate in this case, and also a violation of the practice standards relating to respecting data contributors and data provenance. We can change option a) from "do nothing" to "do not proceed with any use of the data until all data provenance can be established". Then option b) would become, "consult or confer with a peer (ASA) or a supervisor (ACM) as to how best to stop the collection and any use of the data until provenance can be established or data obtained without consent can be removed." Option c) must also be modified, because simply reporting the fact that the data collection violates the GLs (ASA), or refusing to implement the system (that collects data with unknown provenance) (ACM) might not ensure that the data with unknown provenance is both *not used* and also, **its collection ceases**. So, option c) needs to change to "report violations of policy, procedure, ethical guidelines, or law to a peer (ASA) or a supervisor (ACM), and stop the collection and any use of the data until provenance can be established or data obtained without consent can be removed."

5. Make and justify a decision.

Decision: *Note that all options include* "do not proceed with any use of the data". Like the previous cases, the core of response to this situation is to **follow the practice standards** and to utilize only data for which consent of the data contributors was obtained appropriately. Option a) says "follow the ethical practice standards, and do not use inappropriately-obtained data". This would be the *least desirable option* because **both** the data collection *and* its use need to stop, because both are unethical; however, option a) only addresses one of these. So, option a) prevents *you* from unethical practice (i.e., you refuse to work with data that has unknown provenance) but it may have no effect on others, and doesn't make that data collection stop. Moreover, the method(s) by which the data are being collected may also be inappropriate and possibly unlawful. Data with unknown or mixed provenance may be commonly collected and utilized in your work context; it might be impossible for you to do anything ethical in this case/context except refuse to analyse or work with data of unknown provenance and also, seek another team, other colleagues, or a different job. However, since this leaves the data collection mechanism(s) in place, option a) is least desirable.

Option b) would be the next *most desirable option* - not the most desirable one in this case - in that consulting with a colleague/supervisor will ensure some notification of leadership or your colleagues (preferably, both) that the organization may be collecting and using data without appropriate consent – which is frankly unethical, and may also be illegal. Option b) is preferable to option a) because you would be notifying *someone* about the problems with your data collection (and/or its documentation) *and* the fact that individuals who are directed to analyze it "anyway" are actually being directed to violate their ethical practice standards.

There may not be policies in your workplace that prevent violation of the ASA GLs or ACM CE, but *reporting* (option c) is actually *the most desirable option* in this case. The key in this case that differentiates this option from others in this chapter is that the organization appears to have a system that collects data in an unethical and possibly illegal manner. This creates both a professional and an individual/civic obligation to report. You (the practitioner) can – and should - still demonstrate your professional competence and follow the relevant practice standards by not collecting or working with data with uncertain provenance. With its "report" feature, option c) has the key benefit of permitting "educating" people or your company/organization about the practice standards but also notifying others that data collection – and any work on unethically sourced data – violate ethical practice standards and possibly (in terms of illegal data collection) the law. It is critical *to the profession* to support the next practitioner who ends up in a similar situation (or better, to prevent unethical data collection in the future), which is why option c) with its reporting feature is so important in this case: notifying relevant individuals in your organization that the policy around data collection is in direct violation the ACM CE and the ASA GLs (as well as possible laws). Furthermore, reporting may permit you to point out that you cannot be directed by your organization to violate the CE and GLs. Engaging in discussion of alternative data collection methodologies –that *at a minimum* ensure a known provenance for all data - would also be an opportunity to point out to everyone involved that this is the only way to meet ASA GLs/ACM CE, limit harms/risks of harms, and also enable the team to answer DEFW and DSEC questions. This discussion may also lead to better documentation of other features of the enterprise, enabling ACM as well as DSEC items to be addressable.

6. Reflect on the decision.

Construct your own reflection on this decision.

Summary of six case analyses for Data Collection/munging/wrangling

In the six case analyses of Chapter 3.3 there was a common thread: using data without understanding its provenance violates key principles of ethical practice according to both the ASA and the ACM. Although all the cases had similar structure, the analyses did highlight different "most optimal" responses. You will also have noticed that in some cases, the policy or structure at your organization may play a role in whether the option to "report" (options c) or simply to "consult" (options b) violations of the GLs, CE, or applicable laws – although as a concerned citizen, it is a general responsibility to report when laws are being broken. Both the GLs and CE assume that their ethical guidance is relevant in addition to following all applicable laws (and both practice standards state that ethical practitioners should be familiar with applicable laws and codes).

When discussing how to perform "data collection/munging/wrangling" ethically in Section 2, general principles from the GLs or CE, and specific questions from the DSEC and DEFW, could be used to ensure that our general behavior is ethical. However, in response to specific challenges and ethical violations such as those in this chapter, DSEC and DEFW were *not* helpful in either identifying whether or not an ethical problem existed (or determining what exactly it is), nor were they helpful in figuring out what to do. Even though we considered the same 3 alternative options in every case, these were modified according to the available information and that did not include the DSEC and DEFW.

Each of the cases in this chapter were based on actual events; and most of the case analyses instruct the reader to ignore "standard operating procedure" or the cultural norms when these violate the ethical practice standards or when they exert any influence to cause others to violate those standards. These cases arise as the result of *others* trying to impede professional and ethical practice, either directly (supervisor telling you to ignore it) or indirectly (through documentation or the lack thereof). In these cases relating to data collection (and munging/wrangling), the SHA shows clearly that benefits to ignoring data provenance, or using data collected unethically, are minimal and are offset by great potential for serious harms to others, including the data contributors and unknown individuals but also to the profession and the public trust.

The identification of situations where others, or policies, conflict with (or direct you to violate) ethical practice standards requires a good working understanding of the GLs and CE as well as a firm commitment to follow these standards. As was noted in the previous chapter, it might be useful to share the GLs and CE as widely as possible in your working context – to ensure that people are not surprised when you seek to adhere to professional practice standards. The emphasis in ACM CE Principles 3 and 4 (and ASA Principle H) on ensuring that all professionals are allowed to work in ethical contexts suggests that publicising the GLs and CE in the workplace is an important aspect of ethical statistics and data science practice.

Questions for Discussion:

1. Review at least two cases in this chapter.

A. Do you agree with the decisions in these case analyses? Why or why not? Which parts of the ER process are most and least acceptable for each case? Do these parts differ for different cases in this chapter? Were there different ASA or ACM Principles or elements that you identified, and if so, how do these affect your case analyses? Are the elements and/or Principles you identify represented in your reflections?

B. Discuss your results for the tests of universality, justice, and publicity for any (or all) case(s) in this chapter.

C. Discuss the relevance to each case – including the analysis and your reflection – of federal, state, or local/organizational policies regarding human subjects, live vertebrate animal subjects in research, and safe laboratory practices. How can you incorporate what you know about these policies into any part of the case analysis (prerequisite knowledge; identifying the ethical issue; determining alternative actions; making or justifying your decision; or reflection)? Note that, if there is no alignment between federal, state, or your local/institutional policies and a given case, you can comment on that (e.g., do policies in your organization focus on *misconduct* – and ignore detrimental practices?). Do you feel that these policies could have helped in the reasoning, decision or justification?

D. Discuss the relevance to each case - including the analysis and your reflection – of federal, state, or local/organizational policies relating to data acquisition and laboratory tools; management, sharing and ownership. How can you incorporate what you know about these policies into any part of the case analysis (prerequisite knowledge; identifying the ethical issue; determining alternative actions; making or justifying your decision; or reflection)? If there are no relevant policies, you should comment on that, particularly when a) data are obtained using federal monies (grants); or b) data are collected (acquired) from humans, such that Federal Regulations about the treatment of human subjects and the acquisition of tissues and data from humans are relevant.

2. Review at least two (different) cases in this chapter.

Keeping in mind that doing nothing/not responding are *not plausible responses*, are there other plausible alternatives (KSA 4) that you can think of for any case? If not, discuss that; if so, list them and discuss their evaluation (are they equally consistent with GL/CE, do they lead to similar decisions, etc.).

3. Choose any case in this chapter. Redo the analysis, but feature a different GL or CE principle in your reasoning process. Make sure the *justification* is different from what is given. Is the *decision* also different? Discuss your decision with its justification and the tests of universality, justice, and publicity.

4. Consider all six cases in this chapter.

A. Comment on the role of the stakeholder analysis in these decisions. Were stakeholder impacts featured in justifications? Were they features in your own reflections?

B. Comment on the importance of your reflections on the decisions you were asked for: for others doing a similar task at work; for those using results from this task; and for those who might join the profession (and be learning about planning and design).

C. Consider your professional identity and the profession itself: planning/designing projects and data collection/analysis projects is obviously foundational in statistics and data science. How can these six cases and the decisions reached via the analyses presented earn you (the decider) the trust of the practicing community or the public for the profession and its future?

Chapter 3.4
Analysis (Perform, or Program to Perform)

NB: ASA and ACM incorporate "analysis" differently. While the use of this term for statisticians and more statistically oriented data scientists will be clearer (i.e., relating specifically to the analysis of data), for those who use computation extensively in their statistical practice (or who use statistics a bit in their primarily computational practice, the ACM uses the term "evaluation" instead (and does not mention statistical analysis or refer to ethical considerations in the analysis of data *per se*). The ASA GLs refer to the analysis of data, and in some cases to an analysis of how people use the data or the analysis (evaluation) of others' use of data, results, or methods (e.g., in peer evaluation). The ACM CE refers instead to the analysis of risks, the evaluation (and implied analysis) of systems or their plans/designs, and to the analysis of activities, or the effects of computing on stakeholders, the public, or the public good[17].

Case 13. You are told to implement an analysis plan that you suspect was written by someone else (who does not know it is being used) and for another problem/project.

Case 14. Your supervisor ignores your requests for reviews of your work and tells you that no one else can review it either.

Case 15. You are asked to carry out an analysis you are confident that you do *not know how to do or interpret* (or troubleshoot).

Case 16. You are given code to execute and while the code runs, you discover a mistake in the program.

Case 17. You notice that at least some of the assumptions required for interpretable results, using the code you were asked to implement, are not supportable. The code does run and yield results, but the assumptions underpinning those results are not valid.

Case 18. You are asked to evaluate a new system, and told the results that your evaluation should generate.

---

[17] This divergence in the meaning of the task "analysis" for statistics (ASA) and computing (ACM) was first noted in Chapter 2.4 of the book Tractenberg RE. (2022), Ethical Reasoning for a Data Centered World.

Case 19. Your analysis of a new system suggests that there is an unexpectedly high error rate, but only for a small group of users. Overall, the system's results are exactly as expected; for the subgroup, the results are the opposite of the overall result.

Case 20. You institute an interim check of results and discover that there is bias in the results. The interim check is literally the middle of a multi-part process, so there's no way to immediately pinpoint the source of the bias.

Case 21. You are told that your results with new data must match original results (i.e., you must replicate other results), and your analyses/code are right, but they do not replicate earlier results.

Case 13. You are told to implement an analysis plan that you suspect was written by someone else (who does not know it is being used) and for another problem/project.

1. Identify and 'quantify' prerequisite knowledge:

> Which DEFW and DSEC Principles or Areas seem most relevant to this vignette?

This situation may be inconsistent with DEFW Principle 1 because you are not starting with the clear user need but rather, an analysis plan already created. Principle 3 may or may not be addressable, because executing an analysis plan implies all data were already collected. However, DEFW Principles 4-7 can still be followed, and every practitioner is always obliged to meet Principle 2 (be aware of relevant legislation/codes of practice).

This case begins at "analysis", so all data were already collected; however, it is essential (see previous chapter) that data were obtained legally and ethically (DSEC A1). All other DSEC items should be answered/answerable assuming the data are ethically sourced. If the data provenance cannot be determined, see previous Chapter for alternatives.

> Which ASA and ACM Principles and elements seem most relevant to this vignette?

ASA:

Principle A. Professional integrity and accountability (A2-A6; A8-A9; A11-A12)
Principle B. Integrity of data and methods (B2-B3; B6)
Principle C. Responsibilities to stakeholders (C1; C4; C8)
Principle D. Responsibilities to research subjects, data subjects, or those directly affected by statistical practices (D2-D3; D8; D10; D11)
Principle E. Responsibilities to members of multidisciplinary teams (E1-E4)
Principle F. Responsibilities to fellow statistical practitioners and the Profession (F2-F5)
Principle G. Responsibilities of leaders, supervisors, and mentors in statistical practice (G1; G2; G4)
Principle H. Responsibilities regarding potential misconduct: H2

ACM:

1. General Ethical Principles (1.1)
2. Professional Responsibilities (2.1; 2.4; 2.5; 2.9)
3. Professional Leadership Responsibilities (3.5)

Potential result: Stakeholder:	HARM	BENEFIT	UNKNOWN	UNKNOWABLE
YOU	Figuring out whether the analysis plan you received matches the user need/problem and the data takes time – and, the answer may turn out to be "no", possibly creating a confrontation. You may need to create a new plan (also time and effort consuming).	No benefits to starting a project without a good understand-ing of its purpose, but save time and effort by using a pre-existing plan (assuming it is appropriate) and the author is credited).	If you execute the analysis and it is not appropriate but yields interpretable results, the outcomes will be error-prone and could be used and interpreted incorrectly, or misused. Results from an older/wrong analysis plan on different data may be uninterpretable and client (or other stakeholders) may lose faith in *you* even if you were instructed to utilize the wrong/older plan in the first place.	
Your boss/client	New analysis plans take time; the results may not match previous results; creating a new plan may alienate client if it suggests that the "original" plan wasn't right (and yet the client had to pay for it).	If an inappropriate analysis yields expected results, then we get what we expected. Pre-specified analysis plans are faster to execute, possibly cheaper.	An analysis planned for another purpose may not run at all on different data; results may be uninterpretable, and client (or other stakeholders) may lose faith in you.	

Potential result:	HARM	BENEFIT	UNKNOWN	UNKNOWABLE
Stakeholder:				
Unknown individuals	Ineffective design/ poor planning can create bias or permit privacy breaches, or access/ utilize data that should not be included or accessed. The individual who wrote the analysis plan may not be getting appropriate credit for their work if you simply use it without their knowledge.		Without knowing whether or not the given analysis plan fits the current problem or data, a "solution" that comes from a plan created for a different project could be misused and go undetected.	Not all risks/harms can be foreseen and mitigated, but NONE of them can be if you can't be sure your system/analysis is intended to solve a given problem with given data.
Employer	New analysis plans take time; the results may not match previous results; creating a new plan may alienate client if it suggests that the "original" plan wasn't right (and yet the client had to pay for it).	If an inappropriate analysis yields expected results, then we get what we expected. Pre-specified analysis plans are faster to execute, possibly cheaper.		Identifying that someone in your company tried to "pass off" old work as relevant or useable – particularly when it is not - could undermine trust in your company and limit business/profit.

Potential result: Stakeholder:	HARM	BENEFIT	UNKNOWN	UNKNOWABLE
Colleagues	Your "need" to ensure that the given analysis plan is appropriate for the current project may complicate colleagues' work, particularly if you discover it is not, then they also have to make adjustments.	If an inappropriate analysis (you are running the analysis plan from another project) yields expected results, then colleagues get what they expected.		A "solution" – or results - that come from analyses, and your colleagues' work based on that analysis, could be misused, and go undetected. Your colleagues may end up building in bias, unfairness or other – unauditable– problems because they assumed your analysis was appropriate for the data.
Profession	Simply recycling someone else's work suggests incompetence; the client (or other stakeholders) may lose faith in *the profession* when practitioners prioritize expedience over competence.		If you execute an inappropriate analysis plan but generate "interpretable results", the outcomes will be error-prone and could be misused. Blame for misuse or errors/bias will fall on the profession.	

Potential result:	HARM	BENEFIT	UNKNOWN	UNKNOWABLE
Stakeholder:				
Public/public trust	Ineffective design/poor planning –especially if they lead to misuse/abuse can undermine public trust in systems and the profession		Transparency and focus may be distorted if the plan to be executed isn't known/shown to be appropriate for the given data; misuse could occur and be undetected. When this is eventually detected, public trust will be decremented (further).	Without knowing whether or not the given analysis plan fits the current problem or data, a "solution" that comes from a plan created for a different project could be misused and go undetected

Table 3.4.1. *Stakeholder Analysis: Implement analysis plan someone else wrote for another project*

2. Identify decision-making frameworks.

The ASA GLs' virtue approach supports, *at a minimum*, understanding the analysis plan to ensure that it is correct and appropriate for the data before utilizing it, because the ethical practitioner "Uses methodology and data that are valid, relevant, and appropriate, without favoritism or prejudice, and in a manner intended to produce valid, interpretable, and reproducible results." (ASA A2) The ethical statistical practitioner "Accepts full responsibility for their own work, does not take credit for the work of others, and gives credit to those who contribute. Respects and acknowledges the intellectual property of others." (ASA A5) and "Strives to follow, and encourages all collaborators to follow, an established protocol for authorship. Advocates for recognition commensurate with each person's contribution to the work. Recognizes that inclusion as an author does imply, while acknowledgement may imply, endorsement of the work." (ASA A6) These elements of Principle A mean that the author of the analysis plan might have rights to it that should be considered before executing the plan – as a matter of professional courtesy and to signify your own professional integrity. However, simply executing an existing analysis plan creates potential conflicts with many elements of Principles A, B, C and D. For example, assuring that the plan is fit for purpose (ASA C1) and that its execution will not violate ASA A3, "Opposes efforts to predetermine or influence the results of statistical practices and resists pressure to selectively interpret data." Although it would be more expedient to simply run the analysis according to the plan you were given, Principle E4 dictates that the ethical statistical practitioner "Avoids compromising scientific validity for expediency". Moreover, the ethical practitioner "is transparent about assumptions made in the execution and interpretation of statistical practices including methods used, limitations, possible sources of error, and algorithmic biases." (ASA B2). Thus, you need to identify if whether it is scientifically valid to run the existing plan on the current data, otherwise you may appear to condone professional incompetence (violating ASA H2).

ACM CE's utilitarian perspective would focus on the importance of the potential harms (SHA) and the relative weaknesses of any benefits that accrue to expediently running an existing plan on new data. Prior to using the old plan, the ethical practitioner will identify and anticipate harms and risks of harms; if any -or any in excess of what is acceptable- are discovered, then the old plan cannot be utilized.

3. Identify or recognize the ethical issue.

There are several ethical issues in this case. 1. Both the ASA and ACM standards require that appropriate methods are utilized and that ethical practitioners both a) only accept work for which they are competent; and b) acknowledge the intellectual contributions of others in their work. In this case, it is not clear that the analyst *is competent* to have created the given analysis plan; if there are any problems in the plan's execution, we don't know if the analyst will be capable of identifying them and fixing them. 2. The analyst in the case is handed an existing plan – without information about the creator of the plan. This means the analyst will not be able to respect the intellectual contributions of others; take full responsibility for the work; or confirm that the methods are appropriate for the data and the current problem. All of these are violations of both the ASA and ACM standards. 3. The SHA suggests that harms accrue for all stakeholders when there is an incomplete understanding of a project, and in this case, you do not know if the existing plan is appropriate for the current project/problem or the data. Thus, harms and risks of harms of the results of executing the plan cannot be anticipated (contrary to ACM CE).

4. Identify and evaluate alternative actions (on the ethical issue).

Although we identified three different ethical issues in this case, the same three decisions we usually use are applicable: a) do nothing; b) consult or confer with a peer (ASA) or a supervisor (ACM); and c) report violations of policy, procedure, ethical guidelines, or law (ASA) or refuse to implement the system (ACM). In this case we could change option a) from "do nothing" to "execute the plan as directed" – so that "do nothing" refers to "do nothing *else*, including determining whether or not the plan is actually appropriate for the data and problem". This option requires an assumption that you would not be directed to do something unethical – which, as you might surmise given the fact that most of the 47 vignettes in this book arose from this exact direction, is not a sound assumption. Thus, although it is a clear violation of ASA H2 ("Avoids condoning or appearing to condone statistical, scientific, or professional misconduct"), option a) is still an evaluable alternative with this modification. Note that option a) is an evaluable decision for all 3 of the issues identified. Options b) and c) need to provide alternatives to simply executing the plan as directed – conferring and reporting, while executing the plan as directed, are not viable. Thus, option b) can be modified to "confer with a peer or supervisor on how to determine if the plan is appropriate for the problem and data, *and* on what to do if it is not". As we have seen in other case analyses, whether or not "reporting" the fact that you are being directed to ignore your obligation to ensure you are competent for the task and that you are applying appropriate methodology will depend on whether your organization has policies in place either respecting the ACM CE or your rights and responsibilities as a competent and ethical practitioner. However, if you feel that the plan you are given is the intellectual property of someone else, and you feel, or find evidence, that you are unable to credit that other person with the plan, then this could be an illegal infringement of their intellectual property rights. The case does not specify whether you are unable to respect intellectual property, but it clearly does make it impossible for you to take full responsibility for the work. Again, if your organization has policies against this, the option to report being directed to perform something for which you cannot take full responsibility is an important one. So option c) should be modified to be, "determine if the plan is appropriate for the problem and data, *and* determine what to do *if it is not,* and report it if you are directed to ignore intellectual property rights or execute an inappropriate analysis plan (or both)." Note that, like option a), both options b) and c) are also evaluable decisions for all 3 of the issues identified.

5. Make and justify a decision.

Decision: *Option a) is inconsistent with the practice standards.* Just as "do nothing" in every case is never ethical practice, "doing what you are told"- in cases like this vignette outlines - is essentially unethical. Both options b) and c) include the same action, "determine if the plan is appropriate for the problem and data, *and* <make/have a plan for> what to do if it is not". The differences in b) and c) revolve around whether or not you should confer with others to achieve this action, or whether you should simply execute the action and report anyone who tries to direct you to follow option a). Note that if you select option b) and do confer with others, and they also direct you to "just do what you are told", you may need to choose option c). However, if you do determine that the plan *is* in fact appropriate for the problem and the data, then option a) will ultimately work – as long as you are able to acknowledge the contributions of others, make the roles of all contributors known, and take full responsibility for your work.

6. Reflect on the decision.

In this case, as will all the others, "do nothing" is obviously not ethical and is also inconsistent with ethical practice standards. However, "do as you are told" may, in some cases, devolve to "do nothing (about your perceived ethical problem) and just get on with your work/do as you're told". The ACM CE (Principle 4, and Principle 3.4) stipulates that policies and behaviors that are, or direct you to be, in violation of the CEs should not be followed. ASL H2 also states that the ethical statistical practitioner "avoids condoning or appearing to condone statistical, scientific, or professional misconduct" – and following unethical instructions is in conflict with ASA H2. You can see from the SHA that apparent benefits to recycling plans that others created for other problems and data – apparent efficiency and time/effort savings – are minor, and that the potential for incorrect, and undetectably inappropriate, results are much worse, and accrue to many other stakeholders than any benefits may accrue to.

The SHA shows plainly that option a) fails a *test of justice*, and in the SHA the harms that can accrue to the profession and the public trust – as well as to you - may mean that option a) fails the *test of publicity*, because you would not want it publicized that you simply follow directions without knowing (or apparently caring) whether or not the directions are appropriate for the problem and the data. Because of the SHA, and the substantial harms (with minimal benefits) that accrue in this case if you decided on option a), it clearly fails the *test of universality* as well.

Professional integrity and accountability (ASA Principle A) demand that the ethical quantitative practitioner is able to take full responsibility for their work. Simply following directions ("doing as you are told") makes this impossible. The real core of the ethical issues (KSA 3) is that it is unknown whether or not the plan you've been directed to utilize is appropriate. Simply confirming that it is – and having the opportunity to create a new analysis plan if it is not – essentially makes the ethical issue disappear; if there were concerns about intellectual property and contributions but you have to create a new plan, those concerns are addressed.

Case 14. Your supervisor ignores your requests for reviews of your work and tells you that no one else can review it either.

1. Identify and 'quantify' prerequisite knowledge:

Which DEFW and DSEC Principles or Areas seem most relevant to this vignette?

Principle 5 may be limited if you are unable to confirm your approach is the most robust one possible. However, DEFW Principle 6, ensure your work is transparent and be accountable, can and should be followed in this case because your work will, at some point, be reviewed.

By limiting your recourse to reviews, DSEC items relating to considering sources of bias may not be addressable. While *you* may have considered sources, other perspectives (C1; C2; D1; D5) may not be. Most other DSEC items are unaffected by this supervisor's behavior.

Which ASA and ACM Principles seem most relevant to this vignette?

ASA:

Principle A. Professional integrity and accountability
Principle C. Responsibilities to stakeholders
Principle F. Responsibilities to fellow statistical practitioners and the Profession
Principle G. Responsibilities of leaders, supervisors, and mentors in statistical practice

ACM:

2. Professional Responsibilities
3. Professional Leadership Responsibilities

Potential result: Stakeholder:	HARM	BENEFIT	UNKNOWN	UNKNOWABLE
YOU	You may have missed an opportunity to innovate, or to apply a new method. You may have made an error. You are solely responsible for the accuracy of your work.	You are solely responsible for the accuracy of your work – meaning that you can take full responsibility for it, as the independent practitioner that you are.		
Your boss/client	Peer review takes time, and if errors are found, they will take even more time to fix. Key harms or risks of harms may be missed. Independent confirmation that the system/analysis will operate as intended supports valid interpretation and limits misuse opportunities –those safeguards are missing without peer review.	Trusting all practitioners to take full responsibility for independent work is faster and possibly cheaper than requiring peer review.		
Unknown individuals	System and analysis planning can inadvertently create harms or risks of harms; without peer review, these may be missed.			

Potential result: Stakeholder:	HARM	BENEFIT	UNKNOWN	UNKNOWABLE
Employer	Peer review takes time, and if errors are found, they will take even more time to fix. Key harms or risks of harms may be missed. Independent confirmation that the system/analysis will operate as intended supports valid interpretation and limits misuse opportunities –those safeguards are missing without peer review.	Trusting all practitioners to take full responsibility for independent work is faster and possibly cheaper than requiring peer review.		
Colleagues	Peer review takes time, and if errors are found, they will take even more time to fix. Key harms or risks of harms may be missed. Independent confirmation that the system/analysis will operate as intended supports valid interpretation and limits misuse opportunities –those safeguards are missing without peer review. Colleagues' work is also unreviewed, possibly compounding problems.	Trusting all practitioners to take full responsibility for independent work is faster and possibly cheaper than requiring peer review.		A "solution" – or results - that come from unreviewed analyses, and your colleagues' work based on those analyses, could be misused, and go undetected if the original solution was not correct or made unsupported assumptions. Your colleagues may end up building in bias, unfairness or other – unauditable-problems because they assumed your work did not need review (i.e., was perfect).

Potential result: / Stakeholder:	HARM	BENEFIT	UNKNOWN	UNKNOWABLE
Profession	Key harms or risks of harms may be missed in any work, such that when discovered later, the profession is blamed. Independent confirmation that the system/analysis will operate as intended supports valid interpretation and limits misuse opportunities –those safeguards are *missing* without peer review.	Trusting all practitioners to take full responsibility for independent work – i.e., eliminating peer review – suggests that all practitioners are fully competent and possibly strengthens the perception of the profession.	If a statistician or data scientist designs or employs inappropriate methodology, but generates "interpretable results", the outcomes will be error-prone and could be misused. Blame for misuse or errors/bias will fall on the profession, particularly since peer review – per practice standards – could easily have been implemented.	
Public/public trust	If peer review identifies weaknesses or limitations – especially if they did lead, or could have led, to misuse/abuse it can undermine public trust in systems and the profession.	If peer review identifies weaknesses or limitations – especially if they could have led to misuse/abuse – then those will be fixed before those harms/risks of harms can accrue to the public. This can *bolster* public trust in systems and the profession.		

Table 3.4.2. *Stakeholder Analysis: No one can/will review your work*

2. Identify decision-making frameworks.

ASA GL Principle F describes the responsibilities that all statistical practitioners have towards others in the profession and to those who are practicing. Specifically, the ethical statistical practitioner "Helps strengthen, and does not undermine, the work of others through appropriate peer review or consultation. Provides feedback or advice that is impartial, constructive, and objective" (ASA F2). This means that a supervisor who is also a statistics practitioner should fulfill their ethical obligation to strengthen your work through peer review. Moreover, those with leadership roles should "G1 Ensure appropriate statistical practice that is consistent with these Guidelines. Protect the statistical practitioners who comply with these Guidelines, and advocate for a culture that supports ethical statistical practice." (ASA G1) and "Promote a respectful, safe, and productive work environment. Encourage constructive engagement to improve statistical practice." (ASA G2). Leaders who don't are not supportive of Principle F (or the profession). In keeping with ASA Principle A1 ("Takes responsibility for evaluating potential tasks, assessing whether they have (or can attain) sufficient competence to execute each task and that the work and timeline are feasible. Does not solicit or deliver work for which they are not qualified or that they would not be willing to have peer reviewed.") and ASA C1 ("Seeks to establish what stakeholders hope to obtain from any specific project. Strives to obtain sufficient subject-matter knowledge to conduct meaningful and relevant statistical practice."), you should be able to determine your qualification to design and execute an analysis.

The utilitarian perspective is also consistent in its support for your request for input and review of your work, for example ACM CE Principle 2.5 ("accept and provide appropriate professional review"). Your request for review is also consistent with the ACM CE's overall utilitarian perspective to identify and try to mitigate potential harms (ACM 1.1). It is your supervisor whose behavior is inconsistent with the ACM CE, especially Principle 3.

3. Identify or recognize the ethical issue.

This case is complicated by the fact that, while you as an ethical practitioner can follow the GLs and CE, and only accept work you are competent to complete, and do so in a transparent and accountable manner, your supervisor (and/or your work context) are failing to follow the specific practice standards that exist simply to promote everyone in the profession doing their best work. The ethical issues the limitation on peer review creates are: 1. ethical practitioners have an obligation to *provide* competent peer review to help others in their profession (and this is being blocked); and 2. ethical practitioners have an obligation to *seek* competent peer review to help them(selves) maintain currency and expertise in the profession (and this is being blocked). Thus, the act of limiting peer review is itself contrary to the GLs and CE because it effectively limits the development or maintenance of expertise. The lack of competent peer review, at worst, may render identifiable mistakes invisible, and lead to harms or risks of harms that could have been prevented – that is the real harm the policy (or this supervisor) creates. However, every other aspect of ethical practice is unimpeded in this case.

4. Identify and evaluate alternative actions (on the ethical issue).

The same three decisions we usually use are applicable in this case: a) do nothing; b) consult or confer with a peer (ASA) or a supervisor (ACM); and c) report violations of policy, procedure, ethical guidelines, or law (ASA) or refuse to implement the system (ACM). In this case –like every other - "do nothing" is not a viable option. Since stewardship of the profession is threatened by the refusal to give or let you obtain peer review, option a) is: "do the "peer review" *yourself*" (i.e., do not continue to request independent review). While it seems oxymoronic, you can simply adapt an existing rubric for peer review – e.g., for reviewers of grants (e.g., https://grants.nih.gov/grants/peer/critiques/rpg.htm), or of manuscripts (https://publons.com/blog/how-to-write-a-peer-review-12-things-you-need-to-know/), or of any other type of document that is relevant for your particular work. Then, because you are following the GLs and CE, your work is complete and transparent so you should be able to document whether and how your work meets "peer review" criteria that you can also be transparent about. The other options need to be modified, but not because you need to consult on how to formulate a response (b) or to report this behavior (c). Instead, you might modify option b) to "consult with a peer or supervisor to create a mechanism by which peer review can be effectively (and lawfully) implemented in your workplace or team; and meanwhile, do the "peer review" yourself". It may be that no peer review is possible due to security concerns; this makes consultation essential. Finally, option c) can be modified from "report" to "inquire", because impeding professional development may be contrary to the GLs and CE, but it is possible that your organization has important (possibly legal) reasons to do so with respect to peer review. Thus, option c) could become, "inquire about policies that could support –or be modified to support – the creation of a mechanism by which peer review can be effectively (and lawfully) implemented in your workplace or team; and meanwhile, do the "peer review" yourself".

5. Make and justify a decision.

Decision: Note that option a) is the essential response, whether or not you are able to accomplish options b or c. All three alternatives include doing the peer review yourself. As an independent practitioner, you are most likely already reviewing your work before you turn it in/hand it on to the next person on your team or in the workflow. The practitioner in this case may have been asking specifically for help with professional development or to generate some kind of independent documentation of their professional status/expertise. By selecting a peer reviewing tool and utilizing it on your own work, you are documenting your commitment to your own professional growth and development. After figuring out how to do this and doing it, in a new context (out of the situation or team where peer review is not available), you would be able to document the growth as well as initiative in furthering your own skill set, to any future employer or team leader.

6. Reflect on the decision.

Construct your own reflection on this decision.

Case 15. You are asked to carry out an analysis you are confident that you do *not know how to do or interpret* **(or troubleshoot).**

1. Identify and 'quantify' prerequisite knowledge:

Which DEFW and DSEC Principles or Areas seem most relevant to this vignette?

This situation puts you in conflict with DEFW Principle 5 because you would not be working within your skillset. Also, the practice standards ("codes of practice") both articulate a responsibility to only accept work for which you are qualified, so you would be in conflict with Principle 2. However, DEFW Principle 6, ensure your work is transparent and be accountable, can and should be followed in this case because your work will, at some point, be reviewed.

There is no reference to working outside your skillset in DSEC. However, many DSEC items may not be addressable by you, because you do not know how to interpret or troubleshoot the analysis that you're directed to carry out. You cannot determine if the analysis is appropriate, further limiting your ability to answer DSEC items.

Which ASA and ACM Principles seem most relevant to this vignette?

ASA:

Principle A. Professional integrity and accountability
Principle B. Integrity of data and methods
Principle C. Responsibilities to stakeholders
Principle D. Responsibilities to research subjects, data subjects, or those directly affected by statistical practices
Principle E. Responsibilities to members of multidisciplinary teams

ACM:

2. Professional Responsibilities

Potential result: Stakeholder:	HARM	BENEFIT	UNKNOWN	UNKNOWABLE
YOU	You may do the analysis wrong/use wrong seeds or starting values, or make (m)any other errors. You are unable to take responsibility for the accuracy of your work.	This could be a great opportunity to develop a new skillset professionally – as long as you ask for help.		
Your boss/client	Errors may be missed when someone who cannot interpret or troubleshoot is tasked with methods they are not familiar with. Independent confirmation that the system/ analysis will operate as intended supports valid interpretation and limits misuse opportunities –those safeguards are missing without expertise and oversight (and the ability to troubleshoot).	Trusting all practitioners to take full responsibility for independent work is faster and possibly cheaper than requiring expertise and oversight. Instituting peer review can help offset the lack of familiarity with the methods.		

Potential result: Stakeholder:	HARM	BENEFIT	UNKNOWN	UNKNOWABLE
Unknown individuals	Inappropriate analyses can inadvertently create harms or risks of harms; without oversight and lacking the ability to interpret/ troubleshoot, these may be missed.			
Employer	Errors may be missed when someone who cannot interpret or troubleshoot is tasked with methods they are not familiar with. Independent confirmation that the system/analysis will operate as intended supports valid interpretation and limits misuse opportunities –those safeguards are missing without expertise and oversight (and the ability to troubleshoot).	Trusting all practitioners to take full responsibility for independent work is faster and possibly cheaper than requiring expertise and oversight. Instituting peer review can help offset the lack of familiarity with the methods.		

Potential result:	HARM	BENEFIT	UNKNOWN	UNKNOWABLE
Stakeholder:				
Colleagues	Key harms or risks of harms may be missed. Independent confirmation that the system/analysis will operate as intended supports valid interpretation and limits misuse opportunities –those safeguards are missing without expertise and oversight. If colleagues' work is also outside of their skill set, it will compound problems.	Trusting all practitioners to take full responsibility for independent work is faster and possibly cheaper than requiring expertise and oversight.		A "solution" – or results - that come from un-evaluated analyses, and your colleagues' work based on those analyses, could be misused, and go undetected if the original solution was not correct or made unsupported assumptions. Your colleagues may end up building in bias, unfairness or other – unauditable-problems because they assumed your work did not need interpre-tation or trouble-shooting (i.e., was perfect).

Potential result: Stakeholder:	HARM	BENEFIT	UNKNOWN	UNKNOWABLE
Profession	Key harms or risks of harms may be missed in any work, such that when discovered later, the profession is blamed. Independent confirmation that the system/analysis will operate as intended supports valid interpretation and limits misuse opportunities – those safeguards are *missing* without peer review.	Trusting all practitioners to take full responsibility for independent work – i.e., eliminating peer review – suggests that all practitioners are fully competent and possibly strengthens the perception of the profession.	If a statistician or data scientist designs or employs inappropriate methodology, but generates "interpretable results", the outcomes will be error-prone and could be misused. Blame for misuse or errors/bias will fall on the profession, particularly since peer review – per practice standards – could easily have been implemented.	
Public/public trust	If weaknesses or limitations go unnoticed, especially if they led to misuse/abuse of results, this can undermine public trust in systems and the profession.	If a statistician or data scientist employs inappropriate methodology, but generates "interpretable results", the outcomes will be error-prone and could be misused. Blame for misuse or errors/bias will fall on the profession, particularly since peer review – per practice standards – could easily have been implemented.		

Table 3.4.3. *Stakeholder Analysis: Doing an analysis you do not know how to do/interpret*

2. Identify decision-making frameworks.

ASA GL Principle A describes the responsibilities to "take responsibility for evaluating potential tasks, assessing whether they have (or can attain) sufficient competence to execute each task and that the work and timeline are feasible. Does not solicit or deliver work for which they are not qualified or that they would not be willing to have peer reviewed." (ASA A1). Clearly in this case, ASA A1 will be difficult to follow, because while you would be taking responsibility for your work, you do so knowing that it is beyond your capability or experience – so, you would not want it to be peer reviewed. (While you might actually learn a lot from constructive peer review – pointing out what went/was done wrong, that is not part of this vignette!). In carrying out the analysis, the practitioner would be violating many other GL Principles, and will also be unable to follow others (e.g., in ASA Principles B, C, D and E).

ACM CE Principle 2.5 ("A computing professional's ethical judgment should be the final guide in deciding whether to work on the assignment") suggests that this case constitutes a real ethical issue for the computing professional. Just as with the virtue perspective, the computing professional would be violating other CE Principles if they comply with the request.

3. Identify or recognize the ethical issue.

The ethical issue in this case is clear from the GLs and CE: if you are not capable of doing an analysis or completing a work assignment, then you cannot accept that assignment. A secondary ethical issue is that a person seems to be trying to direct this practitioner to do their job not-competently, so that person (the requestor) is behaving contrary to the GLs and CE by trying to get the practitioner to violate the practice standards.

Importantly, *even if you are transparent* about your concern that you did not do the assignment correctly or interpret its results correctly, those concerns might not get passed along to the next – or the last - person in the workflow. (Given the vignette, the requestor is unlikely to follow any of the elements in ASA Principle G, or the Appendix). Thus, ensuring that analyses are done (or systems that do analyses are designed) competently, and with appropriate interpretation and troubleshooting, is essential. The issue here is the fundamental aspect of competent performance is being ignored by the requestor.

4. Identify and evaluate alternative actions (on the ethical issue).

The three decisions that we consider for any circumstance are: a) do nothing. b) consult or confer with a peer (ASA) or a supervisor (ACM); and c) report violations of policy, procedure, ethical guidelines, or law (ASA) or refuse to implement the system (ACM). Clearly, "do nothing" does not make sense in this case. We can change option a) from "do nothing" to "do not accept/refuse the request to do this analysis – but do not consult colleagues/report this behavior". Then option b) would become, "consult or confer with a peer (ASA) or a supervisor (ACM) as to how best to refuse the request from this person who wants you to violate the practice standards and accept an assignment you're not prepared to do". Option c) must also be modified, because simply reporting the requestor (ASA) or refusing the request (ACM) could mean that your time on the project would end, and another person who has no problem executing whatever methods the client requests-irrespective of their competency with those methods - will be brought on. So, option c) needs to change to "report violations of policy, procedure, ethical guidelines, or law to a peer (ASA) or a supervisor (ACM), and refuse the request."

5. Make and justify a decision.

Decision: *Note that all options include* "refuse the request". That is because the core of the response to this situation is to follow the practice standards – *work within your area of competence* – thus following what must be considered "best practices". In this case the analyst is simply "requested" – not directed by a supervisor – to do something contrary to the practice standards. The requestor may have a pre-defined outcome in mind (contrary to several ASA GL Principles) or may simply not understand statistics and data science well enough to recognize the problem they've created with this request. Because we don't know much about the requestor's motivations, what else to do (apart from **not what is requested**!) is less clear. Option a) says "follow best practices, and beyond that, do nothing (just refuse the request to violate practice standards)". This would be the *least desirable option* because the person who is requesting that you ignore your professional practice standards needs to understand that this is inappropriate, and they need to stop that behavior. However, if such behavior is common in your work context, it might be impossible for you to do anything except seek another team, other colleagues, or a different job. Option b) would be the *most desirable option* in that consulting with a colleague/supervisor will ensure that this kind of frankly unethical behavior by the requestor – inappropriate interference with your professional practice - is documented. There may not be policies in your workplace against such interference/violation of the ASA GLs or ACM CE, so reporting (option c) may not be an option. However, even in contexts where such policies do exist, it may still be difficult to report such behaviors; but you (the practitioner) can demonstrate your professional competence and follow the relevant practice standards.

You should keep in mind that the domains of statistics, data science, and "statistics and data science" comprise a huge and growing multitude of methods and no one can be expected to be competent in all of them. If requests to carry out analyses with which you would like to become competent crop up, then you can use those to request support from your team or organization to develop your professional skill set. This case is in no way intended to suggest that growing new skills is not ethical - in fact, continuing professional development is an obligation for statistics practitioners (ASA A10) and those who lead, supervise, or mentor them (ASA G3), as well as computing professionals (ACM 2.2) and those who lead them (ACM 3.5).

6. Reflect on the decision.

Construct your own reflection on this decision.

Case 16. You are given code to execute and while the code runs, you discover a mistake in the program.

1. Identify and 'quantify' prerequisite knowledge:

Which DEFW and DSEC Principles or Areas seem most relevant to this vignette?

This situation interferes with DEFW Principles 5 and 6: robust practice (5) requires that errors be corrected, and 6 requires transparent and accountable work. Results of a system or analysis with an identified error conflict with these Principles.

There is no reference to what to do about errors in analyses in DSEC. However, the existence of the error means C3, C5, D4, E3 cannot be addressed. Depending on the nature of the error, other DSEC items cannot be answered confidently.

Which ASA and ACM Principles seem most relevant to this vignette?

ASA:

Principle A. Professional integrity and accountability
Principle B. Integrity of data and methods
Principle C. Responsibilities to stakeholders
Principle E. Responsibilities to members of multidisciplinary teams
Principle F. Responsibilities to fellow statistical practitioners and the Profession
Principle G. Responsibilities of leaders, supervisors, and mentors in statistical practice
Principle H. Responsibilities regarding potential misconduct

ACM:

2. Professional Responsibilities

Potential result: Stakeholder:	HARM	BENEFIT	UNKNOWN	UNKNOWABLE
YOU	Identifying, documenting, and then fixing the error will take time, adds effort; may identify other problems that require additional time/effort. Failing to fix the error may save time but lead to unpredictable propagation of that error throughout the workflow.	Correction of the error means future work with the code will be facilitated –and correct, and interpretable.	Competent analysis (ASA) supports reproducible results; effective analysis (ACM) promotes public good.	
Your boss/client	Identifying, documenting, and then fixing the error will take time, adds effort; may identify other problems that require additional time/effort. This error may be discovered in a 'product' that the client has paid for in the past, identifying past errors and undermining this and other projects	Correction of the error means future work with the code will be facilitated –and correct, and interpretable.	Identification of error may require resources to fix/address. *Appropriate analysis* –once the error is fixed - may not yield desired results.	

Potential result: Stakeholder:	HARM	BENEFIT	UNKNOWN	UNKNOWABLE
Unknown individuals	Analysis required or ACM CE violated; weak statistical analysis undermines reproducibility and rigor.	Identification and correction of the error allows mitigation of risks of harms (ACM); supports valid inferences (ASA), and can limit the unpredictable propagation of the error to other parts of the workflow.		
Employer	Identifying, documenting, and then fixing the error will take time, adds effort; may identify other problems that require additional time/effort. This error may be discovered in a 'product' that the client has paid for in the past, identifying past errors and undermining this and other projects	Correction of the error means future work with the code will be facilitated –and correct, and interpretable.	Identification of error may require resources to fix/address. *Appropriate* analysis –once the error is fixed - may not yield desired results.	Identification of the error may highlight past payments or purchases/sales of incorrect/error-prone work. Liability may be incurred.

Potential result: / Stakeholder:	HARM	BENEFIT	UNKNOWN	UNKNOWABLE
Colleagues	The discovered error may complicate colleagues' work.	Identification and correction of the error allows mitigation of risks of harms (ACM); supports valid inferences (ASA), and can limit the unpredictable propagation of the error to other parts of the workflow.		Project could be delayed until risks of additional errors, and resulting harms/ potential harms are addressed.
Profession	Identifying, documenting, and then fixing the error will take time, adds effort; may identify other problems that require additional time/effort. Failing to fix the error may save time but lead to unpredictable propagation of that error throughout the workflow.	Correction of the error means future work with the code will be facilitated —and correct, and interpretable. This supports ethical and competent practice by future practitioners.	Competent analysis (ASA) supports reproducible results; effective analysis (ACM) promotes public good.	Commitment to identification and correction of the error, consistent with GL/CE, might strengthen trust in profession. The error itself can enable innovation (to fix, avoid, or detect similar errors in future), as well as conversations about ethical practice.

Potential result:	HARM	BENEFIT	UNKNOWN	UNKNOWABLE
Stakeholder:				
Public/public trust	Even if errors are noted so they can be fixed, weak or error-prone analysis undermines public trust in reproducibility and rigor.	Identification and correction of the error allows mitigation of risks of harms (ACM); supports valid inferences (ASA), and can limit the unpredictable propagation of the error to other parts of the workflow and decisions based on them.	Transparency supports the public trust, even if risk/harms are identified; can tend to inform the public about limitations inherent in the profession.	

Table 3.4.4. *Stakeholder Analysis: you discover an error in a program you received/are running*

2. Identify decision-making frameworks.

As noted, ASA GLs tend to support a virtue approach, which supports fixing the code so that the analysis is correct, because the ethical statistical practitioner "uses methodology and data that are valid, relevant, and appropriate, without favoritism or prejudice, and in a manner intended to produce valid, interpretable, and reproducible results." (ASA A2) The ethical statistical practitioner "avoids compromising scientific validity for expediency" (ASA E4), so although it would be more expedient to let the system run on with the known error, Principle B dictates that the ethical statistical practitioner: "Strives to promptly correct substantive errors discovered after publication or implementation. As appropriate, disseminates the correction publicly and/or to others relying on the results." (ASA B5) The virtue perspective (via the ASA Guidelines) emphasizes the practitioner recognizing their responsibility to themselves (ASA A), stakeholders (ASA C), data donors and those affected by statistical practice (ASA D), others on their team (ASA E), and the profession (ASA F).

In the ACM CE utilitarian perspective, the computing professional has a responsibility to analyze the work they do/are asked to do so that decisions made on the basis of the system or its subparts are *justifiable*. If an error is identified in the analysis part of a system being evaluated, then results from that system may be biased or incorrect, potentially causing predictable (once the error is identified) but fixable harms. More specifically, "1.3 A computing professional should be transparent and provide full disclosure of all pertinent system capabilities, limitations, and potential problems to the appropriate parties." This code was given to you, and you have identified an error with that code; as long as that code has been used, it has been creating a problem.

3. Identify or recognize the ethical issue.

Both the ASA and ACM standards support the identification and fixing of the error, with the ASA GLs stipulating that corrections should be disseminated, and the ACM CE specifying similarly that limitations of the system should be disclosed (unless the error is fixed, in which it is no longer a system limitation). Since the code may have been used in (many) prior analyses, there could be longstanding problems that the error you discovered has created. As articulated, this case simply states that you found an error: *an ethical issue only arises if you do not fix the error and notify any stakeholders who rely on the code or its results.*

4. Identify and evaluate alternative actions (on the ethical issue).

The three decisions that can be made in any circumstance are: a) do nothing. b) consult or confer with a peer (ASA) or a supervisor (ACM); and c) report violations of policy, procedure, ethical guidelines, or law (ASA) or refuse to implement the system (ACM). In this case, however, only two options are viable because conferring with your supervisor or a peer (option b) will be tantamount to "reporting" (option c) that an error in the code was found. As noted, there is only an ethical issue if you do not fix the error and notify all stakeholders that there is an error. Thus, if you fix and report the error, and your supervisor(s) or organization then seek to suppress this information – particularly, not notifying clients or other stakeholders who rely on the results, then option c) ("report the violation of CE, GLs, and relevant laws") *is* differentiated from option b).

Note that the case simply describes you identifying the error. Thus, option a) must be modified to "correct the program (and re-run all data that have been analyzed with the incorrect system), notify all stakeholders, but do not confer"; and the only other option, b), becomes, "correct the program, re-run all data, and confer with peers or supervisors about how best to notify all stakeholders".

5. Make and justify a decision.

Decision: *option b*: "correct the program, re-run all data, and confer with peers or supervisors about how best to notify all stakeholders" is supported by both ASA and ACM standards. This is the heart of Professional integrity and accountability (ASA Principle A) and general ethical principles (ACM Principle 1). Consulting a peer or supervisor is not necessarily an essential feature of your response, unless the decision to correct the program (common to both options a and b) requires approval. Option b) is preferred over option a) in this case because the notification aspect of the solution may very well require approval or supervisory consultation, if the error you discovered has been in place long enough to have featured in actual decisions already having been made. Note that, once you have identified an error in code you are using, if you continue to use that code without correcting it (even though this is easier), you are violating multiple practice standards – not being transparent or honest in your reporting (unless you also include the error in your report); not using the most appropriate methods; and not ensuring that your work is robust and that you are accountable for it.

6. Reflect on the decision.

Construct your own reflection.

Case 17. You notice that at least some of the assumptions required for interpretable results, using the code you were asked to implement, are not supportable. The code does run and yield results, but the assumptions underpinning those results are not valid.

1. Identify and 'quantify' prerequisite knowledge:

Which DEFW and DSEC Principles or Areas seem most relevant to this vignette?

This situation interferes with DEFW Principles 5 and 6: robust practice (5) requires that methods are appropriate, and 6 requires accountable work. Results of a system or analysis with identified errors conflict with these Principles.

There is no reference to what to do about errors in analyses in DSEC. However, the existence of the error(s) means C3, C5, D4, E3 cannot be addressed. Depending on the nature and impact of the assumption violations on the output of the system, other DSEC items cannot be answered confidently.

Which ASA and ACM Principles seem most relevant to this vignette?

ASA:

Principle A. Professional integrity and accountability
Principle B. Integrity of data and methods
Principle C. Responsibilities to stakeholders
Principle E. Responsibilities to members of multidisciplinary teams
Principle F. Responsibilities to fellow statistical practitioners and the Profession
Principle G. Responsibilities of leaders, supervisors, and mentors in statistical practice
Principle H. Responsibilities regarding potential misconduct

ACM:

1. General Ethical Principles
2. Professional Responsibilities

Potential result: Stakeholder:	HARM	BENEFIT	UNKNOWN	UNKNOWABLE
YOU	Identifying, documenting, and then fixing the error will take time, adds effort; may identify other problems that require additional time/effort. Failing to fix the error may save time but lead to unpredictable propagation of that error throughout the workflow.	Correction of the error means future work with the code will be facilitated –and correct, and interpretable.	Competent analysis (ASA) supports reproducible results; effective analysis (ACM) promotes public good.	
Your boss/client	Identifying, documenting, and then fixing the error will take time, adds effort; may identify other problems that require additional time/effort. This error may be discovered in a 'product' that the client has paid for in the past, identifying past errors and undermining this and other projects	Correction of the error means future work with the code will be facilitated –and correct, and interpretable.	Identification of error may require resources to fix/address. *Appropriate* analysis – once the error is fixed - may not yield desired results.	

Potential result: Stakeholder:	HARM	BENEFIT	UNKNOWN	UNKNOWABLE
Unknown individuals	Analysis required or ACM CE violated; weak statistical analysis undermines reproducibility and rigor.	Identification and correction of the error allows mitigation of risks of harms (ACM); supports valid inferences (ASA), and can limit the unpredictable propagation of the error to other parts of the workflow.		
Employer	Identifying, documenting, and then fixing the error will take time, adds effort; may identify other problems that require additional time/effort. This error may be discovered in a 'product' that the client has paid for in the past, identifying past errors and undermining this and other projects	Correction of the error means future work with the code will be facilitated –and correct, and interpretable.	Identification of error may require resources to fix/address. *Appropriate* analysis – once the error is fixed - may not yield desired results.	Identification of the error may highlight past payments or purchases/sales of incorrect/error-prone work. Liability may be incurred.

Potential result: Stakeholder:	HARM	BENEFIT	UNKNOWN	UNKNOWABLE
Colleagues	The discovered error may complicate colleagues' work.	Identification and correction of the error allows mitigation of risks of harms (ACM); supports valid inferences (ASA), and can limit the unpredictable propagation of the error to other parts of the workflow.		Project could be delayed until risks of additional errors, and resulting harms/ potential harms are addressed.
Profession	Identifying, documenting, and then fixing the error will take time, adds effort; may identify other problems that require additional time/effort. Failing to fix the error may save time but lead to unpredictable propagation of that error throughout the workflow.	Correction of the error means future work with the code will be facilitated –and correct, and interpretable. This supports ethical and competent practice by future practitioners.	Competent analysis (ASA) supports reproducible results; effective analysis (ACM) promotes public good.	Commitment to identification and correction of the error, consistent with GL/CE, might strengthen trust in profession. The error itself can enable innovation (to fix, avoid, or detect similar errors in future), as well as conversations about ethical practice.

Potential result: Stakeholder:	HARM	BENEFIT	UNKNOWN	UNKNOWABLE
Public/public trust	Even if errors are noted so they can be fixed, weak or error prone analysis undermines public trust in reproducibility and rigor.	Identification and correction of the error allows mitigation of risks of harms (ACM); supports valid inferences (ASA), and can limit the unpredictable propagation of the error to other parts of the workflow and decisions based on them.	Transparency supports the public trust, even if risk/harms are identified; can tend to inform the public about limitations inherent in the profession.	

Table 3.4.5. *Stakeholder Analysis: the program you ran used inappropriate methods*

2. Identify decision-making frameworks.

As noted, ASA GLs tend to support a virtue approach, which supports fixing the code so that it represents analyses where all assumptions are met – i.e., the method is appropriate for the data and the objective (ASA A2; B1; B2). Even if an analysis yields results, it does not mean that those results are correct. So, just as in case 16, the fact that assumptions are observed to not be met, i.e., a method is inappropriate, it constitutes an error in that code. "The ethical statistical practitioner Uses methodology and data that are valid, relevant, and appropriate, without favoritism or prejudice, and in a manner intended to produce valid, interpretable, and reproducible results." The ethical statistical practitioner also "avoids compromising scientific validity for expediency" (ASA E4), so although it would be more expedient to let the system run on with the known errors, obligations to stakeholders (ASA C) and data donors (ASA D), as well as team members (ASA E) and other practitioners (ASA F) dictate that the ethical statistical practitioner: "strives to promptly correct substantive errors discovered after publication or implementation. As appropriate, disseminates the correction publicly and/or to others relying on the results." (ASA B5)

The ACM CE articulates that the computing professional has a responsibility to analyze the work they do/are asked to do so that decisions made on the basis of the system or its subparts are *justifiable*. Clearly, violated assumptions can invalidate any decisions based on results based on inappropriate methods; this is a (possibly atypical) error in the analysis part of a system being evaluated. However, a system that was designed according to inappropriate assumptions will yield results that are (mathematically or socially) biased or simply incorrect. Importantly, the violation of assumptions means that it isn't a simple error in the code or system that needs fixing, but rather, possibly different methods altogether. Unsupported assumptions can cause predictable (once the error is identified) but fixable harms. Moreover, if you have identified the fact that assumptions are violated, your work must then address those/the fix to comply with ACM 1.3, "A computing professional should be transparent and provide full disclosure of all pertinent system capabilities, limitations, and potential problems to the appropriate parties."

3. Identify or recognize the ethical issue.

Both the ASA and ACM standards support the identification and fixing of the code so that assumptions are met, and not violated. As with any error, the ASA GLs (B5) stipulate that corrections should be disseminated, and the ACM CE specify similarly that limitations of the system should be disclosed (unless the error is fixed, in which it is no longer a system limitation). Since the code may have been used in (many) prior analyses, there could be longstanding problems that the error you discovered has created. As articulated, this case simply states that you found the error(s) resulting in assumptions being violated: *an ethical issue only arises if you do not fix the code to correct the violations/errors, but also if you fail to notify any stakeholders who rely on the code or its results.*

4. Identify and evaluate alternative actions (on the ethical issue).

The three decisions that can be made in any circumstance are: a) do nothing. b) consult or confer with a peer (ASA) or a supervisor (ACM); and c) report violations of policy, procedure, ethical guidelines, or law (ASA) or refuse to implement the system (ACM). In this case, like case 16, only two options are viable because conferring with your supervisor or a peer (option b) will be tantamount to "reporting" (option c) that the error(s) in the code were found. As noted, there is only an ethical issue if you do not fix the errors and notify all stakeholders that there was an error (which could have underpinned decisions in the past). Thus, if you fix and report the error, and your supervisor(s) or organization then seek to suppress this information – particularly, not notifying clients or other stakeholders who rely on the results, then option c) ("report the violation of CE, GLs, and relevant laws") *is* differentiated from option b).

Note that this case – like case 16 - simply describes you identifying the fact that assumptions are violated. Thus, option a) must be modified to "correct the program and re-run all data that have been analyzed with the incorrect system, and notify all stakeholders, but do not confer"; and the only other option, b), becomes, "correct the program, re-run all data, and confer with peers or supervisors about how best to notify all stakeholders". The difference here is only in the extent of collaboration – since assumptions *are* violated, then methods or the data (or both) need to be rethought. The complexity of the system, your role in the organization or team, or some combination of these will meant that you must (option b) or do not need to (option a) consult or confer with others to make the necessary corrections. The notification of stakeholders will be similarly complicated depending on the extent of the corrections, and conferring or consulting may also help map out the strategy for dissemination of the correction, but both the corrections and the notifications are still essential to the response in the case.

5. Make and justify a decision.

Decision: *option b*: "correct the program, re-run all data, and confer with peers or supervisors about how best to notify all stakeholders" is supported by both ASA and ACM standards. This is the heart of Professional integrity and accountability (ASA Principle A) -among other aspects of the ASA GLs - and general ethical principles (ACM Principle 1). Consulting a peer or supervisor is not necessarily an essential feature of your response, unless the decision to correct the program (common to both options a and b) requires approval. Option b) is preferred over option a) in this case because the notification aspect of the solution may very well require approval or supervisory consultation, if the error you discovered has been in place long enough to have featured in actual decisions already having been made. Given the potential for the complexity of revising the code, peers or supervisors may also be important to help ensure that all aspects of the system or analysis do meet all assumptions, and methods are appropriate for the problem and the data. This makes the case for option b) even stronger in this case than the previous one. Just as with the previous case, once you have identified an error in code you are using, if you continue to use that code without correcting it (even though this is easier), you are violating multiple practice standards – not being transparent or honest in your reporting (unless you also include the error in your report); not using the most appropriate methods; and not ensuring that your work is robust and that you are accountable for it.

6. Reflect on the decision.

Construct your own reflection.

Case 18. You are asked to evaluate a new system or analyze a data set, and told the results that your evaluation or analysis should generate.

1. Identify and 'quantify' prerequisite knowledge:

   Which DEFW and DSEC Principles or Areas seem most relevant to this vignette?

This situation is not aligned with DEFW Principle 1, and is inconsistent with Principle 2 (since using *appropriate* **methodology is the hallmark of competent professional practice). The instructions make it impossible to address any of the other DEFW questions honestly.**

Not specifically addressed in DSEC, but this situation will specifically make DSEC C and D impossible to address. This situation technically constitutes DSEC E4 ("Have we taken steps to identify and prevent unintended uses and abuse of the model.") – and the answer here is "no", because this situation is technically an abuse of the model.

   Which ASA and ACM Principles seem most relevant to this vignette?

ASA:

Principle A. Professional integrity and accountability
Principle B. Integrity of data and methods
Principle C. Responsibilities to stakeholders
Principle E. Responsibilities to members of multidisciplinary teams
Principle F. Responsibilities to fellow statistical practitioners and the Profession
Principle H. Responsibilities regarding potential misconduct

ACM:

1. General Ethical Principles
2. Professional Responsibilities

Potential result: Stakeholder:	HARM	BENEFIT	UNKNOWN	UNKNOWABLE
YOU	You will appear incompetent if you design a system or analysis using inappropriate methodology – *especially* if it appears to work as intended.	There is **no benefit** to following directions that direct you to violate professional practice standards of competence.	Your work could lead to unpredicted harms, bias, or unfair results, and although you were told to design the system/analysis incorrectly, *only you* will appear at fault.	Incorrect/ inappropriate results may be taken up into other applications, or policy and yield additional inappropriate and incompetent work as well as unpredictable/ unpredicted harms/bias/unfair results.
Your boss/client	Inappropriate methodology may lead to "desired" results, but they will not be reproducible or reliable, or correct.	Inappropriate methodology may lead to "desired" results (but they will not be reproducible or reliable, or correct.)		Incorrect/ inappropriate results may be taken up into other applications, or policy and yield additional inappropriate and incompetent work as well as unpredictable/ unpredicted harms/bias/unfair results.

Potential result: Stakeholder:	HARM	BENEFIT	UNKNOWN	UNKNOWABLE
Unknown individuals	Ineffective design/ poor planning can create bias or permit privacy breaches. A weak or biased system may be erroneously reported to be strong/fair. Results may be unfair, insecure, or both – and decisions made by incorrectly-designed systems may be unauditable. Untested – known to be incorrect – assumptions can lead to incorrect decisions and policy.			Incorrect/inappropriate results may be taken up into other applications, or policy and yield additional inappropriate and incompetent work as well as unpredictable/ unpredicted harms/bias/ unfair results.

Potential result: Stakeholder:	HARM	BENEFIT	UNKNOWN	UNKNOWABLE
Employer	Ineffective design/ poor planning can create bias or permit privacy breaches. The company will appear to employ incompetent workers if it is discovered that inappropriate methods were used. Results (or the system you evaluate) may be unfair, insecure, or both – and decisions made by incorrectly-designed systems may be unauditable. Untested – known to be incorrect – assumptions can lead to incorrect decisions and policy.			The company will appear to employ incompetent workers- and supervisors - if it is discovered that inappropriate methods were directed to be used.
Colleagues	Colleagues may assume the work was done competently and incorrectly build on inappropriate, biased, or otherwise unfair foundations.	Colleagues who are *not* instructed to use inappropriate methods may carry out tests that demonstrate the system or analysis cannot move forward with this design.		

Potential result: Stakeholder:	HARM	BENEFIT	UNKNOWN	UNKNOWABLE
Profession	The profession may appear incompetent when it is discovered that practitioners design analyses or evaluate systems or analyses using inappropriate methodology to obtain pre-determined results.	There is **no benefit** to the profession that accrues from following directions that violate professional practice standards of competence.	Inappropriate work could lead to unpredicted harms, bias, or unfair results, and although you were told to design the system/analysis incorrectly, *only the practitioner* will appear at fault.	Incorrect/inappropriate results may be taken up into other applications, or policy and yield additional inappropriate and incompetent work as well as unpredictable/ unpredicted harms/bias/unfair results. Trust in the profession will be undermined (why not use amateurs – they're cheaper and would be expected to make the same mistakes – for less!)
Public/public trust	Inappropriate methodology can lead to conflicting results (best case) or misuse/abuse (worst case). Even if it is only one small part of a system, in inappropriate component can lead to many harms and under-mine public trust in systems and the profession.			Aiming for predetermined results undermine public trust in the profession and in computing, statistics, and data science generally. Trust in the profession will be undermined (it doesn't matter if you have a degree in that field, "those people" cannot be trusted to perform competently).

Table 3.4.6. *Stakeholder Analysis: An evaluation/analysis with pre-specified results instead of honest/correct ones*

2. Identify decision-making frameworks.

The ethical statistical practitioner has responsibilities to "Use methodology and data that are valid, relevant, and appropriate, without favoritism or prejudice, and in a manner intended to produce valid, interpretable, and reproducible results." (ASA A2); "oppose efforts to predetermine or influence the results of statistical practices and resists pressure to selectively interpret data." (ASA A4); "regardless of personal or institutional interests or external pressures, does not use statistical practices to mislead any stakeholder." (ASA C2) Also, the ethical practitioner "is transparent about assumptions made in the execution and interpretation of statistical practices including methods used, limitations, possible sources of error, and algorithmic biases. Conveys results or applications of statistical practices in ways that are honest and meaningful." (ASA B2) Apart from being able to take full responsibility for your work, these two principles in particular allow the ethical practitioner to "Avoid condoning or appearing to condone statistical, scientific, or professional misconduct. Encourage other practitioners to avoid misconduct or the appearance of misconduct." (ASA H2) Following the directions to generate desired and pre-determined results is fully inconsistent with the ASA GLs. Making this request is also inconsistent with Principle G and the ASA GL Appendix, particularly Appendix items 8 and 9 (APP 8: "Recognizing that it is contrary to these Guidelines to report or follow only those results that conform to expectations without explicitly acknowledging competing findings and the basis for choices regarding which results to report, use, and/or cite." and APP 9: "Recognizing that the results of valid statistical studies cannot be guaranteed to conform to the expectations or desires of those commissioning the study or employing/supervising the statistical practitioner(s).")

The ACM CE, with its utilitarian perspective, typically focuses on limiting harms but the tradeoff of harms and benefits must also be considered (as it is, implicitly, in DEFW Principle 1). Agreeing to the request will generate results that cannot be reliable or robust, thus harming virtually all stakeholders. The SHA shows that there are no legitimate benefits to following directions to use inappropriate methodology or yield a report that conforms to desires rather than honestly reflecting the expert's evaluation of the system, and the harms are significant – to the practitioner as well as other stakeholders. I.e., there are no benefits to trade off against the harms that following this request will or might create.

3. Identify or recognize the ethical issue.

Both the ASA and ACM practice standards state explicitly and repeatedly that professionals perform only those tasks for which they are competent, and that they do not yield to pressures to behave contrary to their respective practice standards (**ASA**: A; A1, A2, A4, A9, A12; C1, C2; E4; G1; H2; **ACM**: 1.2; 1.3; 2.1; 2.2; 2.3; 3.4; 4). Moreover, both the ASA (ASA Principle G) and ACM CE (Principle 3) also suggest that the practitioner and those in leadership positions have an obligation to ignore the person(s) who want you to violate the ethical practice standards. Both practice standards also articulate a responsibility to encourage others to comply with the standards, if possible.

Leaders should also ensure that the system they and other computing professionals design or contribute to is correct and appropriate. The primary ethical issue is that a person is trying to prevent a professional from doing their job competently; the individual directing you to generate pre-determined results is behaving contrary to the GLs and CE. Moreover, that person is preventing others (or trying to) from following the GLs and CE. It is important to consider whether pressure or intimidation is being utilized to ensure that a predetermined result is obtained; in this case, not only are the GLs and CE violated by the requestor, but also, applicable laws that establish your right to a safe work environment. This case does not suggest that there is undue pressure, but this frankly unethical and illegal behavior *has happened* in many cases (where, if the desired results are not obtained, then the practitioner is fired or replaced).

4. Identify and evaluate alternative actions (on the ethical issue).

As usual, the three decisions that can be made in any circumstance are: a) do nothing. b) consult or confer with a peer (ASA) or a supervisor (ACM); and c) report violations of policy, procedure, ethical guidelines, or law (ASA) or refuse to implement the system (ACM). Clearly, "do nothing" is both totally inappropriate in this case, and also a violation of the practice standard. We can change option a) from "do nothing" to "ignore the request and design the system/analysis appropriately – but do not consult colleagues/report this behavior". Then option b) would become, "consult or confer with a peer (ASA) or a supervisor (ACM) as to how best to respond to a person who wants you to do your job incompetently, while ignoring that request and designing the system/analysis appropriately". Option c) must also be modified, because simply reporting the individual who directs you to violate the GLs (ASA) or refusing to implement the system (to collect an inappropriate sample size) (ACM) could mean that your time on the project would end, and another person who has no problem designing a system or analysis to do whatever is requested irrespective of the violation of the ASA and ACM practice standards will be brought on. Option c) needs to change to "report violations of policy, procedure, ethical guidelines, or law to a peer (ASA) or a supervisor (ACM), and ignore that request and design the system/analysis appropriately." Note again that, if pressure or intimidation is being utilized to ensure that a predetermined result is obtained, then option c) is the best (albeit possibly the most uncomfortable) option. In that event, you should consult applicable laws that establish your right to a safe work environment and protect whistleblowers.

5. Make and justify a decision.

Decision: *Note that all options include* "ignore that request and design the system/analysis appropriately". That is because the ethical response to this situation is to **follow the practice standards** – do your job competently - and ignore the resistance to, or direction to violate, both the guidelines and what must be considered "best practices". What is less clear is what *else* to do. Option a) says "follow best practices, and beyond that, do nothing (besides ignore the encouragement to violate practice standards)". This would be the *least desirable option* because the person/people who are encouraging you to ignore your professional practice standards need to understand that this is inappropriate, and they should stop that behavior. However, if such behavior is common in your work context, it might be impossible for you to do anything except seek another team, other colleagues, or a different job. Option b) would be the *most desirable option* in that consulting with a colleague/supervisor will ensure that this kind of frankly unethical behavior by a collaborator (or boss) – inappropriate interference with your professional practice - is documented. There may not be policies in your workplace against such interference/violation of the ASA GLs or ACM CE, so reporting (option c) may not be an option, making it a slightly less desirable option than b. However, in contexts where such policies do exist, it may still be difficult to report such behaviors; the person directing you to violate your professional practice standards may not be aware of them/this fact. So, option c may not be as much about "reporting" that person as it is about "educating" them. However, is important *to the profession* to support the next practitioner who ends up in a similar situation (or better, to prevent such unethical interference in the future), so it is definitely worth considering politely notifying the individual that you (or anyone) fulfilling their request represents a violation of the ACM CE, and you being directed to violate the CE and GLs is impossible. Finally, if pressure or intimidation are at play in the "request" for a predetermined result, option c) is the best option for you, for your colleagues and the organization, as well as for the profession and the public. The SHA shows that only the requestor may benefit – and that, only marginally – from inappropriate work by you; harms accrue to all other stakeholders.

6. Reflect on the decision.

Construct your own reflection on this decision.

Case 19. Your analysis of your new system suggests that there is an unexpectedly high error rate, but only for a small subgroup of users. *Overall,* **your system's results are exactly as expected;** *for the subgroup,* **the results are the opposite of the overall result.**

1. Identify and 'quantify' prerequisite knowledge:

> Which DEFW and DSEC Principles or Areas seem most relevant to this vignette?

The results can help you fulfill DEFW Principle 4 (and should contribute to Principles 5 and 6).

This case generates "no" answers to multiple items on DSEC (A2; C1; C2; D1; D4), but generates "yes" answers to two: D2; D3. DSEC does not offer suggestions for changing the "noes" to "yes", or how to understand/reconcile some yeses and some noes.

> Which ASA and ACM Principles seem most relevant to this vignette?

ASA:

Principle A. Professional integrity and accountability
Principle B. Integrity of data and methods
Principle C. Responsibilities to stakeholders
Principle D. Responsibilities to research subjects, data subjects, or those directly affected by statistical practices

ACM:

1. General Ethical Principles

Potential result: Stakeholder:	HARM	BENEFIT	UNKNOWN	UNKNOWABLE
YOU	You may "get in trouble" if you demonstrate that the system does not function/results are not as expected.	The sensitivity results are performing as intended, showing your competence! Addressing the results means the system will be less biased.		
Your boss/client	Taking longer to edit or change the system (to address the bias) could translate to missed deadlines or milestones. New methods may be risky even if they produce less biased results.	Using less data (e.g., excluding the subgroup with error-prone results) may limit liability and exposure to risk (and may cost less for data access/storage).		
Unknown individuals	Continuing to use existing methods means risks will not change (will not decrease) relative to "now". Excluding the subgroup from future models may create other bias.	Sensitivity analysis (if not the system itself) is functioning as intended. Addressing the results means the system will be less biased.		

Potential result: Stakeholder:	HARM	BENEFIT	UNKNOWN	UNKNOWABLE
Employer	Taking longer to edit or change the system (to address the bias) could translate to missed deadlines or milestones. New methods may be risky even if they produce less biased results.	Using less data (e.g., excluding the subgroup with error-prone results) may limit liability and exposure to risk (and may cost less for data access/storage).		
Colleagues	Your results may create a requirement for adaptations (requiring extra time and effort from everyone).	Your sensitivity analysis results may create opportunities for innovations from everyone.		
Profession	The profession or professionals may appear too "in the weeds" if it focuses on results from small subgroups.	The sensitivity results are performing as intended, showing their importance to those in and outside of the profession.		

Potential result:	HARM	BENEFIT	UNKNOWN	UNKNOWABLE
Stakeholder:				
Public/public trust	Excluding the subgroup from future models may create other bias.	Continuing to ignore the errors for a small subgroup means risks will not change (will not decrease) for "most people". Addressing the results of the analysis may mean the system will be less biased.		

Table 3.4.7. *Stakeholder Analysis: Subgroup analysis shows unexpected result*

2. Identify decision-making frameworks.

Note that the practitioner in this case has followed the GLs and CE: they used appropriate methodology scientifically and discovered a difficulty with an existing method. Thus, the virtue perspective has been followed in this case. The question that the SHA implies is, what will be done as a result of the identification of the inequity or bias in this system?

The ethical statistical practitioner has responsibilities to "Uses methodology and data that are valid, relevant, and appropriate, without favoritism or prejudice, and in a manner intended to produce valid, interpretable, and reproducible results." (ASA A2), and the sensitivity analyses have demonstrated that A2 cannot be met until the issue with the subgroup is resolved. (ASA A3 may also be at risk here, "Does not knowingly conduct statistical practices that exploit vulnerable populations or create or perpetuate unfair outcomes."). Moreover, the ethical practitioner "considers the impact of statistical practice on society, groups, and individuals. Recognizes that statistical practice could adversely affect groups or the public perception of groups, including marginalized groups. Considers approaches to minimize negative impacts in applications or in framing results in reporting." (ASA D6) Thus, the virtue approach dictates that the ethical practitioner will seek to incorporate sensitivity analyses in systems like the one in this case, and then they will act to ensure that disproportionate harm is not a result of that system or analysis – especially not after it has been identified.

The utilitarian perspective of the ACM CE alerts the ethical computational practitioner to the responsibility to consider cases like this one: "The use of information and technology may cause new, or enhance existing, inequities." (ACM 1.4). The SHA shows that many potential real harms accrue if the bias is ignored, with real benefits accruing if you address the bias.

3. Identify or recognize the ethical issue.

Both the ASA and ACM standards support the identification and fixing of the error, with the ASA GLs stipulating that corrections should be disseminated, and the ACM CE specifying similarly that limitations of the system should be disclosed (unless the error is fixed, in which it is no longer a system limitation). Since the system may have been used in (many) prior analyses and supported prior decisions, which would have adversely affected the specified subgroup, there could be longstanding problems created by the error you discovered. As articulated, this case simply states that you found a source of bias (error really): *an ethical issue only arises if you do not fix the error and notify any stakeholders who rely on the code or its results.*

4. Identify and evaluate alternative actions (on the ethical issue).

The three decisions that can be made in any circumstance are: a) do nothing. b) consult or confer with a peer (ASA) or a supervisor (ACM); and c) report violations of policy, procedure, ethical guidelines, or law (ASA) or refuse to implement the system (ACM). In this case, however, only two options are viable because conferring with your supervisor or a peer (option b) will be tantamount to "reporting" (option c) that a source of bias and error in the system was found. As noted, there is only an ethical issue if you do not fix the error and notify all stakeholders that there is an error, because it may have already adversely affected decisions. Thus, if you fix and report the error, and your supervisor(s) or organization then seek to *suppress* this information – particularly, not notifying clients or other stakeholders who rely on the results, then option c) ("report the violation of CE, GLs, and relevant laws") *is* differentiated from option b).

Note that the case simply describes you identifying the source of error. Thus, option a) must be modified to "correct the program (and re-run all data that have been analyzed with the incorrect system), notify all stakeholders, but do not confer"; and the only other option, b), becomes, "correct the program, re-run all data, and confer with peers or supervisors about how best to notify all stakeholders".

5. Make and justify a decision.

Decision: *option b*: "correct the program, re-run all data, and confer with peers or supervisors about how best to notify all stakeholders" is supported by both ASA and ACM standards, as well as the SHA. This is the heart of Professional integrity and accountability (ASA Principle A) and general ethical principles (ACM Principle 1). Consulting a peer or supervisor is not necessarily a critical feature of your response, unless the decision to correct the program (common to both options a and b) requires approval and/or collaboration with others. Option b) is preferred over option a) in this case because the notification aspect of the solution may very well require approval or supervisory consultation, if the bias you discovered has been in place long enough to have featured in actual decisions already having been made. Note that, once you have identified a source of bias in a system or method you are using, if you continue to use that method/approach without correcting it (even though this is easier), you are violating multiple practice standards – not being transparent or honest in your reporting (unless you also include the error in your report); not using the most appropriate methods; and not ensuring that your work is robust and that you are accountable for it. Finally, if it is the case that you suggest option b) but your supervisor(s) or organization seek to *suppress* this information and to prevent stakeholders who rely on the results from learning about the bias in the decisions they make, then option c) ("report the violation of CE, GLs, and relevant laws") is the optimal one. As with the previous case, you may need to consult whistleblower laws to ensure that you are protected as you pursue option c).

6. Reflect on the decision.

Construct your own reflection.

Case 20. You institute an interim check of results and discover that there is bias in the results. The interim check is literally the middle of a multi-part process that you are working on with several colleagues, so there's no way to immediately pinpoint the source of the bias.

1. Identify and 'quantify' prerequisite knowledge:

Which DEFW and DSEC Principles or Areas seem most relevant to this vignette?

The results can help you fulfill DEFW Principles 5 and 6.

This case generates "no" answers to multiple items on DSEC (A2; C1; C2; D1; D4), but generates "yes" answers to two: D2; D3. DSEC does not offer suggestions for changing the "noes" to "yes", or how to understand/reconcile some yeses and some noes.

Which ASA and ACM Principles seem most relevant to this vignette?

ASA:

Principle A. Professional integrity and accountability
Principle B. Integrity of data and methods
Principle C. Responsibilities to stakeholders
Principle D. Responsibilities to research subjects, data subjects, or those directly affected by statistical practices

ACM:

1. General Ethical Principles

Potential result: Stakeholder:	HARM	BENEFIT	UNKNOWN	UNKNOWABLE
YOU	You may "get in trouble" if you demonstrate that the system does not function/results are not as expected. Addressing this source of bias will take time and effort.	The sensitivity analysis is performing as intended, showing your competence! Addressing the interim results means the system will ultimately be less biased.		
Your boss/client	Taking longer to edit or change the system (to address the bias) could translate to missed deadlines or milestones. If the bias cannot be found and fixed, the whole system may fail.	Addressing the interim results means the system will ultimately be less biased.		
Unknown individuals	Failure to correct bias will undermine public trust in the profession and may further limit enthusiasm for sharing data.	Sensitivity analysis (if not the system itself) is functioning as intended. Addressing the results means the system will ultimately be less biased, or if the bias cannot be corrected, the system may not be deployed.		

Potential result: Stakeholder:	HARM	BENEFIT	UNKNOWN	UNKNOWABLE
Employer	Taking longer to edit or change the system (to address the bias) could translate to missed deadlines or milestones. If the bias cannot be found and fixed, the whole system may fail.	Addressing the interim results means the system will ultimately be less biased. Addressing the bias may ultimately limit liability/need for redress.		
Colleagues	Your results may create a requirement for adaptations (requiring extra time and effort from everyone). Addressing this source of bias will take collaborative time and effort.	Your sensitivity analysis results may create opportunities for innovations from everyone.		
Profession	The profession or professionals may appear too "in the weeds" if it focuses on results from small subgroups.	The sensitivity results are performing as intended, showing their importance to those in and outside of the profession.		
Public/public trust	Failure to correct bias will undermine public trust in the profession and may further limit enthusiasm for sharing data.	Sensitivity analysis (if not the system itself) is functioning as intended. Addressing the results means the system will ultimately be less biased, or if the bias cannot be corrected, the system may not be deployed.		

Table 3.4.8. *Stakeholder Analysis: Interim check shows bias that can't be immediately corrected*

2. Identify decision-making frameworks.

Note that the practitioner in this case has followed the GLs and CE: they used appropriate methodology scientifically and discovered a difficulty with an existing method. Thus, the virtue perspective has been followed in this case. The question that the SHA implies is, what will be done as a result of the identification of the inequity or bias in this system? We can assume that this is a different challenge than the previous case because the interim analysis is literally in the middle of the overall process – so, making changes will require a much greater effort to identify exactly the source of the bias, and then ensure that further bias doesn't get added in when the first source is addressed.

The ethical statistical practitioner has responsibilities to "use methodology and data that are valid, relevant, and appropriate, without favoritism or prejudice, and in a manner intended to produce valid, interpretable, and reproducible results" (ASA A2) and "does not knowingly conduct statistical practices that exploit vulnerable populations or create or perpetuate unfair outcomes." (ASA A3) The interim analyses have demonstrated that ASA A2 cannot be met until the cause of the bias is identified and addressed; in this case we don't know if ASA A3 will be an issue (yet). Also important, the ethical practitioner "strives to promptly correct substantive errors discovered after publication or implementation. As appropriate, disseminates the correction publicly and/or to others relying on the results." (ASA B5) – and the other team members are stakeholders in the process where the bias is being created (as we have seen in all the SHAs). Thus, the virtue approach dictates that the ethical practitioner will continue to use interim checks and sensitivity analyses in systems like the one in this case, and then they will act to ensure that disproportionate harm is not a result of that system or analysis – especially not after bias has been identified.

The utilitarian perspective of the ACM CE alerts the ethical computational practitioner to the responsibility to consider cases like this one: "The use of information and technology may cause new, or enhance existing, inequities." (ACM 1.4). The SHA shows that potential harms accrue if the bias is ignored, with real benefits accruing if you address the bias.

3. Identify or recognize the ethical issue.

Both the ASA and ACM standards support the identification and fixing of the error, with the ASA GLs stipulating that corrections should be disseminated, and the ACM CE specifying similarly that limitations of the system should be disclosed (unless the error is fixed, in which it is no longer a system limitation). Since the system may have been used in (many) prior analyses and supported prior decisions, which would have adversely affected the specified subgroup, there could be longstanding problems created by the error you discovered. As articulated, this case simply states that you found a source of bias (error really): *an ethical issue only arises if you do not find and fix the error and notify any stakeholders who rely on the system or its results.*

4. Identify and evaluate alternative actions (on the ethical issue).

The three decisions that can be made in any circumstance are: a) do nothing. b) consult or confer with a peer (ASA) or a supervisor (ACM); and c) report violations of policy, procedure, ethical guidelines, or law (ASA) or refuse to implement the system (ACM). In this case, however, only two options are viable because conferring with your supervisor or a peer (option b) will be tantamount to "reporting" (option c) that a source of bias and error in the system was found. As noted, there is only an ethical issue if you do not fix the error and notify all stakeholders that there is an error, because it may have already adversely affected decisions. Thus, if you fix and report the error, and your supervisor(s) or organization then seek to *suppress* this information – particularly, not notifying clients or other stakeholders who rely on the results, then option c) ("report the violation of CE, GLs, and relevant laws") *is* differentiated from option b).

Note that, like the previous case, the case here simply describes you identifying that there is a source of bias somewhere in the system. Thus, option a) must be modified to "correct the program (and re-run all data that have been analyzed with the incorrect system), notify all stakeholders, but do not confer"; and the only other option, b), becomes, "correct the program, re-run all data, and confer with peers or supervisors about how best to notify all stakeholders". Unlike the previous case, in this case there may be many more practitioners involved in "correct the program" and multiple iterations of "re-run all data", because the source of bias must first be pinpointed. The fact that the practitioner in this case is *not* the individual who is responsible for the whole system does not relieve them of the burden of pinpointing the source of bias and acting to address it.

5. Make and justify a decision.

Decision: *option b*: "correct the program, re-run all data, and confer with peers or supervisors about how best to notify all stakeholders" is supported by both ASA and ACM standards, as well as the SHA. As noted in the previous case, this is the heart of Professional integrity and accountability (ASA Principle A) and general ethical principles (ACM Principle 1). Note again that the fact that the practitioner in this case is *not* the individual who is responsible for the whole system does mean that there is no responsibility for pinpointing the source of bias that was detected by this practitioner, and acting to address it.

Consulting a peer or supervisor becomes a more critical feature of your response because with the larger system in this case (compared to the last one), a decision to make changes/corrections may require approval and/or collaboration with others. Like the previous case, option b) is preferred over option a) because the notification aspect of the solution may require approval or supervisory consultation, if the bias you discovered has been in place long enough to have featured in actual decisions already having been made. Like the previous case, because you have identified a source of bias in a system or method you are using, if you continue to use that method/approach without correcting it (even though this would be much easier), you are violating multiple practice standards – not being transparent or honest in your reporting (unless you also include the error in your report); not using the most appropriate methods; and not ensuring that your work is robust and that you are accountable for it. Because the system is multi-part and there are at least several others working on it with you, you would then be affecting their ability to practice ethically as well. Finally, just as with the prior case, if it is the case that you suggest option b) but your supervisor(s) or organization seek to *suppress* this information and to prevent stakeholders who rely on the results from learning about the bias in the decisions they make, then option c) ("report the violation of CE, GLs, and relevant laws") is the optimal one. As with the previous cases, you may need to consult whistleblower laws to ensure that you are protected as you pursue option c).

6. Reflect on the decision.

Construct your own reflection on this decision.

Case 21. You are told that your results with new data must match original results (i.e., you must replicate other results), and your analyses/code are right, but they do not replicate earlier results.

1. Identify and 'quantify' prerequisite knowledge:

Which DEFW and DSEC Principles or Areas seem most relevant to this vignette?

This situation is *consistent* with DEFW Principles 5 and 6.

There is no reference to what to do about errors in analyses (or results) in DSEC. However, the situation means that DSEC areas C, D, and E cannot be addressed – until the conflict is resolved.

Which ASA and ACM Principles seem most relevant to this vignette?

ASA:

Principle A. Professional integrity and accountability
Principle B. Integrity of data and methods
Principle C. Responsibilities to stakeholders
Principle E. Responsibilities to members of multidisciplinary teams
Principle F. Responsibilities to fellow statistical practitioners and the Profession
Principle H. Responsibilities regarding potential misconduct

ACM:

2. Professional Responsibilities

Potential result: Stakeholder:	HARM	BENEFIT	UNKNOWN	UNKNOWABLE
YOU	Figuring out whether your - or the predetermined results - are wrong will take time, add effort; may identify other problems that require additional time/effort. Failing to fix the error – whether in yours or the predetermined results - may save time but lead to unpredictable propagation of that error throughout the workflow.	Correction of the error means future work with the code will be facilitated –and correct, and interpretable.	Competent analysis (ASA) supports reproducible results; effective analysis (ACM) promotes public good.	
Your boss/client	Identifying, documenting, and then fixing the error – whether in yours or the predetermined results will take time, adds effort; may identify other problems that require additional time/effort. This error may be discovered in a 'product' that the client has paid for in the past, identifying past errors and undermining this and other projects	Correction of the error means future work with the code will be facilitated –and correct, and interpretable.	Identification of error may require resources to fix/address. *Appropriate* analysis – once the error is found and fixed - may not yield desired results.	

Potential result: Stakeholder:	HARM	BENEFIT	UNKNOWN	UNKNOWABLE
Unknown individuals	Incorrect statistical analysis undermines reproducibility and rigor. Predetermined results negatively affect all stakeholders in the long run.	Identification and correction of the error allows mitigation of risks of harms (ACM); supports valid inferences (ASA), and can limit the unpredictable propagation of the error to other parts of the workflow.		
Employer	Identifying, documenting, and then fixing the error will take time, adds effort; may identify other problems that require additional time/effort. This error may be discovered in a 'product' that a client has/other clients have paid for in the past, identifying past errors and undermining this and other projects	Correction of the error means future work with the code will be facilitated – and correct, and interpretable.	Identification of error may require resources to fix/address. *Appropriate* analysis – once the error is fixed - may not yield desired results.	Identification of the error may highlight past payments or purchases/sales of incorrect/error-prone work. Liability may be incurred.

Potential result: Stakeholder:	HARM	BENEFIT	UNKNOWN	UNKNOWABLE
Colleagues	The discovered error may complicate colleagues' work.	Identification and correction of the error allows mitigation of risks of harms (ACM); supports valid inferences (ASA), and can limit the unpredictable propagation of the error to other parts of the workflow.		Project could be delayed until risks of additional errors, and resulting harms/ potential harms are addressed.
Profession	Failing to fix the error may save time but lead to unpredictable propagation of that error throughout the workflow, undermining the practitioner's credibility –as well as the profession's.	Correction of the error means future work with the code will be facilitated –and correct, and interpretable. This supports ethical and competent practice by future practitioners.	Competent analysis (ASA) supports reproducible results; effective analysis (ACM) promotes public good.	Commitment to identification and correction of the error, consistent with GL/CE, might strengthen trust in profession. The error itself can enable innovation (to fix, avoid, or detect similar errors in future), as well as conversations about ethical practice.

Potential result: Stakeholder:	HARM	BENEFIT	UNKNOWN	UNKNOWABLE
Public/public trust	Even if errors are noted so they can be fixed, weak or error-prone analysis undermines public trust in reproducibility and rigor. An insistence on predetermined results being "correct" undermines public trust as well as reproducibility and rigor.	Identification and correction of the error allows mitigation of risks of harms (ACM); supports valid inferences (ASA), and can limit the unpredictable propagation of the error to other parts of the workflow and decisions based on them.	Transparency supports the public trust, even if risk/harms are identified; can tend to inform the public about limitations inherent in the profession.	

Table 3.4.9. *Stakeholder Analysis: Your results are right, but don't match predetermined results.*

2. Identify decision-making frameworks.

As noted, ASA GLs tend to support a virtue approach, which prioritizes the analytic method over predetermined results. The ethical statistical practitioner "uses methodology and data that are valid, relevant, and appropriate, without favoritism or prejudice, and in a manner intended to produce valid, interpretable, and reproducible results." (ASA A2), and "opposes efforts to predetermine or influence the results of statistical practices and resists pressure to selectively interpret data." (ASA A4). It is not clear what the original results are, the ethical practitioner "does not knowingly conduct statistical practices that exploit vulnerable populations or create or perpetuate unfair outcomes." (ASA A2). The obligation to resist efforts to influence or predetermine outcomes (whether unfair or not) are to oneself (ASA Principle A) but also to stakeholders (ASA C2: "Regardless of personal or institutional interests or external pressures, does not use statistical practices to mislead any stakeholder.") Also, the ethical statistical practitioner "Avoids compromising scientific validity for expediency" (ASA E4). Since the code is right, and the disagreement is with predetermined (unethical!) results, it is likely that it is the predetermined results and not the practitioner's work where the error lies. Principle B dictates that the ethical statistical practitioner: "Strives to promptly correct substantive errors discovered after publication or implementation. As appropriate, disseminates the correction publicly and/or to others relying on the results." (ASA B5) Not only are there many ASA GL elements that are informative about the virtue perspective on this case, ASA B5 offers the actual solution.

In the ACM CE utilitarian perspective, the computing professional has a responsibility to analyze the work they do/are asked to do so that decisions made on the basis of the system or its subparts are *justifiable*. In this case, no error has been identified in the analysis part of a system being evaluated – so, results from that system are not the concern in this case. Instead, in this case it is the predetermined results that may be biased or otherwise incorrect. Insisting on the predetermined results (rather than the correct ones) potentially causes predictable but fixable harms. "A computing professional should be transparent and provide full disclosure of all pertinent system capabilities, limitations, and potential problems to the appropriate parties." In this case, you have identified an error with the previous or pre-determined results, and that is what needs to be transparently reported. Moreover, it is possible that as long as that predetermined result has been "requested" from practitioners, it may have been creating problems.

3. Identify or recognize the ethical issue.

Both the ASA and ACM standards support the identification and fixing of the error, with the ASA GLs stipulating that corrections should be disseminated, and the ACM CE specifying similarly that limitations of the system should be disclosed. As this case is presented, the practitioner has applied their expertise to carry out an analysis appropriately. The ethical issue is that someone told them what their results needed to be; it is possible that those results were obtained in prior analyses using error-prone or frankly incorrect methodology or code. However, it is important for this practitioner – or anyone in this situation – to ensure that if methodological differences –and not errors - led to the differences in results, then that must be reported. Although the "your analyses/code are right", this case does not specify that the same method had been used on this data/in this case as was used on other data. So the ethical issue will be identified by determining if the discrepancy between new and old results arises from a simple methodological difference, from true differences in the data and not to an error in the previous methodology (because we are told the current analysis/code is correct), or from an actual error in the previous analyses. Until the source of the discrepancy is identified, the nature and existence of an ethical issue cannot be determined. Since you cannot determine if there is an ethical challenge without that, nor conceptualize alternative actions, the source of the discrepancy becomes part of your prerequisite knowledge and is not any kind of alternative action for you to consider.

Note that in this case, there is only an ethical issue if you do identify an error in the prior methodology, and you fail to notify all stakeholders that there was an error underlying the original results (and any decisions based on them). It is not unethical to apply different methods than what was previously used to solve a problem; nor is it unethical to discover that correct methods yield different-than-expected results when applied to new data. It *is* unethical to decide that the predetermined results "must be right" and falsify your own results so they match previous/expected results.

As in earlier cases, if there is an error in the method used previously (but you have fixed it now), then that erroneous method or error-prone code may have been used in (many) prior analyses; thus, there could be longstanding problems that the error thus discovered has created. As articulated, this case simply states that you found an error: *an ethical issue arises if you do not fix the error and notify any stakeholders who rely on the code or its results, or if you falsify your results so they match the expected results.* Since the case does not describe falsification of results, we assume the analysis should proceed on the basis of the identification of an error in prior work.

4. Identify and evaluate alternative actions (on the ethical issue).

Recall that the ethical issue will be identified by determining if the discrepancy between new and old results arises from: *i*) a simple methodological difference; *ii*) from true differences in the data and not to an error in the previous methodology (because we are told the current analysis/code is correct); or *iii*) from an actual error in the previous analyses.

The three decisions that can be made in any circumstance are: a) do nothing. b) consult or confer with a peer (ASA) or a supervisor (ACM); and c) report violations of policy, procedure, ethical guidelines, or law (ASA) or refuse to implement the system (ACM). Obviously the options must depend on the source of the discrepancy; if it is either *i* or *ii* then the only option is to report this – that report would be optimally useful to stakeholders if the original data could be analyzed using the new method (*i*) or if a plausible explanation for different results from the same method on different data (e.g., concept drift per DSEC E3) can be tested (*ii*).

One outcome of the discrepancy determination –that an error was made in previous analyses (*iii*) –creates the alternative actions we should evaluate and from which we can choose. Only two of our three standard options are viable because conferring with your supervisor or a peer (option b) will be tantamount to "reporting" (option c) that an error in the code was found if in fact that is what you determine.

As noted, assuming that you do not falsify your results to match what was expected (*not* a plausible alternative action!), there is only an ethical issue if you do not fix the error and notify all stakeholders that there is/was an error. As we saw in other cases, if you fix and report the error internally, and your supervisor(s) or organization then seek to suppress this information – particularly, not notifying clients or other stakeholders who rely on the results, then option c) ("report the violation of CE, GLs, and relevant laws") *is* differentiated from option b).

Note that the case simply describes you identifying that there is a discrepancy; there may be no ethical issue (if it is due to either *i* or *ii*). Thus, alternative actions will relate to the identification of an error. Then, option a) must be modified to "correct the program (and re-run all data that have been analyzed with the incorrect system), notify all stakeholders, but do not confer"; and the only other option, b), becomes, "correct the program, re-run all data, and confer with peers or supervisors about how best to notify all stakeholders".

5. Make and justify a decision.

Decision: *option b*: "correct the program, re-run all data, and confer with peers or supervisors about how best to notify all stakeholders" is supported by both ASA and ACM standards. This ensures you demonstrate Professional integrity and accountability (ASA Principle A) and follow general ethical principles (ACM Principle 1). Consulting a peer or supervisor is not necessarily an essential feature of your response, unless the decision to correct the original results requires approval. Option b) is preferred over option a) in this case because the notification aspect of the solution may very well require approval or supervisory consultation, particularly if the error you discovered has been in place long enough to have featured in actual decisions already having been made. Going forward we assume that your correct code will be utilized, enabling you to meet other practice standards, e.g., being transparent and honest in your reporting; using the most appropriate methods; and ensuring that your work is robust and that you are accountable for it.

6. Reflect on the decision.

Construct your own reflect

Summary of nine case analyses for Analysis

The nine case analyses of Chapter 3.4 were focused on statistical analysis and the analysis of computing systems. As we discussed at the start of the Chapter, the ACM and the ASA define and use the term slightly differently. The cases were designed to balance the two definitions so that readers from either perspective could follow the logic and practice the reasoning.

Although all the cases had similar structure, the analyses did highlight different "most optimal" responses. Like with Chapter 3.3, in some cases, the policy or structure at your organization can play a role in whether the option to "report" violations (options c) or simply to "consult" (options b) about violations of the GLs, CE, or applicable laws is preferred – although as a concerned citizen, it is a general responsibility to report when laws are being broken. We saw in Chapter 3.4 that reporting violations of law may imply a need to understand whistleblower protections and laws. Many organizations have policies that protect whistleblowers, and both the GLs (Principle G1-G2; Appendix 1, 11) and CE (1.2, 2.3, 2.5, 2.9, 3.4, 4.2) offer some guidance about blowing the whistle.

When discussing how to perform "analysis" ethically in Section 2, general principles from the GLs or CE, and specific questions from the DSEC and DEFW, could be used to ensure that our general behavior is ethical. However, in response to specific challenges and ethical violations such as those in this chapter, DSEC and DEFW were *only* helpful in some cases, but like in Chapter 3.3, neither of these instruments offers support for making or justifying ethical decisions.

Each of the cases in this chapter were based on actual events; and most of the case analyses instruct the reader to ignore requests when these violate the ethical practice standards or when requestors seek to exert any influence to cause others to violate those standards. Some of these cases arise as the result of *regular* work – e.g., the identification of bias after sensitivity analyses are implemented. In the other cases, individuals trying to impede professional and ethical practice by making requests that generally reflect a disinterest in methodological propriety and a strong interest in obtaining the expedient results they desire. In these cases relating to analysis or the evaluation of systems, the SHA shows clearly that benefits that accrue when practice standards are ignored or violated are minimal – and are specific to the violators, but accrue for no one else. As you may have come to expect, these minimal benefits are offset by the potential for serious harms to most other stakeholders.

Questions for Discussion:

1. Review at least two cases in this chapter.

A. Do you agree with the decisions in these case analyses? Why or why not? Which parts of the ER process are most and least acceptable for each case? Do these parts differ for different cases in this chapter? Were there different ASA or ACM Principles or elements that you identified, and if so, how do these affect your case analyses? Are the elements and/or Principles you identify represented in your reflections?

B. Discuss your results for the tests of universality, justice, and publicity for any (or all) case(s) in this chapter.

C. Discuss the relevance to each case – including the analysis and your reflection – of federal, state, or local/organizational policies regarding conflict of interest. How can you incorporate what you know about these policies into any part of the case analysis (prerequisite knowledge; identifying the ethical issue; determining alternative actions; making or justifying your decision; or reflection)? Note that, if there is no alignment between federal, state, or your local/institutional policies and a given case, you can comment on that (e.g., do policies in your organization focus on *misconduct* – and ignore detrimental practices?). Do you feel that these policies could have helped in the reasoning, decision or justification?

D. Discuss the relevance to each case - including the analysis and your reflection – of the statistician/ data scientist as a responsible member of society. How can you incorporate what you know about these policies into any part of the case analysis (prerequisite knowledge; identifying the ethical issue; determining alternative actions; making or justifying your decision; or reflection)? If there are no relevant policies, you should comment on that, particularly when a) data are obtained using federal monies (grants); or b) data are collected (acquired) from humans, such that Federal Regulations about the treatment of human subjects and the acquisition of tissues and data from humans are relevant.

2. Review at least two (different) cases in this chapter.

Keeping in mind that doing nothing/not responding are *not plausible responses*, are there other plausible alternatives (KSA 4) that you can think of for any case? If not, discuss that; if so, list them and discuss their evaluation (are they equally consistent with GL/CE, do they lead to similar decisions, etc.).

3. Choose any case in this chapter. Redo the analysis, but feature a different GL or CE principle in your reasoning process. Make sure the *justification* is different

from what is given. Is the *decision* also different? Discuss your decision with its justification and the tests of universality, justice, and publicity.

4. Consider all nine cases in this chapter.

A. Comment on the role of the stakeholder analysis in these decisions. Were stakeholder impacts featured in justifications? Were they features in your own reflections?

B. Comment on the importance of your reflections on the decisions you were asked for: for others doing a similar task at work; for those using results from this task; and for those who might join the profession (and be learning about planning and design).

C. Consider your professional identity and the profession itself: planning/designing projects and data collection/analysis projects is obviously foundational in statistics and data science. How can these nine cases and the decisions reached via the analyses presented earn you (the decider) the trust of the practicing community or the public for the profession and its future?

Chapter 3.5
Interpretation

NB: There is no "interpretation" guidance in the ACM CE.

As we saw in the previous chapter, ASA and ACM consider "analysis" differently. Since the ACM is focused on "evaluation" rather than statistical or numerical type analysis, and the interpretation of these types of analyses as the ASA is, it makes sense that interpretation is not separately addressed in the ACM CE. However, for the ethical statistical practitioner, the *interpretation* of statistical analysis is a critical feature of professional practice. Interpretation focuses on contextualizing, summarizing, or characterizing the results, because statistical inference test results are typically only some summary of the estimate (e.g., sample mean or maximum *a posteriori* estimate) or test score (e.g., z score), its associated degrees of freedom, and the p-value (or a confidence or credibility interval). That is, results of frequentist or Bayesian analysis have to be interpreted or qualified in some way. Thus, interpretation has an important role in both statistics and data science. In this chapter, the cases all feature prerequisite knowledge from the ASA GLs and considerations of DEFW and DSEC, but not the ACM since there is no specific guidance in the CE relating to interpretation[18].

Case 22. You discover that prior (expected) results cannot be reproduced. Sensitivity analyses strongly suggest that earlier results were spurious; reading the team's report of that analysis confirms this suggestion. The results were improperly interpreted to favor the team's objectives.

Case 23. At the end of a long project, you realize you made an error early on. The results cannot be interpreted in a valid way. Everything has to be redone.

Case 24. At the end of a long project, you realize your supervisor made an error early on. The results cannot be interpreted in a valid way. Everything has to be redone.

---

[18] This absence of the task "interpretation" in practice standards for computing professionals (ACM) was first noted in Chapter 2.5 of the book Tractenberg RE. (2022), Ethical Reasoning for the Data Centered World.

Case 25. You complete a very large set of analyses; one result happens to be "significant". A senior team member highlights this result, interpreting it without considering the context.

Case 26. Your supervisor singles out one "meaningful" result to demonstrate that whatever you've been doing "is working", even after you carry out multiple simulations that show their single, "favorite," result is totally spurious.

Case 22. You discover that prior (expected) results cannot be reproduced. Sensitivity analyses strongly suggest that earlier results were spurious; reading the team's report of that analysis confirms this: the results were improperly interpreted to favour the team's objectives.

1. Identify and 'quantify' prerequisite knowledge:

Which DEFW and DSEC Principles or Areas seem most relevant to this vignette?

This situation (your discovery) is *consistent* with DEFW Principles 5 and 6.

The situation is not addressed in DSEC.

Which ASA Principles and elements seem most relevant to this vignette?

ASA:

Principle A. Professional integrity and accountability (A2; A4; A11)

Principle B. Integrity of data and methods (B1; B2; B3; B5)

Principle C. Responsibilities to stakeholders (C1; C2; C4)

Principle D. Responsibilities to research subjects, data subjects, or those directly affected by statistical practices (D10)

Principle E. Responsibilities to members of multidisciplinary teams (E3; E4)

Principle F. Responsibilities to fellow statistical practitioners and the Profession (F2; F4)

Principle G. Responsibilities of leaders, supervisors, and mentors in statistical practice (G5)

Principle H. Responsibilities regarding potential misconduct (H1; H2)

Appendix. Responsibilities of organizations/institutions (APP 8; APP 9; App 11)

Potential result: Stakeholder:	HARM	BENEFIT	UNKNOWN	UNKNOWABLE
YOU	You will need to notify the previous team that their results are wrong; this discovery may identify other problems with other analyses/results that require additional time/effort. Failing to fix the error and correct prior misinterpretation of results may save time but lead to unpredictable propagation of that error throughout the workflow and into the community.	Correction of the error means future work will be facilitated –and correct, and interpretable. Innovation may be stimulated to devise an appropriate way to obtain the "desired" results (e.g., better experimental design).	Competent analysis (ASA) supports reproducible results; effective analysis (ACM) promotes public good. Perpetuation of false reports and misinterpretation undermine both.	
Your boss/client	Documenting, and then fixing the misinterpretation will take time, adds effort; may identify other problems that require additional time/effort. This error may be discovered in a 'product' that the client has paid for in the past, identifying past errors and undermining this and other projects	Correction of the misinterpretation means future work with the code/in the problem area will be facilitated –and correct, and interpretable.	Identification of misinterpretation will require resources to fix/address. *Appropriate* interpretation may not yield desired results.	

Potential result: Stakeholder:	HARM	BENEFIT	UNKNOWN	UNKNOWABLE
Unknown individuals	Incorrect interpretation undermines reproducibility and rigor. Mis-interpreted results negatively affect all stakeholders in the long run.	Identification and correction of the misinterpretation allows mitigation of risks of harms (ACM); supports valid inferences (ASA), and can limit the unpredictable propagation of the error to other parts of the workflow.		
Employer	Identifying, documenting, and then fixing the error will take time, adds effort; may identify other problems that require additional time/effort. This misinterpretation may be discovered in a 'product' that a client has/other clients have paid for in the past, identifying past errors and undermining this and other projects	Correction of the error means future work with the code will be facilitated –and correct, and interpretable.	Identification of misinterpretation may require resources to fix/address. *Appropriate* analysis –once the error is fixed - may not yield desired results.	Identification of the misinterpretation may highlight past payments or purchases/sales of incorrect/error-prone work. Liability may be incurred.

Potential result: Stakeholder:	HARM	BENEFIT	UNKNOWN	UNKNOWABLE
Colleagues	The discovered misinterpretation may complicate colleagues' work.	Identification and correction of the misinterpretation allows mitigation of risks of harms (ACM); supports valid inferences (ASA), and can limit the unpredictable propagation of the error to other parts of the workflow.		Project could be delayed until risks of additional errors, and resulting harms/ potential harms are addressed.
Profession	Failing to fix the misinterpretation may save time but lead to unpredictable propagation of that error throughout the workflow, undermining the practitioner's credibility —as well as the profession's.	Correction of the error means future work with the code will be facilitated —and correct, and interpretable. This supports ethical and competent practice by future practitioners.	Competent analysis (ASA) supports reproducible results; effective analysis (ACM) promotes public good. Perpetuation of false reports and misinterpretation undermine both.	Commitment to identification and correction of the error, consistent with GL/CE, might strengthen trust in profession. The error itself can enable innovation (to fix, avoid, or detect similar errors in future), as well as conversations about ethical practice.

Potential result: Stakeholder:	HARM	BENEFIT	UNKNOWN	UNKNOWABLE
Public/public trust	Even if episodes of misinterpretation are noted so they can be fixed, weak or error-prone analysis undermines public trust in reproducibility and rigor. An insistence on mis-interpreted results being "correct" undermines public trust as well as reproducibility and rigor.	Identification and correction of the error allows mitigation of risks of harms (ACM); supports valid inferences (ASA), and can limit the unpredictable propagation of the error to other parts of the workflow and decisions based on them.	Transparency supports the public trust, even if misinterpretation is identified this can tend to inform the public about limitations inherent in the profession.	

Table 3.5.1. *Stakeholder Analysis: Prior results appear to have been spurious.*

2. Identify decision-making frameworks.

As noted, ASA GLs tend to support a virtue approach, and the previous result interpreters may not have followed this: "Uses methodology and data that are valid, relevant, and appropriate, without favoritism or prejudice, and in a manner intended to produce valid, interpretable, and reproducible results." (ASA A2). They may also have been unable to acknowledge/follow that the ethical statistical practitioner "Avoids compromising scientific validity for expediency" (ASA E4). Since the code is right, and the results of the analysis and sensitivity analysis suggest spurious prior results, there has been an error in interpretation. Principle B dictates that the ethical statistical practitioner has responsibilities about communication, specifically, about the fitness of data for use – and known biases (ASA B1), and about transparency about assumptions made including limitations and possible sources of error (ASA B2). Also, the ethical practitioner "Communicates the stated purpose and the intended use of statistical practices. Is transparent regarding a priori versus post hoc objectives and planned versus unplanned statistical practices. Discloses when multiple comparisons are conducted, and any relevant adjustments." (ASA B3). ASA B3 is particularly important for this case because failure to correct for multiple comparisons allows spurious results to appear statistically significant (like in this case). Finally, the ethical practitioner "strives to promptly correct substantive errors discovered after publication or implementation. As appropriate, disseminates the correction publicly and/or to others relying on the results." (ASA B5) This situation reflects an error being identified, so ASA B5 is relevant even if the ethical practitioner is not the one who created the error.

In fact, every single Principle (A through H) and part (Appendix) of the ASA Guidelines offers information about this case. This underscores the centrality of interpretation – and the need to assure it happens fully and honestly – throughout statistical practice. ASA Principles C, E, F, G and H (H2) as well as the Appendix all contribute virtue-perspective input about the importance of honest and complete interpretation to all stakeholders; while Principle H1 also raises the possibility that the statistical practitioner may have made an honest error in the previous report (whose results cannot be replicated).

Although the ACM CE does not mention interpretation, "A computing professional should be transparent and provide full disclosure of all pertinent system capabilities, limitations, and potential problems to the appropriate parties." (ACM 1.3) In this case, you have identified an error of interpretation of the previous results, and *that* is what needs to be transparently reported.

3. Identify or recognize the ethical issue.

As this case is presented, the practitioner has applied their expertise to carry out an analysis appropriately. *An ethical issue only arises if the misinterpretation is not corrected.* Both the ASA and ACM standards support fixing/reporting of the error, with the ASA GLs stipulating that corrections should be disseminated, and the ACM CE specifying similarly that limitations of the system should be disclosed. Note that in this case, there is only an ethical issue if you *fail to notify* all stakeholders that there was an error in the interpretation of the original results (and any decisions based on them).

4. Identify and evaluate alternative actions (on the ethical issue).

The three decisions that can be made in any circumstance are: a) do nothing. b) consult or confer with a peer (ASA) or a supervisor (ACM); and c) report violations of policy, procedure, ethical guidelines, or law (ASA) or refuse to implement the system (ACM).

As noted, there is only an ethical issue if you do not address the misinterpretation (e.g., publish corrected results) and notify all stakeholders that there was an error. Thus, option a) must be modified to "document your evidence that prior results are spurious, notify all stakeholders, but do not confer"; and the only other option, b), becomes, "document your evidence that prior results are spurious, and confer with peers or supervisors about how best to notify all stakeholders". As we saw in other cases, if you fix and report the error internally, and your supervisor(s) or organization then seek to suppress this information – particularly, not notifying clients or other stakeholders who rely on the results, then option c) ("report the violation of CE, GLs, and relevant laws") *is* differentiated from option b).

5. Make and justify a decision.

Decision: *option b*: "document your evidence that prior results are spurious, notify all stakeholders, and confer with peers or supervisors about how best to notify all stakeholders" is supported by both ASA and ACM standards. Consulting a peer or supervisor is not necessarily an essential feature of your response, but because there was at least one person (probably more than one) involved in the misinterpretation of the prior results, there probably needs to be some careful thought given to why no one on the team objected to the report of spurious results. Option b) is preferred over option a) in this case because the notification aspect of the solution may require organization-level review of policies or incentives that exist for reporting results that cannot be replicated (i.e., are spurious).

6. Reflect on the decision.

In this case, as will all the others, "do nothing" is obviously not the ethical choice for you to make because it is inconsistent with ethical practice standards. However, misinterpretation of statistical results may be a habit of culture in your institution. This demonstrates a disregard for the scientific method and for statistical validity and reproducibility, and is one of the "detrimental research practices" that the National Academies (National Academies of Sciences and Engineering and the Institute of Medicine (NASEM) called attention to in their 2017 report, "Fostering Integrity",

> Actions such as failing to retain or share data and code supporting published work in accordance with disciplinary standards, practices such as honorary or ghost authorship, and *using inappropriate statistical or other methods of measurement and data presentation to enhance the significance of research findings* are clearly detrimental to the research process and may impose comparable or even greater costs on the research enterprise than those arising from research misconduct." NASEM 2017, p. 206 (emphasis added).

These "detrimental research practices" are widespread. The NASEM report urges greater attention to be placed on training new scientists to avoid these practices, but until that happens, ethical statistical practitioners and data scientists will continue to encounter cases like this one. We don't know from the case whether the misinterpreter knowingly reported spurious results or if they were simply doing "what everyone else does" and reporting just what they thought should be reported. Note that this particular detrimental practice not only damages science (as asserted in the NASEM report), but it is also part of the reason why both the ASA and ACM specifically call out all those who utilize their tools, techniques, and methods to follow their practice standards. Note that detrimental research practices are violations of both the GLs and the CE: if every scientist was trained in the GLs and CE, then detrimental research practices would only be utilized by those who purposefully seek to deceive. Currently we cannot know whether misinterpretation is done out of ignorance, in response to cultural pressure, or on purpose. Thus, while it may seem unfair that your ethical practice is contingent on trying to correct (or notify stakeholders of the detrimental practice of) others, you can hopefully see how important it is to the profession and to the wider scientific community.

The SHA shows plainly that options a) and b) pass a *test of justice*, because your notification of others strengthens the profession and the public trust; while it may seem unjust that you become responsible for this notification because of someone else's detrimental practice, the profession, public, and all practitioners benefit when we accept this responsibility; it leads to more just science. Both options pass the *test of publicity*, but possibly only marginally; while we all want the best science and honest work, you might not want it publicized that you found someone was using detrimental practices! However, you *would* want it publicized that you are an ethical practitioner, even in the face of colleagues or former colleagues who … are not. These options pass the *test of universality* as well, since you definitely want all practitioners to be as vigilant as you would have to be in this case. Statisticians and data scientists need universally follow ethical practice standards, and we need to do what we can to identify and correct misuse and misinterpretation of our work –whether that is unintentional or malicious/purposeful.

Case 23. At the end of a long project, you realize you made an error early on. The results cannot be interpreted in a valid way. Everything has to be redone.

1. Identify and 'quantify' prerequisite knowledge:

   Which DEFW and DSEC Principles or Areas seem most relevant to this vignette?

This situation is *consistent* with DEFW Principles 5 and 6: robust practice (5) requires that errors be corrected, and 6 requires transparent and accountable work. Results of a system or analysis with an identified error conflict with these Principles.

There is no reference to what to do about errors in analyses in DSEC. However, the existence of the error means DSEC C3, C5, D4, and E3 cannot be addressed. Depending on the nature of the error, other DSEC items cannot be answered confidently.

   Which ASA Principles seem most relevant to this vignette?

ASA:

Principle A. Professional integrity and accountability
Principle B. Integrity of data and methods
Principle C. Responsibilities to stakeholders
Principle E. Responsibilities to members of multidisciplinary teams
Principle F. Responsibilities to fellow statistical practitioners and the Profession
Principle G. Responsibilities of leaders, supervisors, and mentors in statistical practice
Principle H. Responsibilities regarding potential misconduct

Potential result: Stakeholder:	HARM	BENEFIT	UNKNOWN	UNKNOWABLE
YOU	Identifying, documenting, and then fixing the error will take time, adds effort; may identify other problems that require additional time/effort. Failing to fix the error may save time but lead to unpredictable propagation of that error throughout the workflow.	Correction of the error means future work with the code will be facilitated –and correct, and interpretable.	Competent analysis (ASA) supports reproducible results; effective analysis (ACM) promotes public good.	
Your boss/client	Identifying, documenting, and then fixing the error will take time, adds effort; may identify other problems that require additional time/effort. This error may be discovered in a 'product' that the client has paid for in the past, identifying past errors and undermining this and other projects	Correction of the error means future work with the code will be facilitated –and correct, and interpretable.	Identification of error may require resources to fix/address. *Appropriate* analysis – once the error is fixed - may not yield desired results.	

Potential result: Stakeholder:	HARM	BENEFIT	UNKNOWN	UNKNOWABLE
Unknown individuals	Analysis required or ACM CE violated; weak statistical analysis undermines reproducibility and rigor.	Identification and correction of the error allows mitigation of risks of harms (ACM); supports valid inferences (ASA), and can limit the unpredictable propagation of the error to other parts of the workflow.		
Employer	Identifying, documenting, and then fixing the error will take time, adds effort; may identify other problems that require additional time/effort. This error may be discovered in a 'product' that the client has paid for in the past, identifying past errors and undermining this and other projects	Correction of the error means future work with the code will be facilitated –and correct, and interpretable.	Identification of error may require resources to fix/address. *Appropriate* analysis – once the error is fixed - may not yield desired results.	Identification of the error may highlight past payments or purchases/sales of incorrect/ error-prone work. Liability may be incurred.

Potential result: Stakeholder:	HARM	BENEFIT	UNKNOWN	UNKNOWABLE
Colleagues	The discovered error may complicate colleagues' work.	Identification and correction of the error allows mitigation of risks of harms (ACM); supports valid inferences (ASA), and can limit the unpredictable propagation of the error to other parts of the workflow.		Project could be delayed until risks of additional errors, and resulting harms/ potential harms are addressed.
Profession	Identifying, documenting, and then fixing the error will take time, adds effort; may identify other problems that require additional time/effort. Failing to fix the error may save time but lead to unpredictable propagation of that error throughout the workflow.	Correction of the error means future work with the code will be facilitated –and correct, and interpretable. This supports ethical and competent practice by future practitioners.	Competent analysis (ASA) supports reproducible results; effective analysis (ACM) promotes public good.	Commitment to identification and correction of the error, consistent with GL/CE, might strengthen trust in profession. The error itself can enable innovation (to fix, avoid, or detect similar errors in future), as well as conversations about ethical practice.

Potential result: Stakeholder:	HARM	BENEFIT	UNKNOWN	UNKNOWABLE
Public/public trust	Even if errors are noted so they can be fixed, weak or error-prone analysis undermines public trust in reproducibility and rigor.	Identification and correction of the error allows mitigation of risks of harms (ACM); supports valid inferences (ASA), and can limit the unpredictable propagation of the error to other parts of the workflow and decisions based on them.	Transparency supports the public trust, even if risk/harms are identified; can tend to inform the public about limitations inherent in the profession.	

Table 3.5.2. *Stakeholder Analysis: You discover you made an error and have to rerun everything.*

2. Identify decision-making frameworks.

As noted, ASA GLs tend to support a virtue approach, which supports fixing the error so that results can be interpreted in a valid way, because the ethical practitioner "uses methodology and data that are valid, relevant, and appropriate, without favoritism or prejudice, and in a manner intended to produce *valid, interpretable, and reproducible results*" (ASA A2, emphasis added). The ethical statistical practitioner "avoids compromising scientific validity for expediency" (ASA E4), so ASA Principle B dictates that the ethical statistical practitioner "strives to promptly correct substantive errors discovered after publication or implementation. As appropriate, disseminates the correction publicly and/or to others relying on the results." (ASA B5). As unpleasant as it might be to notify your boss/supervisor and/or teammates that you made an error, it has to be done so that a strategy for correcting that error and redoing the work can be formulated. Responsibilities to address the error accrue across most ASA Principles (A-C, E-H). The SHA table shows that only harms are caused when errors are not corrected, and benefits only accrue when the error is fixed. The only ASA Principle not directly related to the case is D (Responsibilities to research subjects, data subjects, or those directly affected by statistical practices). If the project is not redone, then responsibilities to data contributors and those affected by the practice accrue under ASA Principle C.

In the ACM CE utilitarian perspective, the computing professional has a responsibility to analyze the work they do/are asked to do so that decisions made on the basis of the system or its subparts are *justifiable*. If an error is identified in the analysis part of a system being evaluated, then results from that system may be biased or incorrect, potentially causing predictable (once the error is identified) but fixable harms. More specifically, "A computing professional should be transparent and provide full disclosure of all pertinent system capabilities, limitations, and potential problems to the appropriate parties." (ACM 1.3) While not specifically about interpretation, the ACM element provides a utilitarian perspective on the importance of addressing the error.

3. Identify or recognize the ethical issue.

Both the ASA and ACM standards support the identification and fixing of the error, with the ASA GLs stipulating that corrections should be disseminated, and the ACM CE specifying similarly that limitations of the system should be disclosed (unless the error is fixed, in which it is no longer a system limitation). As articulated, this case simply states that you found an error: *an ethical issue only arises if you do not fix the error and notify any stakeholders who rely on the code or its results*. Since the case specifies that you're only now at the end of the long project, there may not be other stakeholders besides the others on your team. However, since this was a "long project", rerunning everything will take time and may involve many others/others' time and effort as well. Thus, there might be some incentives not to fix the error, or to fix it but not notify everyone. Both the GLs and CE describe the ethical practitioners' responsibilities to do that notification, however, so ethical issues will arise if you do fix the error but do not notify stakeholders.

4. Identify and evaluate alternative actions (on the ethical issue).

The three decisions that can be made in any circumstance are: a) do nothing. b) consult or confer with a peer (ASA) or a supervisor (ACM); and c) report violations of policy, procedure, ethical guidelines, or law (ASA) or refuse to implement the system (ACM). In this case, however, only two options are viable because conferring with your supervisor or a peer (option b) will be tantamount to "reporting" (option c) that an error in the code was found.

Note that the case simply describes you identifying your error. Thus, option a) must be modified to "correct the program (and re-run all data that have been analyzed with the incorrect system), notify all stakeholders, but do not confer"; and the only other option, b), becomes, "correct the program, re-run all data, and confer with peers or supervisors about how best to notify all stakeholders".

As noted, there is only an ethical issue if you do not fix the error **and** notify all stakeholders that there is an error. Thus, if you fix and report the error, and your supervisor(s) or organization then seek to suppress this information – particularly, not notifying clients or other stakeholders who rely on the results, then option c) ("report the violation of CE, GLs, and relevant laws") *is* differentiated from option b) and should be considered. Note that, while you may have chosen to honour your commitments to practice ethically, and notify your supervisor/organization and other stakeholders, the "long project" may entail costs and resources that cannot be renewed. Again, there are some incentives for the organization either not to fix the error, or to fix it but not notify everyone. Your primary responsibility is to ensure that *you* practice ethically; you cannot cause the organization to do so. So, alternative responses were identified for *you* to take so that you can ethically respond to your discovery of the error (obviously making an error is not unethical!) and you will choose a) or b). Your organization may respond to either option by not choosing to endorse or support your ethical practice. If this constitutes fraud – because the organization does not want to incur the cost of fixing your error and redoing the whole project (Flint Water crisis[19]), for example, then you need to consider whistleblowing (again).

5. Make and justify a decision.

Decision: *option b*: "correct the program, re-run all data, and confer with peers or supervisors about how best to notify all stakeholders" is supported by both ASA and ACM standards. Because the case describes a "long project", there very well may be multiple stakeholders and others to notify (or confer with, or both). The decision to correct the program (common to both options a and b) may require approval; the fact that the long project has to be redone may require resources (e.g., your time to be covered to redo that work).

Note that, once you have identified an error in code you are using, if you continue to use that code without correcting it (even though this is easier), you are violating multiple practice standards – not being transparent or honest in your reporting (unless you also include the error in your report); not using the most appropriate methods; and not ensuring that your work is robust and that you are accountable for it. If others choose to continue using code in which you have identified an error, and they purposely ignore that error, they are violating multiple practice standards, and also misleading stakeholders as well as potentially making work for others in the workplace difficult (because it will be based on erroneous work from your code). Note that decisions can still be made – with made up results or interpretations that are not consistent with the results (because those are incorrect). Keep in mind that, if your identified error is not fixed because the workplace environment, then information or work products that you are given might be similarly flawed. You would have to decide if you wanted to work – and be associated with – that type of workplace and context.

6. Reflect on the decision.

Construct your own reflection on this decision.

---

[19] Read about how failures in an organization to allow for correction of an error and notify stakeholders led to the Flint Water Crisis https://en.wikipedia.org/wiki/Flint_water_crisis

Case 24. At the end of a long project, you realize your supervisor made an error early on. The results cannot be interpreted in a valid way. Everything has to be redone.

1. Identify and 'quantify' prerequisite knowledge:

Which DEFW and DSEC Principles or Areas seem most relevant to this vignette?

This situation is *consistent* with DEFW Principles 5 and 6.

The situation is not addressed in DSEC.

Which ASA Principles seem most relevant to this vignette?

ASA:

Principle A. Professional integrity and accountability

Principle B. Integrity of data and methods

Principle C. Responsibilities to stakeholders

Principle E. Responsibilities to members of multidisciplinary teams

Principle F. Responsibilities to fellow statistical practitioners and the Profession

Principle G. Responsibilities of leaders, supervisors, and mentors in statistical practice

Principle H. Responsibilities regarding potential misconduct

Potential result: Stakeholder:	HARM	BENEFIT	UNKNOWN	UNKNOWABLE
YOU	Identifying, documenting, and then fixing the error will take time, adds effort; may identify other problems that require additional time/effort. Your supervisor may not appreciate your finding their error. Failing to fix the error may save time but lead to unpredictable propagation of that error throughout the workflow.	Correction of the error means future work with the code will be facilitated –and correct, and interpretable.	Competent analysis (ASA) supports reproducible results; effective analysis (ACM) promotes public good. If your supervisor is not competent or effective, this event might document that. You and others may be punished (retaliation) or unburdened (if they are no longer your supervisor).	
Your boss/client	Identifying, documenting, and then fixing the error will take time, adds effort; may identify other problems that require additional time/effort. This error may be discovered in a 'product' that the client has paid for in the past; identifying past errors and undermining this and other projects	Correction of the error means future work with the code will be facilitated –and correct, and interpretable.	Identification of error may require resources to fix/address. *Appropriate* analysis –once the error is fixed - may not yield desired results.	

Potential result: Stakeholder:	HARM	BENEFIT	UNKNOWN	UNKNOWABLE
Unknown individuals	Analysis required or ACM CE violated; weak statistical analysis undermines reproducibility and rigor.	Identification and correction of the error allows mitigation of risks of harms (ACM); supports valid inferences (ASA), and can limit the unpredictable propagation of the error to other parts of the workflow.		
Employer	Identifying, documenting, and then fixing the error will take time, adds effort; may identify other problems that require additional time/effort. This error may be discovered in a 'product' that the client has paid for in the past, identifying past errors and undermining this and other projects	Correction of the error means future work with the code will be facilitated –and correct, and interpretable.	Identification of error may require resources to fix/address. *Appropriate analysis* –once the error is fixed - may not yield desired results.	Identification of the error may highlight past payments or purchases/sales of incorrect/ error-prone work. Liability may be incurred.

Potential result: Stakeholder:	HARM	BENEFIT	UNKNOWN	UNKNOWABLE
Colleagues	The discovered error may complicate colleagues' work.	Identification and correction of the error allows mitigation of risks of harms (ACM); supports valid inferences (ASA), and can limit the unpredictable propagation of the error to other parts of the workflow.		Project could be delayed until risks of additional errors, and resulting harms/ potential harms are addressed.
Profession	Identifying, documenting, and then fixing the error will take time, adds effort; may identify other problems that require additional time/effort. Failing to fix the error may save time but lead to unpredictable propagation of that error throughout the workflow.	Correction of the error means future work with the code will be facilitated—and correct, and interpretable. This supports ethical and competent practice by future practitioners.	Competent analysis (ASA) supports reproducible results; effective analysis (ACM) promotes public good.	Commitment to identification and correction of the error, consistent with GL/CE, might strengthen trust in profession. The error itself can enable innovation (to fix, avoid, or detect similar errors in future), as well as conversations about ethical practice.

Potential result: Stakeholder:	HARM	BENEFIT	UNKNOWN	UNKNOWABLE
Public/public trust	Even if errors are noted so they can be fixed, weak or error-prone analysis undermines public trust in reproducibility and rigor.	Identification and correction of the error allows mitigation of risks of harms (ACM); supports valid inferences (ASA), and can limit the unpredictable propagation of the error to other parts of the workflow and decisions based on them.	Transparency supports the public trust, even if risk/harms are identified; can tend to inform the public about limitations inherent in the profession.	

Table 3.5.3. *Stakeholder Analysis: You discover your supervisor made an error and you have to rerun everything.*

2. Identify decision-making frameworks.

ASA GLs support a virtue approach, which supports fixing the error so that results can be interpreted in a valid way, because the ethical statistical practitioner "uses methodology and data that are valid, relevant, and appropriate, without favoritism or prejudice, and in a manner intended to produce *valid, interpretable, and reproducible results*" (ASA A2, emphasis added). The ethical statistical practitioner "Avoids compromising scientific validity for expediency" (ASA E4) – no matter how much more expedient it would be to ignore this error, without appropriate interpretability, the entire project is invalidated and would be a waste of resources. The results themselves would not be reproducible and would not support valid results or decision making, so Principle B dictates that the ethical statistical practitioner: "Strives to promptly correct substantive errors discovered after publication or implementation. As appropriate, disseminates the correction publicly and/or to others relying on the results." (ASA B5) In this case, ASA Principles A, B, C, E, F and H apply to **your supervisor** – and support a decision by them to start over (or approve that decision), while Principle B5 applies *to you*. This is a good time to notice Appendix item 11, "Those in leadership, supervisory, or managerial positions who oversee statistical practitioners promote ethical statistical practice by following Principle G and, in cases where ethical issues are raised, representing them fairly within the organization's leadership team". Key elements of ASA Principle G (Responsibilities of leaders, supervisors, and mentors in statistical practice) include the expectations that leaders will "promote a respectful, safe, and productive work environment. Encourage constructive engagement to improve statistical practice." (ASA G2) and that they will "establish a culture that values validation of assumptions, and assessment of model/algorithm performance over time and across relevant subgroups, as needed. Communicate with relevant stakeholders regarding model or algorithm maintenance, failure, or actual or proposed modifications." (ASA G5)

In the ACM CE utilitarian perspective, the computing professional has a responsibility to analyze the work they do/are asked to do so that decisions made on the basis of the system or its subparts are *justifiable*. If an error is identified in the analysis part of a system being evaluated, then results from that system may be biased or incorrect, potentially causing predictable (once the error is identified) but fixable harms. More specifically, "A computing professional should be transparent and provide full disclosure of all pertinent system capabilities, limitations, and potential problems to the appropriate parties." (ACM 1.3) In this case, the CE Principle applies *to you* since your assessment of the results is how the limitation was identified. While not specifically about interpretation, the ACM element provides a utilitarian perspective on the importance of addressing the error.

3. Identify or recognize the ethical issue.

Both the ASA and ACM standards support the identification and fixing of the error; however, the most important part of their guidance pertains to *you*: this is different from the previous case. Both standards specify that the error (ASA) and limitations of the system (ACM) should be disclosed. As articulated, this case simply states that you found an error: *an ethical issue only arises if* **you** *do not notify any stakeholders who rely on the code or its results*. Obviously, if your supervisor chooses not to fix the error, you may not be able to fix it yourself; you may also not be able to authorize someone else to fix it so that everything can be redone. Since the case specifies that you're at the end of the long project, rerunning everything will take time and may involve many others/others' time, effort, and resources as well. Thus, there might be some incentives for your supervisor to *not fix the error*. If you are aware of this – that there is an error, the results cannot be validly interpreted, and the supervisor chooses not to fix the error or re-run the project, or to fix it but not re-run it, **you** have a responsibility to notify relevant stakeholders, which includes your boss's boss, as well as your colleagues. Otherwise these stakeholders will not be able to rely on the work (or any interpretations). Both the GLs and CE describe the ethical practitioners' responsibilities to do that notification, however, so ethical issues will also arise if the supervisor does fix the error but does not notify stakeholders.

4. Identify and evaluate alternative actions (on the ethical issue).

The three decisions that can be made in any circumstance are: a) do nothing. b) consult or confer with a peer (ASA) or a supervisor (ACM); and c) report violations of policy, procedure, ethical guidelines, or law (ASA) or refuse to implement the system (ACM). In this case, however, only two options are viable because conferring with your supervisor or a peer (option b) will be tantamount to "reporting" (option c) that an error in the code was found.

Note that the case simply describes you identifying your supervisor's error. Thus, option a) must be modified to "notify all stakeholders of the error and need to redo the project, but do not confer"; and the only other option, b), becomes, "confer with peers or supervisors about how best to notify all stakeholders of the error and need to redo the project".

As noted, there is only an ethical issue if you do not notify all stakeholders that there is an error. Thus, if your supervisor fixes and reports the error, and *their* supervisor(s) or organization then seek to suppress this information – particularly, not notifying clients or other stakeholders who rely on the results, then option c) ("report the violation of CE, GLs, and relevant laws") *is* differentiated from option b) and should be considered. Note that, while you – and perhaps your supervisor - may have chosen to honour your commitments to practice ethically, and notify your organization and other stakeholders, the "long project" may entail costs and resources that cannot be renewed. Again, there are some incentives for the organization either not to fix the error, or to fix it but not notify everyone. Your primary responsibility is to ensure that *you* practice ethically; you cannot cause the organization to do so. So, alternative responses were identified for *you* to take so that you can ethically respond to your discovery of the error, and you would choose a) or b). Your organization may respond to either option by not choosing to endorse or support your ethical practice (violating ASA G1 and Appendix items 1, 4, and possibly 11). If this constitutes fraud – because the organization does not want to incur the cost of fixing your error and redoing the whole project (Theranos crisis[20]) and so decides to mislead stakeholders, for example, then you need to consider whistleblowing (again).

5. Make and justify a decision.

Decision: *option b*: "confer with peers or supervisors about how best to notify all stakeholders of the error and need to redo the project" is supported by both ASA and ACM standards. Because the case describes a "long project", there very well may be multiple stakeholders and others to notify (or confer with, or both). Note that, once you have identified an error in code you are using, if you continue to use that code without correcting it (even though this may be easier and cheaper), you are violating multiple practice standards – not being transparent or honest in your reporting (unless you also include the error in your report); not using the most appropriate methods; and not ensuring that your work is robust and that you are accountable for it. Keep in mind that, if the error you identified is not fixed because of attributes of the workplace environment, then information or work products that you are given might be similarly flawed. If others choose to continue using code in which you have identified an error, and they purposely ignore that error, *they* are violating multiple practice standards. More to the point, in cases where ASA Principle G and the Appendix are ignored, particularly when you try to do the right thing and/or blow the whistle, you would have to decide if you wanted to work – and be associated with – that type of workplace and context.

6. Reflect on the decision.

Construct your own reflection on this decision.

---

20 Read about how failures in an organization to allow for correction of an error and notify stakeholders contributed to the Theranos Crisis: https://www.wsj.com/articles/theranos-has-struggled-with-blood-tests-1444881901

26255

262526252626

Case 25. You complete a very large set of analyses; one result happens to be "significant". A senior team member highlights this result, interpreting it without considering the context.

1. Identify and 'quantify' prerequisite knowledge:

Which DEFW and DSEC Principles or Areas seem most relevant to this vignette?

This team member is _behaving contrary to_ DEFW Principles 4 and 5; if you follow Principles 5 and 6, you may help to address that behavior.

DSEC does not address this situation, but note that DSEC C5 (auditability) and D4 (explainability) are both "no" in this case, and E3 (concept drift) cannot be assessed. These difficulties will exist even if all other answerable DSEC items are "yes".

Which ASA Principles seem most relevant to this vignette?

ASA:

Principle A. Professional integrity and accountability
Principle B. Integrity of data and methods
Principle C. Responsibilities to stakeholders
Principle D. Responsibilities to research subjects, data subjects, or those directly affected by statistical practices

Potential result: Stakeholder:	HARM	BENEFIT	UNKNOWN	UNKNOWABLE
YOU	You need to notify the team – including this member- that their *results are wrong*; this discovery may identify other problems with other analyses/results that require additional time/effort. Failing to correct the misinterpretation of results may save time but *will* lead to unpredictable propagation of that error throughout the workflow and into the community.	Correction of the error means future work will be correct, and interpretable. Innovation may be stimulated to devise an appropriate way to obtain the "desired" results (e.g., better experimental design; collection of different data).	Competent interpretation (ASA) supports reproducible results; effective contextualization of your work (ACM) promotes public good. Perpetuation of false reports and misinterpretation undermine both while also perpetuating detrimental research practices.	Detrimental research practices like cherry-picking (this case) undermine science, the public trust in the scientific enterprise, and the entire scientific community.
Your boss/client	Documenting, and then fixing the misinterpretation will take time, adds effort; may identify other problems that require additional time/effort. This interpretation error may be discovered in a 'product' that the client has paid for in the past, identifying past errors and undermining this and other projects.	Correction of the misinterpretation means future work with the results (if not with this team member) will be facilitated –and correct, and interpretable.	Identification of misinterpretation will require resources to fix/address. *Appropriate* interpretation may not yield desired results.	

Potential result: / Stakeholder:	HARM	BENEFIT	UNKNOWN	UNKNOWABLE
Unknown individuals	Incorrect interpretation undermines reproducibility and rigor. Misinterpreted results negatively affect all stakeholders in the long run.	Correction of the error means future work will be correct, and valid. Innovation may be stimulated to devise an appropriate way to obtain the "desired" results (e.g., better experimental design; collection of different data).	Competent interpretation (ASA) supports reproducible results; effective contextualization of your work (ACM) promotes public good. Perpetuation of false reports and misinterpretation undermine both while also perpetuating detrimental research practices.	Detrimental research practices like cherry-picking (this case) undermine science, the public trust in the scientific enterprise, and the entire scientific community.
Employer	Identifying, documenting, and then fixing the error will take time, adds effort; may identify other problems that require additional time/effort. This misinterpretation may be discovered in a 'product' that a client has/other clients have paid for in the past, identifying past errors and undermining this and other projects	Correction of the interpretation error means future work with this member may be facilitated – and correct, and valid.	Identification of misinterpretation may require resources to fix/address. *Appropriate analysis* –once the error is fixed - may not yield desired results.	Identification of the misinterpretation may highlight past payments or purchases/ sales of incorrect/ error-prone work. Liability may be incurred.

Potential result: Stakeholder:	HARM	BENEFIT	UNKNOWN	UNKNOWABLE
Colleagues	The discovered misinterpretation may complicate colleagues' work. Pointing out and correcting a (senior) team member's detrimental research practices can be discomfiting – and may also highlight other detrimental practices that will then invite more scrutiny.	Identification and correction of the misinterpretation allows mitigation of risks of harms (ACM); supports valid inferences (ASA), and can limit the unpredictable propagation of the error to other parts of the workflow. Other colleagues may be emboldened to call out –and stop – other detrimental practices they observe.	Competent interpretation (ASA) supports reproducible results; effective contextualization of your work (ACM) promotes public good. Perpetuation of false reports and misinterpretation undermine both while also perpetuating detrimental research practices.	Detrimental research practices like cherry-picking (this case) undermine science, the public trust in the scientific enterprise, and the entire scientific community.

Potential result: Stakeholder:	HARM	BENEFIT	UNKNOWN	UNKNOWABLE
Profession	Failing to fix the misinterpretation *may* save time but will lead to unpredictable propagation of that error throughout the workflow and into the scientific community, undermining the practitioner's credibility –as well as the profession's.	Correction of the interpretation error means future work with the member may be facilitated –and correct, and valid. This supports ethical and competent practice by future practitioners.	Competent interpretation (ASA) supports reproducible and valid results; effective analysis (ACM) promotes public good. Perpetuation of false reports and misinterpretation undermine both.	Commitment to identification and correction of the error, consistent with GL/CE, might strengthen trust in profession. The error itself can enable innovation (to fix, avoid, or detect similar errors in future), as well as conversations about ethical practice.

Potential result: Stakeholder:	HARM	BENEFIT	UNKNOWN	UNKNOWABLE
Public/public trust	Even if episodes of misinterpretation are noted so they can be fixed, detrimental practices like cherry-picking undermines public trust in reproducibility and rigor. An insistence on misinterpreted results being "correct" undermines public trust as well as reproducibility and rigor.	Identification and correction of the interpretation error allows mitigation of risks of harms (ACM); supports valid inferences (ASA), and can limit the unpredictable propagation of the error to other parts of the workflow and community, to and decisions based on them.	Competent interpretation (ASA) represents reproducible results; effective contextualization of your work (ACM) promotes public good. Perpetuation of false reports and misinterpretation undermine both while also perpetuating detrimental research practices.	Detrimental research practices like cherry-picking (this case) undermine science, the public trust in the scientific enterprise, and the entire scientific community.

Table 3.5.4. *Stakeholder Analysis: Prior results appear to have been spurious.*

2. Identify decision-making frameworks.

As noted, the virtue approach describes what "the ideal (ethical) person would do". This vignette describes a highly prevalent and extremely detrimental research practice, **cherry-picking**[21]. Defined as a logical[22] or statistical[23] fallacy in the best-case scenario, and as propaganda or lying[24] in the worst case, cherry-picking is *obviously not ethical*. Because it is inherently incomplete – using only some of the available evidence, and usually deceptive, it should always be resisted by any ethical practitioner. This practice violates ASA A2, which states that the ethical statistical practitioner "uses methodology and data that are valid, relevant, and appropriate, without favoritism or prejudice, and in a manner intended to produce valid, interpretable, and reproducible results." In *and outside of* contexts where science is being done, the ethical statistical practitioner "avoids compromising scientific validity for expediency" (ASA E4). In this vignette, both of these principles were followed but one person on the team seeks to subvert these ethical practices and has misinterpreted the results. Principle B dictates that the ethical statistical practitioner: "Strives to promptly correct substantive errors discovered after publication or implementation. As appropriate, disseminates the correction publicly and/or to others relying on the results." (ASA B5) Thus, the GLs include a potential course of action for this case – even if the ethical practitioner is not the one who created the error.

Note that one of the stakeholders relying on (just some) of the results is the cherry-picking team member: ASA B5 outlines the responsibility to correct this error –not only so that false reports are not shared or published, but also to hopefully inform this team member that cherry-picking is unethical and will not be tolerated (as per ASA H2). It must be said that non-statistics practitioners may learn about science or business applications in a way that encourages cherry-picking, in which case the cherry-picker may be making an "honest error" (which isn't misconduct per ASA H1). However, even if it is an honest error, cherry-picking is still a detrimental practice. The ethical practitioner should be able to leverage ASA Principle B to inform the cherry-picker – and other stakeholders – that this detrimental practice has to stop. Recall that the Purpose of the Guidelines statement points out that "If an unexpected ethical

---

[21] https://en.wikipedia.org/wiki/Cherry-picking

[22] http://legacy.earlham.edu/~peters/courses/inflogic/onesided.htm

[23] https://web.archive.org/web/20140325173711/http://www.statlit.org/pdf/2008Klass ASA.pdf

[24] https://scienceornot.net/2012/04/03/devious-deception-in-displaying-data-cherry-picking/

challenge arises, the ethical practitioner seeks guidance, not exceptions, in the guidelines. To justify unethical behaviors, or to exploit gaps in the guidelines, is unprofessional and inconsistent with these guidelines." This language means that the cherry-picker cannot say, "the ASA Ethical Guidelines do not say I *can't* cherry-pick, therefore it's fine!"

Although the ACM CE does not mention interpretation, "A computing professional should be transparent and provide full disclosure of all pertinent system capabilities, limitations, and potential problems to the appropriate parties." In this case, you have identified an error of interpretation, and *that* is what needs to be transparently reported. This detrimental practice needs to be identified and discouraged at every possible opportunity, because it is a perfectly predictable limitation that will invariably cause harm – even if that harm is only to the scientific record.

3. Identify or recognize the ethical issue.

As this case is presented, the practitioner has applied their expertise to carry out an analysis appropriately. *An ethical issue only arises if the misinterpretation is not corrected.* Both the ASA and ACM standards support fixing/reporting of the error, with the ASA GLs stipulating that corrections should be disseminated, and the ACM CE specifying similarly that limitations of the system should be disclosed. Note that in this case, there is only an ethical issue if you *fail to notify* all stakeholders of the cherry-picking – to prevent false and invalid reporting (and any decisions based on cherry-picked results). Further, both the ACM (Principle 4) and ASA (Purpose, Principles A & G) outline specific responsibilities to follow the practice standards *and encourage others to do so as well*. It does not always work to change cherry-picking behavior, but the ethical practitioner will *do what they can to discourage cherry-picking and any other detrimental practices*. Thus, to the extent possible given your role (and job security), an ethical issue for the practitioner also arises when they do not seek to discourage this kind of deceptive and detrimental practice. Recall that the ethical practitioner "Avoids condoning or appearing to condone statistical, scientific, or professional misconduct." While the NASEM 2017 report acknowledges that "scientific misconduct" is currently limited in its scope to fraud, falsification, and fabrication, the identified detrimental research practices (p. 206) clearly recognize cherry-picking as statistical misconduct – and detrimental.

4. Identify and evaluate alternative actions (on the ethical issue).

The three decisions that we discuss for any circumstance are: a) do nothing. b) consult or confer with a peer (ASA) or a supervisor (ACM); and c) report violations of policy, procedure, ethical guidelines, or law (ASA) or refuse to implement the system (ACM).

As noted, there is only an ethical issue if you do not address the misinterpretation (e.g., publish corrected results) and notify all stakeholders that there was an error. Further ethical problems arise if you do not also – to the extent that it is possible for you – act to prevent this kind of unethical practice in the future. Thus, option a) must be modified to "correct the cherry-picking, notify all stakeholders of the potential for cherry-picked results (or their existence if you couldn't stop their publication/dissemination)"; and option b), becomes, "confer with peers or supervisors about how best to correct the cherry-picking, notify all stakeholders of the potential for cherry-picked results (or their existence if you couldn't stop their publication/dissemination), and ensure that cherry-picking and other detrimental practices are recognized as unethical and not tolerated".

As discussed in Case 22, the 2017 National Academies of Sciences and Engineering and the Institute of Medicine (NASEM) report, *"Fostering Integrity"*, called on the scientific community in the US to identify and act to prevent such detrimental practices. Unfortunately, as of 2022, cherry-picking is not highlighted in many policies (including those relating to training in the responsibly conduct of research required by the National Institutes of Health and Office of Research Integrity in the US) as a practice to avoid and eradicate. Thus, an option that includes "reporting" (our standard option c) is not going to be viable –until and unless such policies are put into place. However, as we saw in other cases, if you fix and report the error internally, and your supervisor(s) or organization then seek to suppress this information – particularly, not notifying clients or other stakeholders who rely on the results, then option c) ("report the violation of CE, GLs, and relevant laws") *may* become a viable option. You would have effectively demonstrated that the report with cherry-picked data is false and deceptive; the fact that the organization suppressed this information demonstrates their intent to deceive stakeholders as well. While cherry-picking is unethical but not illegal, deception and withholding evidence *is* illegal – particularly when there are financial conflicts of interest in play. If an organization seeks to suppress your documentation that results are incorrect, it may be perpetrating fraud – and both your commitment to ethical practice and your civic duty will require that you report this. Then option c) effectively becomes, "correct the cherry-picking, notify all stakeholders of the potential for cherry-picked results (or their existence if you couldn't stop their publication/dissemination), and report the cherry-picking and suppression of its existence to the appropriate authorities (and to the ACM)."

5. Make and justify a decision.

Decision: All options include "correct the cherry-picking and notify all stakeholders of the potential for cherry-picked results (or their existence if you couldn't stop their publication/dissemination)". This is the only ethical option, although how it is done – or what else is also done- may require additional consideration. *Options b and c* include sharing (i.e., reporting, option c) or conferring (option b) with others. While option c), reporting, is only viable when policies exist and reporting violations of those policies, or relevant laws, is feasible, it may constitute some reporting when you share the cherry-picking behavior of the team member with your peer or supervisor. Consulting a peer or supervisor may be more relevant for a more junior practitioner. As you grow in experience and seniority, you may be able to use the conferring to give some careful thought given to why no one else on the team objected to the reporting of cherry-picked results. There may be a culture in your team or organization that either tolerates or encourages detrimental practices, and the ethical practitioner is obliged to change this culture. When an otherwise ethical practitioner fails to stop others from unethical practice, they would then violate ASA H2 and condone, or appear to condone, these unethical practices. Option b) is preferred over option a) in this case because the notification aspect of the solution may require organization-level review of policies or incentives that exist for reporting (and pursuing) results that cannot be replicated (i.e., are spurious). Option b) is preferred over option c) because higher-level organizational fraud (suppression of your findings) will hopefully be very rare. Option b) will also be more appropriate to initiate culture change, and to lead those who are ignorant that their behavior is unethical (e.g., if their practice guidelines do not decry deceptive and detrimental practices) to the understanding that *cherry-picking is unethical*.

6. Reflect on the decision.

Construct your own reflection on this decision.

Case 26. Your supervisor singles out one "meaningful" result to demonstrate that whatever you've been doing "is working", even after you carry out multiple simulations that show their single, "favourite," result is totally spurious.

1. Identify and 'quantify' prerequisite knowledge:

> Which DEFW and DSEC Principles or Areas seem most relevant to this vignette?

This team member is *behaving contrary to* **DEFW Principles 4 and 5; if you follow Principles 5 and 6, you may help to address that behavior.**

DSEC does not address this situation, but note that DSEC C5 (auditability) and DSEC D4 (explainability) are both "no" in this case, and DSEC E3 (concept drift) cannot be assessed. These difficulties will exist even if all other answerable DSEC items are "yes".

> Which ASA Principles seem most relevant to this vignette?

ASA:

Principle A. Professional integrity and accountability

Principle B. Integrity of data and methods

Principle C. Responsibilities to stakeholders

Principle E. Responsibilities to members of multidisciplinary teams

Principle F. Responsibilities to fellow statistical practitioners and the Profession

Principle G. Responsibilities of leaders, supervisors, and mentors in statistical practice

Principle H. Responsibilities regarding potential misconduct

Potential result:	HARM	BENEFIT	UNKNOWN	UNKNOWABLE
Stakeholder: YOU	You need to notify the team – beyond your supervisor - that their *results are wrong*; this discovery may identify other problems with other analyses/ results that require additional time/effort. Failing to correct the misinterpretation of results may save time but *will* lead to unpredictable propagation of that error throughout the workflow and into the community.	Correction of the error means future work will be correct, and interpretable. Innovation may be stimulated to devise an appropriate way to obtain the "desired" results (e.g., better experimental design; collection of different data).	Competent interpretation (ASA) supports reproducible results; effective contextualization of your work (ACM) promotes public good. Perpetuation of false reports and misinterpretation undermine both while also perpetuating detrimental research practices.	Detrimental research practices like cherry-picking and suppressing evidence that presented results are false (this case) undermine science, the public trust in the scientific enterprise, and the entire scientific community.

Potential result:	HARM	BENEFIT	UNKNOWN	UNKNOWABLE
Stakeholder:				
Your boss/client	Documenting, and then fixing the misinterpretation will take time, adds effort; may identify other problems that require additional time/effort. This supervisor's deception may be discovered to have affected a 'product' that the client has paid for in the past, possibly identifying past errors and undermining this and other projects.	Presenting evidence that conflicts with a supervisor may be uncomfortable, possibly embarrassing (for one or both of you), but could result in a) a different supervisor that respects ethical practice; or b) this supervisor realizing ethical practice does not include deception.	Identification of misinterpretation will require resources to fix/address. *Appropriate* interpretation may not yield desired results.	
Unknown individuals	Incorrect interpretation undermines reproducebility and rigor. Misinterpreted results negatively affect all stakeholders in the long run.	Correction of the error means future work will be correct, and valid. Innovation may be stimulated to devise an appropriate way to obtain the "desired" results (e.g., better experimental design; collection of different data).	Competent interpretation (ASA) supports reproducible results; effective contextualization of your work (ACM) promotes public good. Perpetuation of false reports and misinterpretation undermine both while also perpetuating detrimental research practices.	Detrimental research practices like cherry-picking and suppressing evidence that presented results are false (this case) undermine science, the public trust in the scientific enterprise, and the entire scientific community.

Potential result:	HARM	BENEFIT	UNKNOWN	UNKNOWABLE
Stakeholder:				
Employer	Identifying, documenting, and then fixing the error will take time; this supervisor is trying to deceive the employer. The fact that your project isn't working as promised may identify other problems that require additional time/effort. This misinterpretation may be discovered in a 'product' that a client has/other clients have paid for in the past, identifying past errors and undermining this and other projects	Presenting evidence that conflicts with a supervisor may be uncomfortable but could result in a) a different supervisor that respects ethical practice; or b) this supervisor realizing ethical practice does not include deception. The simulations showing current approaches are not working could stimulate innovations that will ultimately work better.	Identification of misinterpretation may require resources to fix/address. *Appropriate* interpretation –once the cherry-picking is eliminated - may not yield desired results.	Identification of the misinterpretation may highlight past payments or purchases/sales of incorrect/error-prone work. Liability may be incurred.

Potential result: Stakeholder:	HARM	BENEFIT	UNKNOWN	UNKNOWABLE
Colleagues	Pointing out and correcting a (senior) team member's detrimental research practices can be discomfiting – and may also highlight other detrimental practices that will then invite more scrutiny.	Identification and correction of the misinterpretation allows mitigation of risks of harms (ACM); supports valid inferences (ASA), and can limit the unpredictable propagation of the error to other parts of the workflow. Other colleagues may be emboldened to call out –and stop – other detrimental practices they observe.	Competent interpretation (ASA) supports reproducible results; effective contextualization of your work (ACM) promotes public good. Perpetuation of false reports and misinterpretation undermine both while also perpetuating detrimental research practices.	Detrimental research practices like cherry-picking and suppressing evidence that presented results are false (this case) undermine science, the public trust in the scientific enterprise, and the entire scientific community.

Potential result: Stakeholder:	HARM	BENEFIT	UNKNOWN	UNKNOWABLE
Profession	Failing to fix the misinterpretation *may* save time but will lead to unpredictable propagation of that error throughout the workflow and into the scientific community, undermining the practitioner's credibility—as well as the profession's.	The simulations showing current approaches are not working could stimulate innovations that will ultimately work better.	Competent interpretation (ASA) supports reproducible and valid results; effective analysis (ACM) promotes public good. Perpetuation of false reports and misinterpretation undermine both.	Commitment to identification and correction of the error, consistent with GL/CE, might strengthen trust in profession. The error itself can enable innovation (to fix, avoid, or detect similar errors in future), as well as conversations about ethical practice.
Public/public trust	Even if episodes of misinterpretation are noted so they can be fixed, detrimental practices like cherry-picking undermine public trust in reproducibility and rigor. An insistence on misinter-preted results being "correct" undermines public trust as well as reproducibility and rigor.	Identification and correction of the interpretation error allows mitigation of risks of harms (ACM); supports valid inferences (ASA), and can limit the unpredictable propagation of the error to other parts of the workflow and community, to and decisions based on them.	Competent interpretation (ASA) represents reproducible results; effective contextualization of your work (ACM) promotes public good. Perpetuation of false reports and misinterpretation undermine both while also perpetuating detrimental research practices.	Detrimental research practices like cherry-picking and suppressing evidence that presented results are false (this case) undermine science, the public trust in the scientific enterprise, and the entire scientific community.

Table 3.5.5. *Stakeholder Analysis: A favorite result is documented to be spurious.*

2. Identify decision-making frameworks.

Like Case 25, this vignette describes a highly prevalent and extremely detrimental research practice, **cherry-picking**[25]. As noted, cherry-picking is *obviously not ethical*. However, in this vignette, you have generated additional evidence that the "favoured" result is false and not reproducible – and the supervisor is ignoring, possibly suppressing, that fact. This might actually be how "cherry-picking" happens both in and outside of scientific applications.

The report in this vignette only includes what is desired, and also includes knowingly false information, and is thus both incomplete and deceptive. Reporting like this should be resisted by any ethical practitioner. The ethical statistical practitioner "uses methodology and data that are valid, relevant, and appropriate, without favoritism or prejudice, and in a manner intended to produce valid, interpretable, and reproducible results" (ASA A2) and "avoids compromising scientific validity for expediency" (ASA E4). In this vignette, both of these principles were followed but one person on the team seeks to subvert these ethical practices and has misinterpreted the results. Principle B dictates that the ethical statistical practitioner: "Strives to promptly correct substantive errors discovered after publication or implementation. As appropriate, disseminates the correction publicly and/or to others relying on the results." (ASA B5) In addition to notifying stakeholders of the cherry-picking (and false report) in this case, the ethical practitioner must follow ASA H2, and "Avoid condoning or appearing to condone statistical, scientific, or professional misconduct. Encourage other practitioners to avoid misconduct or the appearance of misconduct." The attempted suppression of your simulation results showing that the favoured result is spurious is obviously statistical and scientific ("falsification"[26]), if not also professional, misconduct.

Although the ACM CE does not mention cherry-picking or falsification, "A computing professional should be transparent and provide full disclosure of all pertinent system capabilities, *limitations, and potential problems* to the appropriate parties." (Emphasis added). In this case, you have identified – and provided evidence of - an error of the supervisor's interpretation, and *that* needs to be transparently reported. As noted in the previous case, detrimental practice needs to be identified and discouraged at every possible opportunity. In the vignette, the favoured result is clearly spurious, so whatever the method or system is that produced it will not function validly or reproducibly. This is a wholly predictable limitation that will invariably cause harm – even if that

---

[25] https://en.wikipedia.org/wiki/Cherry_picking
[26] https://ori.hhs.gov/definition-misconduct

harm is only to the scientific record. However, this vignette could also apply to system tests and other non-science situations. The harms are not limited to science, the scientific record, or the scientific community in this vignette.

3. Identify or recognize the ethical issue.

As this vignette is presented, the practitioner has applied their expertise to carry out an analysis appropriately, but the supervisor seeks to suppress the totality of the appropriate work. There is no way for the ethical practitioner to allow the supervisor to misrepresent their work by favouring any result, and then suppressing the (also appropriate) demonstration by simulation that the favoured result is in fact, irreproducible (spurious). Both the ASA and ACM standards support reporting of the error, with the ASA GLs stipulating that corrections should be disseminated, and the ACM CE specifying similarly that limitations of the system (i.e., that it cannot work and presented evidence that it does is spurious) should be disclosed. Note that in this case, there is an ethical issue if you *fail to notify* all stakeholders of the cherry-picking *and its attempted suppression.* These failures would be unethical because the notification can prevent false and invalid reporting (and any decisions based on cherry-picked results). Moreover, the fact that the supervisor is given evidence that their favoured result is spurious *and ignores that,* seeking to promote their preferred interpretation even though it is shown to be irreproducible, suggests strongly that the supervisor intends to deceive others. While some people simply do not understand computing and statistical principles sufficiently to understand the depth of the impropriety of cherry-picking – but do not cherry-pick results because they are dishonest, it does not make cherry-picking ethical. The ethical practitioner must act to call attention to cherry-picking as the unethical and deceptive, detrimental practice that it is, or else they violate ASA H2.

Both the ACM (Principle 4) and ASA (Purpose, Principle G and the Appendix) outline specific responsibilities of the ethical practitioner to follow the practice standards *and encourage others to do so as well*. It does not always work to change cherry-picking behavior, but the ethical practitioner will *do what they can to discourage cherry-picking and any other detrimental practices*. Documenting for this supervisor that their cherry-picking is unsupportable does not prevent them from continuing this practice. Thus, to the extent possible given your role (and job security), an ethical issue for the practitioner also arises when they do not seek to discourage this *deceptive and detrimental* behavior.

4. Identify and evaluate alternative actions (on the ethical issue).

The three decisions that we discuss for any circumstance are: a) do nothing. b) consult or confer with a peer (ASA) or a supervisor (ACM); and c) report violations of policy, procedure, ethical guidelines, or law (ASA) or refuse to implement the system (ACM).

As usual, option a) must be modified. Given the fact in the vignette that you've already identified that the supervisor is cherry-picking, the ethical alternatives are complicated by the supervisor ignoring (or suppressing) evidence that their favoured result is spurious. Option a) should change to "correct the cherry-picking, notify all stakeholders with the documentation of cherry-picked results, and act to prevent their use or further publication." Option b) becomes, "confer with peers or supervisors about how best to correct the cherry-picking, notify all stakeholders with the documentation of cherry-picked results, and act to prevent their use or further publication, *and* ensure that cherry-picking and other detrimental practices (including suppression of evidence of cherry-picking or spurious results) are recognized as unethical and not tolerated".

As discussed in Case 25, the 2017 National Academies of Sciences and Engineering and the Institute of Medicine (NASEM) report, *"Fostering Integrity"*, called on the scientific community in the US to identify and act to prevent such detrimental practices. In this vignette, you have reported the error internally, and your supervisor then seeks to suppress this information because it conflicts with their preferred result. To the extent that the supervisor ignoring your evidence that their result is spurious includes them preventing clients or other stakeholders, who rely on the results, of the fact, then option c) ("report the violation of CE, GLs, and relevant laws") is a viable option. You have effectively demonstrated that the report with cherry-picked data is false and deceptive; the fact that the supervisor wants to suppress this information demonstrates their intent to deceive stakeholders. While cherry-picking is unethical but not illegal, deception and withholding evidence *is* illegal – particularly when there are financial conflicts of interest [27] in play. If an organization agrees with the supervisor and also seeks to suppress your documentation that results are incorrect, it may be perpetrating fraud[28] – and both your commitment to ethical practice and your civic duty will require that you report this. Then option c) effectively becomes, "correct the cherry-picking, notify all stakeholders with the documentation of cherry-picked results, and act to prevent their use or further publication, and report the cherry-picking and suppression of its existence to the appropriate authorities (and to the ACM)." Note that the ACM does, but the ASA does not (as of May 2022), have

---

[27] https://ori.hhs.gov/education/products/ucla/chapter4/default.htm
[28] https://en.wikipedia.org/wiki/Fraud

a reporting mechanism as well as a specifically articulated responsibility to report violations of the practice standards.

Recall that both the ASA and ACM practice standards state explicitly and repeatedly that professionals do not yield to pressures to behave contrary to their respective practice standards (**ASA**: A; A1, A2, A4, A9, A12; C1, C2; E4; G1; H2; **ACM**: 1.2; 1.3; 2.1; 2.2; 2.3; 3.4; 4). Moreover, both the ASA (ASA Principle G) and ACM CE (Principle 3) also suggest that the practitioner and those in leadership positions have an obligation not to comply with requests from persons who want you to violate the ethical practice standards. Both practice standards also articulate a responsibility to encourage others to comply with the standards, if possible.

5. Make and justify a decision.

Decision: All options include "correct the cherry-picking, notify all stakeholders with the documentation of cherry-picked results, and act to prevent their use or further publication". This is the only ethical response to the supervisor's behavior in this vignette, although how it is done – or what else is also done- may depend on other features of your specific context. *Options b and c* include sharing (i.e., reporting, option c) or conferring (option b) with others. While option c), reporting, is only viable when policies exist and reporting violations of those policies, or relevant laws, is feasible, it may constitute some "reporting" when you share the cherry-picking behavior of the supervisor with other stakeholders. Consulting a peer or supervisor may be more relevant for a more junior practitioner. As you grow in experience and seniority, you may be able to use the conferring to give some careful thought given to why no one else on the team objected to the reporting of cherry-picked results. There may be a culture in your team or organization that either tolerates or encourages detrimental practices, and the ethical practitioner is obliged to change this culture. When an otherwise ethical practitioner fails to stop others' unethical practice, they would then violate ASA H2 and "condone, or appear to condone", these unethical practices. Option b) is preferred over option a) in this case because the notification aspect of the solution may require organization-level review of policies or incentives that exist for reporting (and pursuing) results that cannot be replicated (i.e., are spurious). Option b) will also be more appropriate to initiate culture change, and to lead those who are ignorant that their behavior is unethical (e.g., if their practice guidelines do not decry deceptive and detrimental practices) to the understanding that *cherry-picking is unethical*. However, because the supervisor seems to be intent on suppressing your identification of their cherry-picking, option c) may be the decision that has the greatest impact, promoting ethical practice but also preventing frank deception from propagating from your work.

6. Reflect on the decision.

Construct your own reflection on this decision.

Summary of five case analyses for Interpretation

The five case analyses of Chapter 3.5 were focused on interpretation and utilized the ASA GLs without input from the ACM CE. You can see in each case that, even if *you* follow DEFW Principles 5 (use robust practices) and 6 (make your work transparent & be accountable), it will not help you to address the ethical challenges that are created by incorrect interpretation. DSEC questions also do not help address any of these challenges, although you might be able to augment your documentation of how and why misinterpretation is unethical by pointing out that irreproducible results will not be auditable (DSEC C5) or explainable (DSEC D4), and it will be impossible to detect concept drift (DSEC E3). It was also noted that the misuse of p-values – a common source of misinterpretation of results across disciplines – is effectively an example of DSEC D3: using "statistical significance" as the only metric by which results are judged[29]. As has been noted, others on our work and research teams may not be familiar with, or do not hold themselves to, the GLs or CE. While some of this Chapter's cases arise from a lack of understanding of either what constitutes "robust practice" or best practices in data analysis and interpretation, it is essential to recognize and act to discourage the prioritization of a favoured theory or pre-defined/desired result over reproducibility and validity of results.

Because the majority of individuals who are actually statistical practitioners are *not* formally trained in statistics the policy or structure at your organization may play a role in whether the option to "report" (options c) or simply to "consult" (options b) violations of the GLs, CE, or applicable laws – although as a concerned citizen, it is a general responsibility to report when laws are being broken. Both the GLs and CE assume that their ethical guidance is relevant in addition to following all applicable laws (and both practice standards state that ethical practitioners should be familiar with applicable laws and codes). Each of the cases in this chapter is based on actual events. Willful misinterpretation of results is a widespread problem whenever statistical analyses are employed – again, pointing out the importance of, and justification for, both the ACM and ASA stipulating in their preambles that their ethical practice standards are applicable, and should be followed, by *all* who use these tools, methods, and techniques. In the reflection on case 22 in this Chapter, we considered the 2017 report of the National Academies of

---

[29] Read about how p-values are misunderstood and (more often) misused in this editorial or throughout this special issue (2019) of The American Statistician: https://www.tandfonline.com/doi/full/10.1080/00031305.2019.1583913

Sciences and Engineering and the Institute of Medicine (NASEM), *"Fostering Integrity"*,

> Actions such as failing to retain or share data and code supporting published work in accordance with disciplinary standards, practices such as honorary or ghost authorship, and *using inappropriate statistical or other methods of measurement and data presentation to enhance the significance of research findings* are clearly detrimental to the research process and may impose comparable or even greater costs on the research enterprise than those arising from research misconduct." NASEM 2017, p. 206 (emphasis added).

These "detrimental research practices" are widespread, as discussed throughout the 2019 issue of The American Statistician (see footnote 29). The NASEM report urges greater attention to be placed on training new scientists to avoid these practices, but until that happens, ethical statistical practitioners and data scientists will continue to encounter cases like this one. If every scientist was trained in the GLs and CE, then detrimental research practices would only be utilized by those who purposefully seek to deceive. That might make only a small dent in the number deceptive research practitioners, and we already know that people who would purposefully deceive readers or users of their work would not follow any kind of guidelines or practice standards. However, hopefully these cases show how important it is to the profession and to the wider scientific community that we make every effort to encourage those who would practice ethically to do so. Outside of the research and scientific community, widening the population of ethical practitioners and alerting all stakeholders to the fact that detrimental practices only do harm will have the effect, eventually, of promoting the public good and limiting harm to the extent that ethical practice standards can do.

Recognizing and accepting these responsibilities pass the *tests of publicity, universality,* and *justice.* Statisticians and data scientists need to do what we can to *identify and correct* misuse and misinterpretation of our work –whether that is unintentional or malicious/purposeful. The professions of statistics, data science, and 'statistics and data science', the public, and all practitioners using these techniques/methods will benefit when we accept the responsibilities outlined in the practice standards; their use leads to more just science.

Questions for Discussion:

1. Review at least two cases in this chapter.

A. Do you agree with the decisions in these case analyses? Why or why not? Which parts of the ER process are most and least acceptable for each case? Do these parts differ for different cases in this chapter? Were there different ASA or ACM Principles or elements that you identified, and if so, how do these affect your case analyses? Are the elements and/or Principles you identify represented in your reflections?

B. Discuss your results for the tests of universality, justice, and publicity for any (or all) case(s) in this chapter.

C. Discuss the relevance to each case – including the analysis and your reflection – of federal, state, or local/organizational policies regarding research misconduct and policies for handling misconduct. How can you incorporate what you know about these policies into any part of the case analysis (prerequisite knowledge; identifying the ethical issue; determining alternative actions; making or justifying your decision; or reflection)? Note that, if there is no alignment between federal, state, or your local/ institutional policies and a given case, you can comment on that (e.g., do policies in your organization focus solely on *misconduct* – and ignore detrimental or other unethical practices?). Do you feel that these policies could have helped in the reasoning, decision or justification?

D. Discuss the relevance to each case - including the analysis and your reflection – of responsible authorship and publication. How can you incorporate what you know about these policies into any part of the case analysis (prerequisite knowledge; identifying the ethical issue; determining alternative actions; making or justifying your decision; or reflection)? If there are no relevant policies, you should comment on that, particularly when a) data are obtained using federal monies (grants); or b) data are collected (acquired) from humans, such that Federal Regulations about the treatment of human subjects and the acquisition of tissues and data from humans are relevant.

2. Review at least two (different) cases in this chapter.

Keeping in mind that doing nothing/not responding are *not plausible responses,* are there other plausible alternatives (KSA 4) that you can think of for any case? If not, discuss that; if so, list them and discuss their evaluation (are they equally consistent with GL/CE, do they lead to similar decisions, etc.).

3. Choose any case in this chapter. Redo the analysis, but feature a different GL or CE principle in your reasoning process. Make sure the *justification* is different from what is given. Is the *decision* also different? Discuss your decision with its justification and the tests of universality, justice, and publicity.

4. Consider all five cases in this chapter.

A. Comment on the role of the stakeholder analysis in these decisions. Were stakeholder impacts featured in justifications? Were they features in your own reflections?

B. Comment on the importance of your reflections on the decisions you were asked for: for others doing a similar task at work; for those using results from this task; and for those who might join the profession (and be learning about planning and design).

C. Consider your professional identity and the profession itself: planning/designing projects and data collection/analysis projects is obviously foundational in statistics and data science. How can these five cases and the decisions reached via the analyses presented earn you (the decider) the trust of the practicing community or the public for the profession and its future?

Chapter 3.6
Documenting your work

Case 27. It takes as long to fully and transparently document your work as it does to do the work itself. Since this is just your job, not documenting it will only affect you (for the foreseeable future) –and is faster.

Case 28. You failed to fully document your work a few months ago and now your supervisor is requesting your comprehensive documentation so that another person can replicate your work. You really only have time for minimal documentation.

Case 29. You receive documentation of an ongoing program/analysis that lacks all information about data provenance

Case 30. Prior documentation of a company-wide method is complete and correct. The method development did not include sensitivity analyses. You do a few and identify two important errors in the method.

Case 31. You are given documentation that is not complete: it lacks details about exactly what methods and in what order were used.

Case 32. You provide complete and correct documentation, and this gets "edited" so that it is now no longer complete or correct.

Case 33. The documentation you receive specifies an analysis method that is not appropriate for the specific question that must be addressed.

Case 27. It takes as long to fully and transparently document your work as it does to do the work itself. Since this is just *your* job, not documenting it will only affect you (for the foreseeable future) –and is faster.

1. Identify and 'quantify' prerequisite knowledge:

Which DEFW and DSEC Principles or Areas seem most relevant to this vignette?

Not discussed in DSEC or in DEFW.

Which ASA and ACM Principles and elements seem most relevant to this vignette?

ASA:

Principle A. Professional integrity and accountability (A; A2; A4; A11)
Principle B. Integrity of data and methods (B1; B2; B3)
Principle D. Responsibilities to research subjects, data subjects, or those directly affected by statistical practices (D8; D10)
Principle E. Responsibilities to members of multidisciplinary teams (E3)
Principle F. Responsibilities to fellow statistical practitioners and the Profession (F; F2; F4; F5)
Principle G. Responsibilities of leaders, supervisors, and mentors in statistical practice (G5)
Principle H. Responsibilities regarding potential misconduct (H2)
Appendix. Responsibilities of organizations/institutions (APP 4; APP 8; APP 10; APP 12)

ACM:

1. General Ethical Principles (1.3)
2. Professional Responsibilities (2.1; 2.4; 2.5)

Potential result: Stakeholder:	HARM	BENEFIT	UNKNOWN	UNKNOWABLE
YOU	Documentation takes time, adds effort (to fully document plans/ evaluations/systems); may facilitate use of work that is unauthorized/ does not acknowledge original creator	Simplifies evaluations of the work (ASA) or system, and updates (ACM); adds transparency and accountability.	ACM: New harms/risks may be detectable more quickly (and addressed) with full documentation. ASA: replications are facilitated.	
Your boss/client	Takes time	Creates transparency and accountability of statistician/ computing professional work		Documentation may need IP protection, which could be costly.
Unknown individuals		Creates transparency and accountability		
Employer	Takes time	Demonstrates commitment to transparency & accountability.	ACM: New harms/risks may be detectable more quickly (and addressed) with full documentation. ASA: replications are facilitated.	ACM: Harms, if identified, must be remediated, so could limit business/profit (although ASA and ACM would not consider this ethical justification to forego documentation).

Potential result: Stakeholder:	HARM	BENEFIT	UNKNOWN	UNKNOWABLE
Colleagues	May add time (in reviewing)	Supports transparency, accountability, and replicability. Promotes peer evaluation.		
Profession	Takes time, adds effort (to fully document plans/evaluations/ systems); may facilitate use of work that is unauthorized/ does not acknowledge original creator	Simplifies evaluations of the work (ASA) or system, and updates (ACM); adds transparency and accountability. Promotes and facilitates peer evaluation.	ACM: encourages careful review, so new harms/ risks may be detectable more quickly (and addressed). ASA: transparency and validity of analyses and interpretations are facilitated.	
Public/public trust	Without full documentation of work/thought processes, public trust in systems or their results, and the profession (ACM/ASA) are compromised. Perpetuating black box mentality does not help, and explicitly does not inform, society.	Transparency and accountability – and the potential to understand how decisions are made based on automation and systems - supports public trust in the profession and its work.	Transparency supports trust, even if errors are discovered; their discoverability arises only and specifically from documentation, transparency, and accountability.	

Table 3.6.1. *Stakeholder Analysis template: Not documenting your work*

2. Identify decision-making frameworks.

ASA GLs describe the ethical statistical practitioner as fully documenting their work so as to be transparent and to promote reproducibility in all of their work. As a virtue approach, this suggests that full documentation is the ideal, and even if it is inconvenient (as this vignette suggests it is), documentation is still essential to ethical statistical practice. All but one Principle of the ASA Ethical Guidelines are relevant in documentation (only Principle C. Responsibilities to stakeholders is missing). Documentation is a key factor in transparency as well as in reproducibility, which is why so much of the Guidelines are relevant to this case.

The SHA supports transparent and complete documentation of your work as consistent with the utilitarian perspective of the ACM CE: while "major" (dark grey highlighted) harms could potentially accrue to you and to the profession, other harms are minor, while benefits accrue to all stakeholders. Moreover, transparency and accountability through documentation can streamline peer review, and promote consistent, high-quality work (ACM 2.2) which streamlines the identification of harms or limitations – and can also facilitate correcting those harms. Thus, documentation promotes the profession and by doing so, can limit harms.

3. Identify or recognize the ethical issue.

Both the virtue and utilitarian perspectives, as well as practice standards, support complete and transparent documentation of your work. While inconvenient and time consuming, failure to document your work violates both practice standards, and has the potential to undermine real benefits to all stakeholders (including the profession). *A failure to document fully and transparently is unethical* –even if it appears that you might be the only person affected, you can see from the SHA and practice standards that that is not true. In this vignette, the completeness of coverage for this case plus the SHA are helpful in recognizing the ethical issue.

4. Identify and evaluate alternative actions (on the ethical issue).

Although we examine the same three decisions that can be made in any circumstance: a) do nothing b) consult or confer with a peer (ASA) or a supervisor (ACM); or c) report violations of policy, procedure, ethical guidelines, or law (ASA) or refuse to implement the system (ACM), none of these fit our case. Instead, option a) becomes "document your work"; and option b) becomes "confer with a peer or supervisor about how best to document your work (and then document your work)". Given the harms that can accrue to you and your profession if intellectual piracy is facilitated, option

c) could become "document your work and safeguard that documentation to the extent possible".

5. Make and justify a decision.

Decision: Option a) is clearly consistent with practice standards. Options b) and c) are the most feasible, but option c) is the preferred response to the vignette because it has the greatest chance of realizing all the benefits while minimizing the harms identified in the SHA. How exactly your documentation can be "safeguarded" is something that might require consultation with a colleague or supervisor; options include publishing the documentation in an open-source environment, so that both "black hat" (intending on harm) and "white hat" (intending on good) actors in the community can be engaged in safeguarding the work, and ensuring that it cannot be stolen or someone else cannot take it and use it without others knowing. Applying for a patent, or obtaining peer review and publication, are other ways to ensure that your work is recognized as having originated with you. The objective of the documentation, to be transparent and accountable, as well as supporting replication (ASA) and honest peer review (ACM) as appropriate, can be achieved with either options b) or c); enacting whatever safeguards are feasible under option c) will have the added benefit of minimizing harms, so that should be pursued. Depending on your role in your team or organization, you may need to consult or confer to determine what options are available for safeguarding your work. For example, if your work is proprietary or otherwise protected, then it cannot be peer reviewed and published, but there may be alternative methods to safeguard the work and consultation will help uncover those. Discussion with peers may help you discover open source or other channels or venues for your work if they are appropriate.

6. Reflect on the decision.

This vignette shows how – as is argued in other cases- violations of the practice standards may arise from a sense of inconvenience rather than maliciousness. The only ethical options in this vignette are to document your work (because option a) is unethical!), because full and correct documentation is the responsibility of all ethical practitioners. Keep in mind that this responsibility pertains to everyone who uses tools, techniques, and methods of both statistics and computation, as per the GLs and CE. Recognizing and accepting these responsibilities pass the *tests of publicity, universality,* and *justice.* Statisticians and data scientists – as well as those using these disciplines during the course of other work - need to *ensure that* our work, or work that depends on our effort, is fully and transparently documented. Whether failures of documentation that are described in Cases 29-33 were unintentional or malicious/purposeful, the ethical issues those failures of documentation are real; not only does the lack of documentation lead to actual ethical issues, but also the unethical failures to document can create new ethical issues that later statisticians, data scientists, and other scientists might be forced to react/respond to or correct. The professions of statistics, data science, and 'statistics and data science', the public, and all practitioners using these techniques/methods will benefit when we accept the responsibilities outlined in the practice standards. The use of these standards leads to more just science and better stewardship of resources.

The SHA shows that in this vignette, the main harms that failures to appropriately document can lead to will accrue to you and to the profession. Specifically, these harms could be in the form of permitting misuse, abuse, and/or frank stealing of your work – if you do not document that the work is yours/originated with you, then others are easily able to steal it and take credit for it. The better documented your work is, the lower this risk is. The ASA and ACM ethical practice standards both clearly charge the ethical practitioner to respect the intellectual contributions of others; but unfortunately, people who obtain your documentation may not respect these practice standards. Obviously, the potential misuse of your documentation by those who do not acknowledge your contributions is not just "unethical", it might also be illegal – and of course, people who are willing to break the law will not care that they are also violating ethical practice standards. However, that is not sufficient motivation or justification to ignore your responsibility to be transparent and accountable, and to ensure that your work is auditable and peer evaluable, i.e., fully documented.

Case 28. You failed to fully document your work a few months ago and now your supervisor is requesting your comprehensive documentation so that another person can replicate your work. You really only have time for minimal documentation.

1. Identify and 'quantify' prerequisite knowledge:

Which DEFW and DSEC Principles or Areas seem most relevant to this vignette?

Not discussed in DSEC or in DEFW.

Which ASA and ACM Principles seem most relevant to this vignette?

ASA:

Principle A. Professional integrity and accountability

Principle B. Integrity of data and methods

Principle D. Responsibilities to research subjects, data subjects, or those directly affected by statistical practices

Principle E. Responsibilities to members of multidisciplinary teams

Principle F. Responsibilities to fellow statistical practitioners and the Profession

Principle G. Responsibilities of leaders, supervisors, and mentors in statistical practice

Principle H. Responsibilities regarding potential misconduct

Appendix. Responsibilities of organizations/institutions

ACM:

1. General Ethical Principles

2. Professional Responsibilities

Potential result: Stakeholder:	HARM	BENEFIT	UNKNOWN	UNKNOWABLE
YOU	Takes time, adds effort (to fully document plans/ evaluations/systems); minimal documentation will not support review or replication. The lack of complete documentation may suggest to your supervisor that you are not very good at your job.	Even minimal documentation, if it is correct, simplifies evaluations of the work (ASA) or system, and updates (ACM); adds transparency and accountability.	ACM: New harms/risks may not be detectable or addressed with minimal documentation. ASA: Replications may not be facilitated with minimal documentation.	
Your boss/client	Takes time; may require additional time/resources.	Even minimal documentation, if correct, adds transparency and accountability of statistician/ computing professional work.		
Unknown individuals	Minimal documentation will not support review or replication.	Creates transparency and accountability		

Potential result: Stakeholder:	HARM	BENEFIT	UNKNOWN	UNKNOWABLE
Employer	Takes time; may require additional time/resources.	Even minimal documentation, if correct, adds transparency and accountability of statistician/ computing professional work.	ACM: New harms/risks may not be detectable or addressed with minimal documentation. ASA: Replications may not be facilitated with minimal documentation.	
Colleagues	Incomplete documentation may mislead colleagues into thinking the work itself is also incomplete; review and identification of "what is missing" could take time and effort by others.	Even minimal documentation, if it is correct, simplifies evaluations of the work (ASA) or system, and updates (ACM); adds transparency and accountability.		

Potential result: Stakeholder:	HARM	BENEFIT	UNKNOWN	UNKNOWABLE
Profession	Minimal documentation will not support review or replication. Failure to document work fully may signal a lack of respect for other professions or others on the team. The lack of complete documentation may suggest that professionals like you are not very communicative or ethical practitioners.	Even minimal documentation, if it is correct, simplifies evaluations of the work (ASA) or system, and updates (ACM); adds transparency and accountability.	ACM: New harms/risks may not be detectable or addressed with minimal documentation. ASA: Replications may not be facilitated with minimal documentation.	
Public/ public trust	Without full documentation of work/thought processes, public trust in systems or their results, and the profession (ACM/ASA) are compromised. Perpetuating black box mentality does not help, and explicitly does not inform, society.	Transparency and accountability – and the potential to understand how decisions are made based on automation and systems - supports public trust in the profession and its work, even if the documentation is only minimal.	Transparency supports trust, even if errors are discovered; their discover-ability arises only and specifically from documentation, transparency, and accountability. These can arise from even minimal documentation as long as it is correct.	

Table 3.6.2. *Stakeholder Analysis template: Incompletely/minimally documenting your work*

2. Identify decision-making frameworks.

ASA GLs describe the ethical statistical practitioner as fully documenting their work so as to be transparent and to promote reproducibility in all of their work. As a virtue approach, this suggests that *full* documentation is the ideal, and essential to ethical practice. Thus, since minimal documentation – even assuming it is correct – is not ideal, the practitioner should notify the supervisor that the documentation is incomplete and set up a reasonable time frame within which full documentation will be delivered. Documentation is so important that it cuts across nearly all of the ASA Ethical Guidelines. Asking for more time to carry that out is warranted from the virtue perspective.

The SHA supports transparent and *complete* documentation of your work as consistent with the utilitarian perspective of the ACM CE. "Major" (dark grey highlighted) harms could potentially accrue to you (mostly having to do with time) and to the profession (having to do with the appearance of incompetence or unethical practice). Other harms in the SHA are minor. By contrast, if work is fully and transparently documented, then benefits accrue to **all** stakeholders. Benefits of documentation are minimized when it is incomplete, thus highlighting the harms accruing when documentation is not complete. The utilitarian perspective, even without specific elements from the ACM CE to support them, makes the SHA very informative in this case.

3. Identify or recognize the ethical issue.

Both the virtue and utilitarian perspectives, as well as practice standards of the ASA and ACM, support complete and transparent documentation of your work. While inconvenient and time consuming, failure to document your work violates both practice standards, and has the potential to undermine real benefits to all stakeholders. *A failure to document fully and transparently is unethical* – even if it appears that you might be the only person affected, you can see from the SHA and this vignette that that is not true, because others who need to replicate your work are also negatively impacted. However, both General Ethical Principles (ACM 1.3) and Professional Responsibilities (ACM 2.1; 2.4; 2.5) are clear about the need for transparency and documentation of work.

4. Identify and evaluate alternative actions (on the ethical issue).

Although we examine the same three decisions that can be made in any circumstance: a) do nothing b) consult or confer with a peer (ASA) or a supervisor (ACM); or c) report violations of policy, procedure, ethical guidelines, or law (ASA) or refuse to implement the system (ACM), none of these fit our case. Instead, option a) becomes "document your work"; and option b) becomes "confer with a peer or supervisor about how best to document your work (and then document your work)". Given the harms that can accrue to you and your profession if intellectual piracy is facilitated, option c) could become "document your work and safeguard that documentation to the extent possible". However, this vignette suggests that you must also notify your supervisor that your documentation is not yet complete and set up a reasonable time frame within which full documentation will be delivered. All three of the options must include this feature.

5. Make and justify a decision.

Decision: As indicated, all options include "fully document your work" and "notify the supervisor that the documentation is not yet complete". While option a) is clearly consistent with practice standards, options b) and c) are preferred, but option c) is the most preferred response to the vignette because it has the greatest chance of realizing all the benefits while minimizing the harms identified in the SHA.

As was noted in the previous case, how your documentation can be "safeguarded" may require consultation with a colleague or supervisor. The objective of the documentation, to be transparent and accountable, as well as supporting replication (ASA) and honest peer review (ACM) as appropriate, can be achieved with either options b) or c); enacting whatever safeguards are feasible under option c) will have the added benefit of minimizing harms, so that should be pursued. As with the previous case, depending on your role in your team or organization, you may need to consult or confer to determine what options are available for safeguarding your work. For example, if your work is proprietary or otherwise protected, then it cannot be peer reviewed and published, but there may be alternative methods to safeguard the work and consultation will help uncover those. Discussion with peers may help you discover open source or other channels or venues for your work if they are appropriate.

6. Reflect on the decision.

Construct your own reflection on this decision.

Case 29. You receive documentation of an ongoing program/analysis that lacks all information about data provenance.

1. Identify and 'quantify' prerequisite knowledge:

Which DEFW and DSEC Principles or Areas seem most relevant to this vignette?

This situation is inconsistent with DEFW Principle 4 (understand the limitations of the data) but if some of the data were illegally or unethically derived (leading to missing or incorrect provenance), then DEFW Principle 2 will also not be addressed.

In this case, the answer to the first question in the DSEC set, A1 (have human subjects given informed consent) is "no" – because it cannot be answered confidently in the affirmative. Moreover, because there is no evidence about the data provenance, there might also be inconsistencies in evidence for answering other DSEC questions, including whether a "right to be forgotten" (B2) can even be honoured. Depending on the extent of poor documentation, the data may also be missing key perspectives (C1), be biased (A2, C2), be unauditable (C5), and be both unfair across groups (D2) and un-explainable (D4).

Which ASA and ACM Principles seem most relevant to this vignette?

ASA:

Principle A. Professional integrity and accountability
Principle B. Integrity of data and methods
Principle C. Responsibilities to stakeholders
Principle D. Responsibilities to research subjects, data subjects, or those directly affected by statistical practices
Principle E. Responsibilities to members of multidisciplinary teams
Principle F. Responsibilities to fellow statistical practitioners and the Profession
Principle G. Responsibilities of leaders, supervisors, and mentors in statistical practice
Principle H. Responsibilities regarding potential misconduct
Appendix. Responsibilities of organizations/institutions

ACM:

1. General Ethical Principles
2. Professional Responsibilities

Potential result: Stakeholder:	HARM	BENEFIT	UNKNOWN	UNKNOWABLE
YOU	Using data with unknown provenance, particularly if some of it is known not to have been contributed with informed consent, violates the data contributor rights as well as ethical practice standards.	There are **no benefits** that accrue when data are obtained without appropriate consent. Without understanding the provenance of the data, you cannot be sure you had appropriate consent/the data were collected ethically.	Failing to obtain consent to use data could lead to unpredicted harms, bias, or unfair results, as well as risks to data contributors - with the statistician or data scientist bearing responsibility for misuse, unauthorized access, or losses of that data.	Using data with unknown provenance may mean that data that was not contributed with informed consent could have been included. The use of this type of data may suggest to others that ignoring data provenance is OK, even though this directly violates practice standards.
Your boss/client	Data with unknown provenance can lead to unpredicted harms, bias, or unfair results, and can create risks for the data contributors that **were** foreseeable.	Ignoring the provenance of data, while a violation of multiple specific elements of the professional practice standards, is simpler and cheaper than ensuring data are obtained with proper consent (not a viable benefit).		

Potential result: / Stakeholder:	HARM	BENEFIT	UNKNOWN	UNKNOWABLE
Unknown individuals	Data that is accessed without authorization and consent may expose data contributors to risks, as well as to (further) misuse by others.	There are **no benefits** that accrue when data are utilized in unauthorized ways, and no one stops this type of inappropriate behavior.		Using data that was not contributed with informed consent may suggest to others that data provenance is not important, even though this directly violates practice standards as well as basic human rights.
Employer	Data with unknown provenance can lead to unpredicted harms, bias, or unfair results, and can create risks for the data contributors that were foreseeable, thus incurring liability to the employer.	Ignoring the provenance of data, while a violation of multiple specific elements of the professional practice standards, is simpler and cheaper than ensuring data are obtained with proper consent (not a viable benefit).		
Colleagues	If others on the team are not aware that the data provenance is mixed, colleagues may mistakenly share –i.e., further the misuse of- the data.	There are **no benefits** that accrue when data are utilized in unauthorized ways, and no one stops this type of inappropriate behavior.		

Potential result: Stakeholder:	HARM	BENEFIT	UNKNOWN	UNKNOWABLE
Profession	The profession may appear untrustworthy when it is discovered that practitioners use whatever data they collect, even if no consent was given or even if the collected data was "available" because of a data breach. Using data with unknown provenance is a violation of multiple specific elements of the professional practice standards.	There are **no benefits** that accrue when stolen data are utilized, and no one stops the use of that kind of data.	Failing to obtain consent to use data and using stolen data (the results of breaches) could lead to unpredicted harms, bias, or unfair results, as well as risks to data contributors - with the statistician or data scientist bearing responsibility for misuse, unauthorized access, or losses of that data.	Using stolen data and data that was not contributed with informed consent may suggest to others/other system developers that collecting stolen/breached data is OK, even though this directly violates practice standards. This decrements professional integrity in a concrete way.
Public/public trust	Public sentiment about the procedures by which they do and do not give their consent will continue to worsen, and people will become less inclined to contribute or share their data if public concerns about the lack of honesty in data collection systems, continue.	There are **no benefits** that accrue when consents are not obtained lawfully, and data are inappropriately utilized, and no one stops the use of that kind of data.		

Table 3.6.3. *Stakeholder Analysis: Documentation suggests unknown data provenance*

2. Identify decision-making frameworks.

Some might imagine that, as long as there's data, the statistical practitioner can get their work done. The critical importance of knowing data provenance is reflected in the fact that every element of the ASA Guidelines is invoked in this vignette, because ethical sourcing of data is contingent on knowing the data provenance. Without data provenance, the ethical practitioner cannot ensure they do "not knowingly conduct statistical practices that exploit vulnerable populations or create or perpetuate unfair outcomes." (ASA A3), nor are they able to "Communicate data sources and fitness for use, including data generation and collection processes and known biases. Disclose and manages any conflicts of interest relating to the data sources. Communicate data processing and transformation procedures, including missing data handling." (ASA B1). In fact, Principle B states, "The ethical statistical practitioner seeks to understand and mitigate known or suspected limitations, defects, or biases in the data or methods and communicates potential impacts on the interpretation, conclusions, recommendations, decisions, or other results of statistical practices." This is not possible without knowing the data provenance. The ethical statistical practitioner "informs stakeholders of the potential limitations on use and re-use of statistical practices in different contexts and offers guidance and alternatives, where appropriate, about scope, cost, and precision considerations that affect the utility of the statistical practice." (ASA C4) – also impossible without full knowledge of the data provenance.

Principle D states, "The ethical statistical practitioner does not misuse or condone the misuse of data. They protect and respect the rights and interests of human and animal subjects. These responsibilities extend to those who will be directly affected by statistical practices." This and most of the D elements are impossible without knowing the provenance of the data.

Utilizing data with unknown provenance violates most ASA GL Principles. Obviously, failing to protect basic human rights by using data you do not have consent to use violates both C4 and D5 (among others), but also violates ASA H2 ("Avoid condoning or appearing to condone statistical, scientific, or professional misconduct. Encourage other practitioners to avoid misconduct or the appearance of misconduct.") as well as federal laws in some cases.

The ACM CE, with its utilitarian perspective, focuses on limiting harms and in this case, there are many harms, and no benefits, to using data that was contributed to your analysis without consent. Moreover, a system that collects data without obtaining consent may include methodologies or other features that are insecure or otherwise create risks (in addition to risks of confidentiality and privacy breaches) that are not addressed – because the manner in which data are collected may not be legitimate or sufficiently specified. In this case, you simply do not know what the provenance is of the data.

Note also that, because of the poor documentation of the data and its provenance in this case, there *may* be other key features of the data collection or other aspects of the system/pipeline or workflow that are similarly poorly documented. These weaknesses would impede your ability to evaluate a system effectively to identify strengths and liabilities (i.e., following ACM CE Principles). Similarly, the fact that there is documentation, but it lacks this one feature (data provenance) that is so crucial should serve as a red flag about the documentation that is provided, as well as the source of that documentation.

3. Identify or recognize the ethical issue.

The primary ethical issue in this case is that both GLs and CE require that data be obtained with consent and the contributors' knowledge of what the data will be used for, and in this case, you simply cannot tell if this did happen. In this case we do not know what type of data is being collected– nor what types of expectations of privacy contributors have – but it is clear that at least some people do have expectations – otherwise there would be no cause to consider provenance. Utilizing data that was obtained *without appropriate consent* is unethical. The case suggests that there *is cause for concern*, because you cannot obtain a correct report of data provenance. Although this case does not specify that other documentation also being missing or inconsistent, that would also be a serious challenge to implementing key features of the ASA and ACM practice standards. Finally, we have seen in other vignettes and their analysis that full and transparent documentation of all work is essential to ethical practice. The fact that there is incomplete documentation in this case raises that ethical concern, although for the individual in this case, that does not pertain to them (because they've received the documentation, and are not responsible for it being incomplete).

4. Identify and evaluate alternative actions (on the ethical issue).

This vignette requires us to respond to documentation that suggests data is being collected without consent and used in ongoing analyses. We do not know if the data are stolen or collected appropriately because there is no documentation at all. As usual, the three decisions that can be made in any circumstance are: a) do nothing. b) consult or confer with a peer (ASA) or a supervisor (ACM); and c) report violations of policy, procedure, ethical guidelines, or law (ASA) or refuse to implement the system (ACM). Clearly, "do nothing" is totally inappropriate in this case, and also a violation of the practice standards relating to respecting data contributors and data provenance. However, the alternative actions must relate to what to do to ensure that data of unknown provenance is not utilized – and not collected if it is being collected inappropriately; and also to ensure that full and correct documentation of the data is compiled and maintained going forward.

Thus, we can change option a) from "do nothing" to "do not proceed with any use of the data until all data provenance can be established (and do not act to correct the lack of documentation)". Then option b) would become, "consult or confer with a peer (ASA) or a supervisor (ACM) as to how best to stop the collection and any use of the data until provenance can be established or data obtained without consent can be removed; and act to correct the lack of documentation." Option c) must also be modified, because simply reporting the fact that the data collection violates the GLs (ASA), or refusing to implement the system (that collects data with unknown provenance) (ACM) might not ensure that the data with unknown provenance is both *not used* and also, **its collection ceases**. It also does not impact the documentation, although reporting this system may lead to the creation of some policy that prevents undocumented systems from being deployed in the future. So, option c) needs to change to "report violations of policy, procedure, ethical guidelines, or law to a peer (ASA) or a supervisor (ACM); *and* stop the collection and any use of the data until provenance can be established or data obtained without consent can be removed; *and* act to correct the lack of documentation."

5. Make and justify a decision.

Decision: *Note that all options include* "do not proceed with any use of the data"; only options b) and c) include "and act to correct the lack of documentation" because the person in this vignette just received the documentation – and may not be able to correct it. Thus, options b) and c) require engagement with others in the organization or team to ensure that whoever is responsible for the system or its documentation will correctly and transparently document it. The core of response to this situation is to **follow the practice standards** and to utilize only data for which consent of the data contributors was obtained appropriately, and also to ensure that work is fully and transparently documented. Option a) says "follow the ethical practice standards, and do not use inappropriately-obtained data". This would be the *least desirable option* because **both** the data collection *and* its use need to stop, since both are unethical; but option a) only addresses one of these. So, option a) prevents *you* from unethical practice (i.e., you refuse to work with data that has unknown provenance) but it may have no effect on others, and doesn't make that data collection stop –it also doesn't fix the documentation problem. Moreover, the method(s) by which the data are being collected may also be inappropriate and possibly unlawful. Data with unknown or mixed provenance may be commonly collected and utilized in your work context; it might be impossible for you to do anything ethical in this case/context except refuse to analyse or work with data of unknown provenance and also, seek another team, other colleagues, or a different job. However, since this leaves the data collection mechanism(s) in place and doesn't address the documentation problem, option a) is least desirable.

Option b) would be the next *most desirable option* - not the most desirable one in this case - in that consulting with a colleague/supervisor will ensure some notification of leadership or your colleagues (preferably, both) that the organization may be collecting and using data without appropriate consent – which is frankly unethical, and may also be illegal. Option b) is preferable to option a) because you would be notifying *someone* about the problems with your data collection (and/or its documentation) *and* the fact that individuals who are directed to analyze it "anyway" are actually being directed to violate their ethical practice standards.

There may not be policies in your workplace that prevent violation of the ASA GLs or ACM CE, but *reporting* (option c) is actually *the most desirable option* in this case. The key in this case that differentiates this option from others in this chapter is that the organization appears to have a system that collects data in an unethical and possibly illegal manner. This creates both a professional and an individual/civic obligation to report. You (the practitioner) can – and should - still demonstrate your professional competence and follow the relevant practice standards by not collecting or working with data with uncertain provenance. With its "report" feature, option c) has the key benefit of permitting "educating" people or your company/organization about the practice standards but also notifying others that data collection – and any work on unethically sourced data – violate ethical practice standards and possibly (in terms of illegal data collection) the law. It is critical *to the profession* to support the next practitioner who ends up in a similar situation (or better, to prevent unethical data collection in the future), which is why option c) with its reporting feature is so important in this case: notifying relevant individuals in your organization that the policy around data collection (and possibly documentation) is in direct violation the ACM CE and the ASA GLs (as well as applicable laws).

6. Reflect on the decision.

Construct your own reflection on this decision.

Case 30. Prior documentation of an organization-wide method is complete and correct. The method development did not include sensitivity analyses. You do a few and identify two important errors in the method.

1. Identify and 'quantify' prerequisite knowledge:

Which DEFW and DSEC Principles or Areas seem most relevant to this vignette?

The results can help you fulfill DEFW Principles 5 and 6.

Not discussed in DSEC.

Which ASA and ACM Principles seem most relevant to this vignette?

ASA:

Principle A. Professional integrity and accountability
Principle B. Integrity of data and methods
Principle C. Responsibilities to stakeholders
Principle G. Responsibilities of leaders, supervisors, and mentors in statistical practice
Principle H. Responsibilities regarding potential misconduct
Appendix. Responsibilities of organizations/institutions

ACM:

1. General Ethical Principles

Potential result: Stakeholder:	HARM	BENEFIT	UNKNOWN	UNKNOWABLE
YOU	You may "get in trouble" if you demonstrate that the system does not function/results are not as expected. Addressing the methodological errors will take time and effort.	The documentation was complete! And your sensitivity analysis is performing as intended, showing your competence! Addressing the interim results means the system will ultimately be less biased.		
Your boss/client	Changing the system (to address method errors) could translate to missed deadlines or milestones. If the method cannot be fixed, the whole system may fail. New results from the fixed method may not match old results/client expectations.	Addressing the method errors means the system will ultimately be better.		

Potential result: Stakeholder:	HARM	BENEFIT	UNKNOWN	UNKNOWABLE
Unknown individuals	Failure to correct method errors and decisions based on the old (wrong) method may undermine trust in the profession or organization.	Sensitivity analysis (if not the system itself) is functioning as intended. Addressing the errors means the system will ultimately be more correct.		
Employer	Taking longer to edit or change the system (to address the errors) could translate to missed deadlines or milestones. If the errors cannot be corrected, the system may not be deployed – resources were wasted.	Addressing the interim results means the system will ultimately be less biased. Addressing the bias may ultimately limit liability/need for redress.	New results from the fixed method may not match old results/client expectations. Longstanding use of a method with errors may have led to many decisions that now must be revisited.	
Colleagues	New errors may create a requirement for adaptations (requiring extra time and effort from everyone). Addressing the errors will take collaborative time and effort.	Your sensitivity analysis results may create opportunities for innovations from everyone.		

Potential result: Stakeholder:	HARM	BENEFIT	UNKNOWN	UNKNOWABLE
Profession	The profession or professionals may appear too "in the weeds" if it focuses on sensitivity analyses that were never part of the original system/method's development.	The full documentation and the sensitivity analyses are each performing as intended, showing their importance to those in and outside of the profession.		
Public/public trust	Failure to correct method errors and decisions based on the old (wrong) method may undermine public trust in the profession or organization.	Sensitivity analysis (if not the system itself) is functioning as intended. Addressing the results means the system will ultimately be less biased, or if the bias cannot be corrected, the system may not be deployed.		

Table 3.6.4. *Stakeholder Analysis: Documentation helped to identify errors in a method.*

2. Identify decision-making frameworks.

ASA GLs support a virtue approach, which supports fixing the errors, because the ethical practitioner "uses methodology and data that are valid, relevant, and appropriate, without favoritism or prejudice, and in a manner intended to produce valid, interpretable, and reproducible results" (ASA A2). This is true even for material/work that is done entirely within and for an organization. The ethical statistical practitioner "strives to promptly correct substantive errors discovered after publication or implementation. As appropriate, disseminates the correction publicly and/or to others relying on the results." (ASA B5) – ASA B5 applies even if it is not expedient ("The ethical statistical practitioner avoids compromising scientific validity for expediency." ASA E4). Note that ASA E4 applies even if the work is not scientific or in support of science. In this case, ASA Principles A, C, and E apply to whatever organization-wide decisions might be necessary to address whatever errors in decisions this system has created. Meanwhile Principle B5 applies *to you*.

In the ACM CE utilitarian perspective, the computing professional has a responsibility to analyze the work they do/are asked to do so that decisions made on the basis of the system or its subparts are *justifiable*. If an error is identified in the method, results from that system may be biased or incorrect, potentially causing predictable (once the error is identified) but fixable harms. More specifically, "1.3 A computing professional should be transparent and provide full disclosure of all pertinent system capabilities, limitations, and potential problems to the appropriate parties." In this case, the CE Principle applies *to you* since your assessment of the results is how the errors were identified.

3. Identify or recognize the ethical issue.

Both the ASA and ACM standards support the identification and fixing of the error; however, the most important part of their guidance pertains *to you*: this is different from the previous case. Both standards specify that the error (ASA) and limitations of the system (ACM) should be disclosed. As articulated, this case simply states that you found an error: *an ethical issue only arises if **you** do not notify any stakeholders who rely on the method its results*. Obviously, if your organization chooses not to fix the errors, you may not be able to fix them yourself; you may also not be able to authorize someone else to fix the method so that everything can be redone. There may also be significant costs associated with redoing all analyses that involved this method. Rerunning everything will take time and may involve many others/others' time, effort, and resources as well. Thus, there might be some incentives for your organization to *not fix the error*. If you are aware of this – that there are at least two errors, the results cannot be validly interpreted, and the organization chooses not to fix the errors or re-run the projects/analyses that used the method, or to fix the errors but not re-run all analyses, *you* have a responsibility to notify relevant stakeholders. Both the GLs and CE describe the ethical practitioners' responsibilities to do that notification, however, so ethical issues will arise if the organization does fix the errors but does *not* notify stakeholders.

4. Identify and evaluate alternative actions (on the ethical issue).

The three decisions that can be made in any circumstance are: a) do nothing. b) consult or confer with a peer (ASA) or a supervisor (ACM); and c) report violations of policy, procedure, ethical guidelines, or law (ASA) or refuse to implement the system (ACM). In this case, however, only two options are viable because conferring with your supervisor or a peer (option b) will be tantamount to "reporting" (option c) that errors in the method were found.

Note that the case simply describes you identifying errors in an existing method. Thus, option a) must be modified to "notify all stakeholders of the errors and potential need for them to rerun the method on their data (or whatever they used it for), but do not confer"; and the only other option, b), becomes, "confer with peers or supervisors about how best to notify all stakeholders of the errors and their needs to rerun the method".

As noted, there is only an ethical issue if you do not notify all stakeholders that there are at least two errors. Thus, if your supervisor fixes and reports the errors, and *their* supervisor(s) or the organization then seeks to suppress this information – particularly, not notifying clients or other stakeholders who rely on the results, then option c) ("report the violation of CE, GLs, and relevant laws") *is* differentiated from option b) and should be considered. Note that, while you – and perhaps your supervisor - may have chosen to honour your commitments to practice ethically, and notify your organization and other stakeholders, rerunning this method may entail costs and resources that cannot be renewed. Again, there are some incentives for the organization either not to fix the error, or to fix it but not notify everyone. Your primary responsibility is to ensure that *you* practice ethically; you cannot cause the organization to do so. So, alternative responses were identified for *you* to take so that you can ethically respond to your discovery of the error, and you would choose a) or b). Your organization may respond to either option by not choosing to endorse or support your ethical practice. *If this constitutes fraud* – because the organization does not want to incur the cost of fixing the method's errors and redoing any project (National Football League denying impacts cause concussions[30]), for example - then you need to consider whistleblowing (again).

5. Make and justify a decision.

Decision: *option b*: "confer with peers or supervisors about how best to notify all stakeholders of the error and need to rerun the method" is supported by both ASA and ACM standards. Because the case describes a "long project", there very well may be multiple stakeholders and others to notify (or confer with, or both). Note that, once you have identified an error in code you are using, if you continue to use that code without correcting it (even though this may be easier and cheaper), you are violating multiple practice standards – not being transparent or honest in your reporting (unless you also include the error in your report); not using the most appropriate methods; and not ensuring that your work is robust and that you are accountable for it. If others choose to continue using code in which you have identified an error, and they purposely ignore that error, *they* are violating multiple practice standards. You would have to decide if you wanted to work – and be associated with – that type of workplace and context.

6. Reflect on the decision.

Construct your own reflection on this decision.

---

[30] Read about how the NFL suppressed reports of how impacts from football causes concussions to avoid paying health care costs and potentially incur new liabilities: https://www.pbs.org/wgbh/pages/frontline/sports/league-of-denial/timeline-the-nfls-concussion-crisis/

Case 31. You are given documentation that is not complete: it lacks details about exactly what methods and in what order were used.

1. Identify and 'quantify' prerequisite knowledge:

Which DEFW and DSEC Principles or Areas seem most relevant to this vignette?

DEFW: Principle 6 has been violated so you cannot fulfill Principle 5 (or 6).

DSEC: contrary to DSEC C1 or C2, will prevent C3. May perpetuate DSEC D1 and D2 problems. DSEC D4 and D5 will be incorrect. Publication of this incomplete documentation will mean that DSEC E2 will be "no" and there will be no way for DSEC C3, C5, D4, or E3 to be or become "yes".

Which ASA and ACM Principles seem most relevant to this vignette?

ASA:

Principle A. Professional integrity and accountability
Principle B. Integrity of data and methods
Principle C. Responsibilities to stakeholders
Principle E. Responsibilities to members of multidisciplinary teams
Principle F. Responsibilities to fellow statistical practitioners and the Profession
Principle G. Responsibilities of leaders, supervisors, and mentors in statistical practice
Appendix. Responsibilities of organizations/institutions

ACM:

2. Professional Responsibilities

Potential result: Stakeholder:	HARM	BENEFIT	UNKNOWN	UNKNOWABLE
YOU	Takes time, adds effort (to fully document plans/ evaluations/ systems); minimal and potentially incorrect documentation will not support review or replication.	There is no benefit to using incomplete and potentially incorrect documentation.	ACM: New harms/risks may not be detectable or addressed with minimal and potentially incorrect documentation. ASA: Replications are not facilitated with minimal and potentially incorrect documentation.	
Your boss/client	Takes time; may require additional time/resources. The lack of complete documentation may suggest that employees are not very good at their jobs.			

Potential result: Stakeholder:	HARM	BENEFIT	UNKNOWN	UNKNOWABLE
Unknown individuals	Minimal and potentially incorrect documentation will not support review or replication.			
Employer	Takes time; may require additional time/resources. Future evaluation or use of poorly documented work will be hindered.		ACM: New harms/risks may not be detectable or addressed with minimal documentation. ASA: Replications may not be facilitated with minimal documentation.	
Colleagues	Incomplete and potentially incorrect documentation may mislead colleagues into thinking the work itself is also incomplete; review and identification of "what is missing" could take extra time and effort by others.			

Potential result: Stakeholder:	HARM	BENEFIT	UNKNOWN	UNKNOWABLE
Profession	Minimal and potentially incorrect documentation will not support review or replication. Failure to document work fully may signal a lack of respect for other professions or others on the team.		ACM: New harms/risks may not be detectable or addressed with minimal documentation. ASA: Replications are not facilitated with minimal and potentially incorrect documentation.	
Public/public trust	Without full and correct documentation of work/thought processes, public trust in systems or their results, and the profession (ACM/ASA) are compromised. Perpetuating black box mentality does not help, and explicitly does not inform, society.			

Table 3.6.5. *Stakeholder Analysis template: Documentation is methodologically incomplete*

2. Identify decision-making frameworks.

ASA GLs describe the ethical statistical practitioner as fully documenting their work so as to be transparent and to promote reproducibility in all of their work. As a virtue approach, this suggests that *full* documentation is the ideal, and essential to ethical practice. In this case the documentation is incomplete and may be incorrect (since the order of methods is not specified). This makes it impossible for the practitioner to believe or rely on any output of such a system or analysis. Utilizing the results from a system or method for which the documentation is incomplete and may be incorrect would violate ASA A2, (the ethical practitioner "Uses methodology and data that are valid, relevant, and appropriate, without favoritism or prejudice, and in a manner intended to produce valid, interpretable, and reproducible results.)" Incomplete and possibly incorrect makes it difficult to follow Principle A ("taking responsibility for one's work") and could make Principle B impossible to meet ("The ethical statistical practitioner seeks to understand and mitigate known or suspected limitations, defects, or biases in the data or methods and communicates potential impacts on the interpretation, conclusions, recommendations, decisions, or other results of statistical practices.") Responsibilities to stakeholders are impossible to carry out (e.g., "informs stakeholders of the potential limitations on use and re-use of statistical practices in different contexts and offers guidance and alternatives, where appropriate, about scope, cost, and precision considerations that affect the utility of the statistical practice." (ASA C4), "Explains any expected adverse consequences from failing to follow through on an agreed-upon sampling or analytic plan." (ASA C5), and "Strives to make new methodological knowledge widely available to provide benefits to society at large. Presents relevant findings, when possible, to advance public knowledge." ASA C6). Principle D would not be possible ("The ethical statistical practitioner does not misuse or condone the misuse of data. They protect and respect the rights and interests of human and animal subjects. These responsibilities extend to those who will be directly affected by statistical practices.") The vignette is contrary to ASA E3 ("Ensures that all communications regarding statistical practices are consistent with these Guidelines. Promotes transparency in all statistical practices.") as well as ASA F4 ("Promotes reproducibility and replication, whether results are "significant" or not, by sharing data, methods, and documentation to the extent possible.") Leaders and workplaces that allow incomplete and potentially incorrect documentation violate ASA G1 and Appendix Item 1, both of which relate to promoting a culture that prioritizes and follows the Guidelines because the weaknesses of the documentation in this case undermines the ethical practitioner's ability to follow most of the ASA GLs.

The SHA supports transparent and *complete* documentation of your work as consistent with the utilitarian perspective of the ACM CE: while there are harms to every stakeholder, there are no benefits whatsoever in the SHA. The lack of sufficient documentation makes it impossible to obtain peer review (2.4) and also impossible to fully and honestly evaluate a system (2.5).

3. Identify or recognize the ethical issue.

While *the failure to document fully and transparently is unethical*, that describes the persons responsible for documentation, not you (in the vignette). In fact, this case does not describe anything unethical by the practitioner who received this incomplete documentation. An ethical issue would arise if a practitioner tried to use a method without complete documentation and then make any kind of decision or inference based on results from such a system/method.

4. Identify and evaluate alternative actions (on the ethical issue).

Although we examine the same three decisions that can be made in any circumstance: a) do nothing b) consult or confer with a peer (ASA) or a supervisor (ACM); or c) report violations of policy, procedure, ethical guidelines, or law (ASA) or refuse to implement the system (ACM), none of these fit this vignette. Instead, option a) becomes "do not use the system/methods until documentation is complete and correct and notify any stakeholders that their results from this method cannot be characterized as robust, interpretable, or valid"; and option b) becomes "confer with a peer or supervisor about how to notify any stakeholders that their results from this method cannot be characterized as robust, interpretable, or valid and prevent anyone from using the system/methods until documentation is complete and correct". These are the only two ethical responses to this vignette; using the method or approach that is so poorly documented, or continuing with the project would mean failure on much of the ASA Guidelines for the individual and the leadership and workplace.

Option c) needs to change to "do not use the system/methods until documentation is complete and correct, *and* notify any stakeholders that their results from this method cannot be characterized as robust, interpretable, or valid; and report violations of policy, procedure, ethical guidelines, or law to a peer (ASA) or a supervisor (ACM) if anyone seeks to prevent you from stopping the use of the system or notifying stakeholders of its weaknesses/vulnerabilities."

5. Make and justify a decision.

Decision: As indicated, all options include "do not use the system/methods until documentation is complete and correct and notify any stakeholders that their results from this method cannot be characterized as robust, interpretable, or valid". While option a) is clearly consistent with practice standards, option b) is preferred. Since options b) and c) only differ "if anyone seeks to prevent you from stopping the use of the system or notifying stakeholders of its weaknesses/vulnerabilities", we simply hope that does not happen but plan a response in case it does.

6. Reflect on the decision.

Construct your own reflection on this decision.

Case 32. You provide complete and correct documentation, and this gets "edited" by a supervisor so that it is now no longer complete or correct.

1. Identify and 'quantify' prerequisite knowledge:

Which DEFW and DSEC Principles or Areas seem most relevant to this vignette?

DEFW: *You* fulfilled Principle 6, but the documentation no longer meets that standard.

DSEC: This vignette describes dishonest representation (DSEC C3) and an unintended use (DSEC E4) of your documentation. The editing will mean that DSEC E2 will be "no" and there will be no way to ensure DSEC C3, C5, D4, or E3 are or can become "yes".

Neither DEFW nor DSEC discusses how to *respond*.

Which ASA and ACM Principles seem most relevant to this vignette?

ASA:

Principle A. Professional integrity and accountability
Principle B. Integrity of data and methods
Principle C. Responsibilities to stakeholders
Principle E. Responsibilities to members of multidisciplinary teams
Principle F. Responsibilities to fellow statistical practitioners and the Profession
Principle G. Responsibilities of leaders, supervisors, and mentors in statistical practice
Principle H. Responsibilities regarding potential misconduct

ACM:

1. General Ethical Principles
2. Professional Responsibilities

Potential result: Stakeholder:	HARM	BENEFIT	UNKNOWN	UNKNOWABLE
YOU	Incorrect and incomplete documentation means the work can no longer be used for valid or interpretable results. Your effort on the documentation has been wasted, *and* now neither the documentation nor what it describes can be used with confidence. Failing to correct the documentation may save time but *will* lead to unpredictable propagation of error throughout the workflow and into the community.	Correction of the misuse/editing means future work will be correct, reproducible, and interpretable.	Perpetuation of false reports and misinterpretation undermine ethical practice as well as any work based on the misleading documentation, while also perpetuating deceptive and detrimental research practices.	Deceptive and detrimental research practices like cherry-picking (what the editing in this case represents) undermine science, the public trust in the scientific enterprise, and the entire scientific community.
Your boss/client	Incorrect and incomplete documentation means the work can no longer be used for valid or interpretable results. All effort on the documentation has been wasted, *and* now neither the documentation nor what it describes can be used with confidence.	Correction of the misuse/editing means future work will be correct, reproducible, and interpretable.	Correction of this misuse of the documentation will require resources to fix/address. *Appropriate* documentation may not yield desired results.	

Potential result: Stakeholder:	HARM	BENEFIT	UNKNOWN	UNKNOWABLE
Unknown individuals	Incorrect interpretation undermines reproducibility and rigor. Misleading documentation negatively affects all stakeholders in the long run.	Correction of the error means future work will be correct, and valid.	Competent interpretation (ASA) supports reproducible results; effective contextualization of your work (ACM) promotes public good. Perpetuation of false reports and misinterpretation undermine both while also perpetuating detrimental research practices.	Deceptive and detrimental research practices like cherry-picking (what the editing in this case represents) undermine science, the public trust in the scientific enterprise, and the entire scientific community.

Potential result: Stakeholder:	HARM	BENEFIT	UNKNOWN	UNKNOWABLE
Employer	Incorrect and incomplete documentation means the work can no longer be used for valid or interpretable results. All effort on the documentation has been wasted, *and* now neither the documentation nor what it describes can be used with confidence. The incorrect documentation may be discovered to have been used by a client, or by other clients who paid for honest work, identifying past errors and undermining this and other projects	Correction of the documentation errors means future work with the product it describes (if not the employee who cherry-picked the documentation) may be facilitated –and correct, and valid.	Correction of the documentation errors may require resources to fix/address. *Appropriate* analysis –once the error is fixed - may not yield desired results.	Correction of the documentation errors may highlight past payments or purchases/sales of incorrect/error-prone work. Liability may be incurred.

Potential result: Stakeholder:	HARM	BENEFIT	UNKNOWN	UNKNOWABLE
Colleagues	Having to respond to this misuse of your documentation may complicate colleagues' work. Pointing out and correcting a team member's deceptive and detrimental research practices can be discomfiting – and may also highlight other detrimental practices that will then invite more scrutiny.	Identification and correction of the cherry-picking allows mitigation of risks of harms (ACM); supports valid inferences (ASA), and can limit the unpredictable propagation of the error to other parts of the workflow. Other colleagues may be emboldened to call out – and stop – other deceptive and detrimental practices they observe.	Competent interpretation (ASA) supports reproducible results; effective contextualization of your work (ACM) promotes public good. Perpetuation of false reports and misinterpretation undermine both while also perpetuating detrimental research practices.	Deceptive and detrimental research practices like cherry-picking (what the editing in this case represents) undermine science, the public trust in the scientific enterprise, and the entire scientific community.

Potential result: Stakeholder:	HARM	BENEFIT	UNKNOWN	UNKNOWABLE
Profession	Incorrect and incomplete documentation means the work can no longer be used for valid or interpretable results. If the documentation is shared, this will create invalidity in others' work. Failing to correct the documentation may save time but *will* lead to unpredictable propagation of error throughout the workflow and into the community.	Correction of the misuse/editing means future work will be correct, reproducible, and interpretable. Identification of the deceptive and detrimental research practice of cherry-picking (what the editing in this case represents) may lead to more in the profession preventing it, and fewer people trying it.	Perpetuation of false reports and cherry-picking undermine ethical practice as well as any work based on the misleading documentation, while also perpetuating deceptive and detrimental research practices.	Deceptive and detrimental research practices like cherry-picking (what the editing in this case represents) undermine science, the public trust in the scientific enterprise, and the entire scientific community.

Potential result: Stakeholder:	HARM	BENEFIT	UNKNOWN	UNKNOWABLE
Public/public trust	Deceptive and detrimental practices like cherry-picking undermine public trust in reproducibility and rigor. An insistence (by the editor) on cherry-picked documentation being "correct" –resistance to its correction - undermines public trust as well as reproducibility and rigor.	Identification and correction of the misleading documentation allows mitigation of risks of harms (ACM); supports valid inferences (ASA), and can limit the unpredictable propagation of the error to other parts of the workflow and community, to and decisions based on them.	Competent interpretation (ASA) represents reproducible results; effective contextualization of your work (ACM) promotes public good. Perpetuation of false reports and misrepresentation in documentation undermine both while also perpetuating deceptive and detrimental research practices.	Deceptive and detrimental research practices like cherry-picking (what the editing in this case represents) undermine science, the public trust in the scientific enterprise, and the entire scientific community.

Table 3.6.6. *Stakeholder Analysis: Your documentation was correct and was edited to be incomplete and incorrect.*

2. Identify decision-making frameworks.

Like Cases 25 and 26, this vignette describes a highly prevalent, deceptive, and extremely detrimental research practice, **cherry-picking**[31]. Cherry-picking, a clearly deceptive, as well as detrimental, practice, is *obviously not ethical*. However, in this vignette, you took the time and effort to fully and transparently document your work – and someone else (possibly more than one person) is ignoring, possibly suppressing, the honest representation of your work. Because the editing that was done renders the documentation both incomplete and incorrect in this vignette, it should be resisted, and corrected, by an ethical practitioner.

In the last case (Case 31) it was not clear if the incomplete documentation was also incorrect. In this case, it is clearly both. Since the documentation that you originally produced was correct, you or your name may be associated with the documentation at this point, even though someone else modified it - possibly to mislead. The failure to respect your original work could even represent statistical and scientific ("falsification"[32]), if not also professional, misconduct.

Incomplete and incorrect makes it difficult for you to follow ASA Principle A ("taking responsibility for one's work") but makes the rest of the Guidelines challenging for others. As we saw in Case 31, incomplete and incorrect documentation make ASA Principle B impossible to meet ("The ethical statistical practitioner seeks to understand and mitigate known or suspected limitations, defects, or biases in the data or methods and communicates potential impacts on the interpretation, conclusions, recommendations, decisions, or other results of statistical practices.") Responsibilities to stakeholders are impossible to carry out (e.g., "informs stakeholders of the potential limitations on use and re-use of statistical practices in different contexts and offers guidance and alternatives, where appropriate, about scope, cost, and precision considerations that affect the utility of the statistical practice." (ASA C4), "Explains any expected adverse consequences from failing to follow through on an agreed-upon sampling or analytic plan." (ASA C5), and "Strives to make new methodological knowledge widely available to provide benefits to society at large. Presents relevant findings, when possible, to advance public knowledge." (ASA C6). Principle D would not be possible ("The ethical statistical practitioner does not misuse or condone the misuse of data. They protect and respect the rights and interests of human and animal

---

[31] https://en.wikipedia.org/wiki/Cherry_picking
[32] https://ori.hhs.gov/definition-misconduct

subjects. These responsibilities extend to those who will be directly affected by statistical practices.")

The vignette describes a situation (and someone's behavior) that is contrary to ASA E3 ("Ensures that all communications regarding statistical practices are consistent with these Guidelines. Promotes transparency in all statistical practices.") as well as ASA F4 ("Promotes reproducibility and replication, whether results are "significant" or not, by sharing data, methods, and documentation to the extent possible.") The problem is, it was someone else's activities (editing your documentation badly) that violates ASA E3 and ASA F4 – and will cause everyone else to violate those as well. Leaders and workplaces that allow incomplete and incorrect documentation violate ASA G1 and Appendix Item 1, both of which relate to promoting a culture that prioritizes and follows the Guidelines because the weaknesses of the documentation in this case undermines the ethical practitioner's ability to follow most of the ASA GLs. Even more worrisome is whether the workplace encourages the editing that was done to create the incomplete and incorrect record of your work.

Although the ACM CE does not mention cherry-picking or falsification, "A computing professional should be transparent and provide full disclosure of all pertinent system capabilities, *limitations, and potential problems* to the appropriate parties." (1.3, emphasis added). It also specifies that "professionals should be cognizant of any serious negative consequences affecting any stakeholder that may result from poor quality work and should resist inducements to neglect this responsibility" (2.1). In this case, documentation that would support reproducible and interpretable, valid results, has been modified and the new (edited) documentation cannot be said to meet these standards - and *that* needs to be transparently reported. As noted in other cases, detrimental practices need to be identified and discouraged at every possible opportunity.

In the vignette, the documentation was complete and correct, but someone changed it so now it is incorrect and incomplete. This means that whatever comes out of the method or system the documentation describes will not function validly or reproducibly. This is a predictable limitation that will invariably cause harm (so should be avoided, according to ACM CE)– even if that harm is only to the scientific record. However, in *and outside of* contexts where science is being done, the ethical statistical practitioner "avoids compromising scientific validity for expediency" (ASA E4).

3. Identify or recognize the ethical issue.

As this vignette is presented, the practitioner has applied their expertise to document their work appropriately, but someone else has edited this work so that it is no longer fit for purpose. There is no way for the ethical practitioner to allow the misrepresentation of their work. Both the ASA and ACM standards support reporting of the error, with both the ASA GLs and the ACM CE specifying that the errors (ASA) and limitations of the system (ACM, i.e., that it cannot yield valid results) should be disclosed. Note that in this case, while the inappropriate editing is unethical, that is *not* the practitioner's behavior. The ethical practitioner must respond – ethically – to either correct the inappropriate editing or notify stakeholders, preferably both. That is, an ethical issue arises if you *fail to notify* all stakeholders of the cherry-picking. This failure to notify is unethical because the notification can prevent inappropriate use of the work (that was poorly documented), false and invalid reporting based on the incorrect use of a misleadingly documented system or method, and any decisions based on the incorrect documentation or the method/system it describes. While some people simply do not understand computing and statistical principles sufficiently to understand the how essential correct and complete documentation is to the reproducibility and validity of results, and do not cherry-pick in their documentation or reporting because they are dishonest, it does not make cherry-picking ethical. The ethical practitioner has a responsibility to call attention to cherry-picking as the unethical and deceptive, detrimental practice that it is, or else they violate ASA H2 (the ethical practitioner "avoids condoning or appearing to condone statistical, scientific, or professional misconduct. Encourages other practitioners to avoid misconduct or the appearance of misconduct"), and failing in the virtue perspective.

Both the ACM (Principle 4) and ASA (Purpose, Principles A & G) outline specific responsibilities of the ethical practitioner to follow the practice standards *and encourage others to do so as well*. It does not always work to change cherry-picking behavior, but the ethical practitioner will *do what they can to discourage cherry-picking and any other detrimental practices*. Documenting for this supervisor that their cherry-picking is unsupportable does not prevent them from continuing this practice. Thus, to the extent possible given your role (and job security), an ethical issue for the practitioner also arises when they do not actively seek to discourage this *deceptive and detrimental* behavior. The harms that accrue when incorrect and incomplete documentation are permitted – especially when this is the result of possibly intentional misuse of appropriate documentation – violates ACM 1.3 and 2.1, and fails in the utilitarian perspective.

4. Identify and evaluate alternative actions (on the ethical issue).

The three decisions that we discuss for any circumstance are: a) do nothing. b) consult or confer with a peer (ASA) or a supervisor (ACM); and c) report violations of policy, procedure, ethical guidelines, or law (ASA) or refuse to implement the system (ACM).

As usual, option a) must be modified: the practitioner *was* ethical in their original documentation but to ignore the editing, and "do nothing", is clearly in violation of ethical practice standards. Option a) should change to "correct the erroneous documentation, notify all stakeholders that the edited documentation is incorrect and incomplete, and act to prevent the use or further publication of the false documentation." Option b) becomes, "confer with peers or supervisors about how best to correct the erroneous documentation, notify all stakeholders that the edited documentation is incorrect and incomplete, and act to prevent the use or further publication of the false documentation, *and* ensure that cherry-picking and other detrimental practices (including suppression of evidence of cherry-picking or spurious results) are recognized as unethical and not tolerated".

As discussed in Cases 25 and 26, the 2017 National Academies of Sciences and Engineering and the Institute of Medicine (NASEM) report, *"Fostering Integrity",* called on the scientific community in the US to identify and act to prevent such deceptive and detrimental practices. In this vignette, to the extent that anyone tries to prevent notification of stakeholders of the misleading edits that prevent clients or other stakeholders, who rely on the results, from using the method or system to generate valid results, then option c) ("report the violation of CE, GLs, and relevant laws") is a viable option. You have effectively demonstrated that the edited documentation is cherry-picked – and is now false and deceptive; if a supervisor or the organization wants to suppress this information, it *demonstrates their intent to deceive* stakeholders. While cherry-picking is unethical but not illegal, deception and withholding evidence *is* illegal – particularly when there are financial conflicts of interest[33]in play. If an organization or supervisor seeks to suppress the correction of misleading and incorrect documentation, it may be perpetrating fraud[34] – and both your commitment to ethical practice and your civic duty will require that you report this. Then option c) effectively becomes, "correct the erroneous documentation, *and* notify all stakeholders that the edited documentation is incorrect and incomplete, *and* act to prevent the use or further publication of

---

[33] https://ori.hhs.gov/education/products/ucla/chapter4/default.htm
[34] https://en.wikipedia.org/wiki/Fraud

the false documentation, *and* ensure that cherry-picking and other detrimental practices (including suppression of evidence of cherry-picking or spurious results) are recognized as unethical and not tolerated, *and* report the cherry-picking and suppression of its existence to the appropriate authorities (and to the ACM)." Note that the ACM does, but the ASA does not (as of May 2022), have a reporting mechanism as well as a specifically articulated responsibility to report violations of the practice standards.

5. Make and justify a decision.

Decision: All options include "correct the erroneous documentation, notify all stakeholders that the edited documentation is incorrect and incomplete, and act to prevent the use or further publication of the false documentation". This is the only ethical response to the unethical, deceptive, and detrimental behavior in this vignette, although how it is done – or what else is also done– may depend on other features of your specific context. *Options b and c* include sharing (i.e., reporting, option c) or conferring (option b) with others. While option c), reporting, may only be relevant if your efforts to notify stakeholders of the erroneousness of the documentation are resisted, it may constitute some "reporting" when you share the cherry-picking behavior of the supervisor with other stakeholders in option b). Also, consulting a peer or supervisor as to how best to proceed (option b) may be more relevant for a more junior practitioner. As you grow in experience and seniority, you may be able to use the conferring to give some careful thought given to why no one else on the team objected to the reporting of cherry-picked results. There may be a culture in your team or organization that either tolerates or encourages detrimental practices, and the ethical practitioner is obliged to do what they can change this culture; "what you can do to change the culture from one that supports or even encourages detrimental practices to an ethical one" will naturally change with seniority and experience – but it will only become more important over time and will never be less important! When an otherwise ethical practitioner fails to stop others' unethical practice, they would then violate ASA H2 and "condone, or appear to condone", these unethical practices – meaning that a team or organizational culture or context where you are encouraged or directed to violate your practice guidelines is an *unethical* culture. Option b) is preferred over option a) because the notification aspect of the solution may require organization-level review of policies or incentives that exist for reporting unethical modifications to documentation that will lead to results that are invalid, or that cannot be replicated. Option b) will also be more appropriate to initiate culture change, and to lead those who are ignorant that their behavior is unethical (e.g., if their practice guidelines do not decry deceptive and detrimental practices) to the understanding that *cherry-picking is unethical.*

However, option c) may be the decision that has the greatest impact, preventing frank deception from propagating from your work, and also promoting ethical practice.

6. Reflect on the decision.

Construct your own reflection on this decision.

Case 33. The documentation you receive specifies an analysis method that is not appropriate for the specific question that must be addressed.

1. Identify and 'quantify' prerequisite knowledge:

Which DEFW and DSEC Principles or Areas seem most relevant to this vignette?

This situation is contrary to DEFW Principle 5 and is inconsistent with Principle 6: robust practice (5) requires that methods are appropriate, and 6 requires accountable work. Results of a system or analysis with identified errors conflict with these Principles.

There is no suggestion of what to do about errors in analyses in DSEC. However, the existence of the error(s) means DSEC C3, C5, D4, E3 cannot be addressed. Depending on the nature and impact of the assumption violations on the output of the system, other DSEC items cannot be answered confidently.

Which ASA and ACM Principles seem most relevant to this vignette?

ASA:

Principle A. Professional integrity and accountability
Principle B. Integrity of data and methods
Principle C. Responsibilities to stakeholders
Principle D. Responsibilities to research subjects, data subjects, or those directly affected by statistical practices
Principle E. Responsibilities to members of multidisciplinary teams
Principle F. Responsibilities to fellow statistical practitioners and the Profession

ACM:

1. General Ethical Principles
2. Professional Responsibilities

Potential result: Stakeholder:	HARM	BENEFIT	UNKNOWN	UNKNOWABLE
YOU	Identifying, documenting, and then fixing the error will take time, adds effort; may identify other problems that require additional time/effort. Failing to fix the error may save time but lead to unpredictable propagation of that error throughout the workflow.	Correction of the error means future work with the code will be facilitated –and correct, and interpretable.	Competent analysis (ASA) supports reproducible results; effective analysis (ACM) promotes public good.	
Your boss/client	Identifying, documenting, and then fixing the error will take time, adds effort; may identify other problems that require additional time/effort. This error may be discovered in a 'product' that the client has paid for in the past, identifying past errors and undermining this and other projects	Correction of the error means future work with the code will be facilitated –and correct, and interpretable.	Identification of error may require resources to fix/address. *Appropriate* analysis – once the error is fixed - may not yield desired results.	

Potential result:	HARM	BENEFIT	UNKNOWN	UNKNOWABLE
Stakeholder:				
Unknown individuals	Analysis required or ACM CE violated; weak statistical analysis undermines reproducibility and rigor.	Identification and correction of the error allows mitigation of risks of harms (ACM); supports valid inferences (ASA), and can limit the unpredictable propagation of the error to other parts of the workflow.		
Employer	Identifying, documenting, and then fixing the error will take time, adds effort; may identify other problems that require additional time/effort. This error may be discovered in a 'product' that the client has paid for in the past, identifying past errors and undermining this and other projects	Correction of the error means future work with the code will be facilitated –and correct, and interpretable.	Identification of error may require resources to fix/address. *Appropriate* analysis – once the error is fixed - may not yield desired results.	Identification of the error may highlight past payments or purchases/sales of incorrect/error-prone work. Liability may be incurred.

Potential result: Stakeholder:	HARM	BENEFIT	UNKNOWN	UNKNOWABLE
Colleagues	The discovered error may complicate colleagues' work.	Identification and correction of the error allows mitigation of risks of harms (ACM); supports valid inferences (ASA), and can limit the unpredictable propagation of the error to other parts of the workflow.		Project could be delayed until risks of additional errors, and resulting harms/ potential harms are addressed.
Profession	Identifying, documenting, and then fixing the error will take time, adds effort; may identify other problems that require additional time/effort. Failing to fix the error may save time but lead to unpredictable propagation of that error throughout the workflow.	Correction of the error means future work with the code will be facilitated –and correct, and interpretable. This supports ethical and competent practice by future practitioners.	Competent analysis (ASA) supports reproducible results; effective analysis (ACM) promotes public good.	Commitment to identification and correction of the error, consistent with GL/CE, might strengthen trust in profession. The error itself can enable innovation (to fix, avoid, or detect similar errors in future), as well as conversations about ethical practice.

Potential result:	HARM	BENEFIT	UNKNOWN	UNKNOWABLE
Stakeholder: Public/public trust	Even if errors are noted *so they can be fixed*, weak or error-prone analysis undermines public trust in reproducibility and rigor.	Identification and correction of the error allows mitigation of risks of harms (ACM); supports valid inferences (ASA), and can limit the unpredictable propagation of the error to other parts of the workflow and decisions based on them.	Transparency supports the public trust, even if risk/harms are identified; can tend to inform the public about limitations inherent in the profession.	

Table 3.6.7. *Stakeholder Analysis: The documentation specifies a program that uses inappropriate methods*

2. Identify decision-making frameworks.

As noted, ASA GLs tend to support a virtue approach, which supports fixing this problem so that the method is appropriate for the data and the objective. The vignette does not specify if the analysis method has already been run, so "fixing" the problem might entail correcting the documentation (but nothing else, if the method has not been applied yet), or it might mean correcting reports that are based on incorrect methods that have been correctly documented. Ensuring that the documentation is actually correct, if the method has already been run, must be the first objective to determine what else needs to be done.

Even if an analysis has been done and yielded results, it does not mean that those results are correct. So, just as in Cases 16 and 17, the fact that the method is inappropriate constitutes an error in that code. Recall that the ethical statistical practitioner "uses *methodology and data that are valid, relevant, and appropriate*, without favoritism or prejudice, and in a manner intended to produce valid, interpretable, and reproducible results." (ASA A2, emphasis added). That was not done in this case, according to the documentation – which itself *may be* correct! The ethical statistical practitioner "Avoids compromising scientific validity for expediency" (ASA E4), so although it would be more expedient to let the method be employed even though it is inappropriate for the data and the question, Principle B dictates that the ethical statistical practitioner "strives to promptly correct substantive errors discovered after publication or implementation. As appropriate, disseminates the correction publicly and/or to others relying on the results." (ASA B5). Note that the **use** of the inappropriate method may have been an honest error (or may have been the best option at the time the method was created), or it may have been purposefully utilized to ensure that prior or desired results were obtained. In this vignette, we do not know why the method is incorrect, but since that is what the documentation describes, the challenge here is to respond to this existing, incorrect, methodology. The virtue perspective is unequivocal that a response must be made. Responsibilities to stakeholders (ASA Principle C), data donors and those affected by statistical practice (ASA Principle D), others on the team (ASA Principle E) and other practitioners (ASA Principle F) all accrue to the individual who recognizes the error in the documentation.

The ACM CE articulates that the computing professional has a responsibility to analyze the work they do/are asked to do so that decisions made on the basis of the system or its subparts are *justifiable*. Clearly, inappropriate methods can invalidate any decisions based on results that they generate. However, a system that was designed according to inappropriate methods (or assumptions) will yield results that are (mathematically or socially) biased or simply incorrect. Importantly, the violation of assumptions means that it isn't a simple error in the code or system that needs fixing, but rather, possibly different methods altogether. These correct methods may need to be run and stakeholders notified. Unsupported or frankly violated assumptions can cause predictable (once the error is identified) but fixable harms. Moreover, if you have identified the fact that assumptions are violated, your work must then address those/the fix to comply with ACM 1.3, "a computing professional should be transparent and provide full disclosure of all pertinent system capabilities, limitations, and potential problems to the appropriate parties."

3. Identify or recognize the ethical issue.

Both the ASA and ACM standards support the identification and fixing of the code (or system) so that appropriate methods are used, and assumptions are met – and then, the documentation will be corrected and reflect the right design. As with any error, the ASA GLs stipulate that corrections should be disseminated, and the ACM CE specify similarly that limitations of the system should be disclosed (unless the error in the documentation/with the method being, or to be, used is fixed, in which it is no longer a system limitation). Since the method that the documentation describes may have been used in (many) prior analyses, there could be longstanding problems that the error you discovered has created. As articulated, this case simply states that you found the error(s): *an ethical issue only arises for you if you do not fix the method/documentation to correct future deployments, and notify any stakeholders who rely on the code or its results up to now*. You are not responsible for the error, only for addressing it and notifying stakeholders.

4. Identify and evaluate alternative actions (on the ethical issue).

The three decisions that can be made in any circumstance are: a) do nothing. b) consult or confer with a peer (ASA) or a supervisor (ACM); and c) report violations of policy, procedure, ethical guidelines, or law (ASA) or refuse to implement the system (ACM). In this case, only two options are viable because conferring with your supervisor or a peer (option b) will be tantamount to "reporting" (option c) that the error(s) in the code were found. As noted, there is only an ethical issue if you do not fix the errors and notify all stakeholders that there was an error (which could have been used to make decisions, or build other systems, in the past). Thus, if you fix and report the error, and your team, supervisor(s), or organization then seek to suppress this information – particularly, not notifying clients or other stakeholders who rely on the results, then option c) ("report the violation of CE, GLs, and relevant laws") *is* differentiated from option b) and becomes important to consider. In this vignette, only two options are viable because conferring with your supervisor or a peer (option b) will be tantamount to "reporting" (option c) that an error in the methods utilized (or planned) was found.

Note that this vignette simply describes you identifying the fact that inappropriate methods were used. Thus, option a) must be modified to "correct the method (and any system that uses it) (and re-run all data that have been analyzed with the incorrect method), notify all stakeholders, but do not confer"; and the only other option, b), becomes, "correct method (and any system that uses it) and re-run all data that have been analyzed with the incorrect method, and confer with peers or supervisors about how best to notify all stakeholders". The difference between options a) and b) is only in the extent of collaboration – since the method is inappropriate, then methods or the data (or both) need to be rethought. The notification of stakeholders will be more complex because of this, but it is still essential to the response in the case. You may need to confer with others to fix the method (and any system that uses it) simply because it is too complex to fix everything that needs fixing by yourself.

As noted, there is only an ethical issue if you do not notify all stakeholders that there is an error. Thus, if your supervisor fixes and reports the error, and *their* supervisor(s) or organization then seek to suppress this information – particularly, not notifying clients or other stakeholders who rely on the results, then option c) ("report the violation of CE, GLs, and relevant laws") *is* differentiated from option b) and should be considered. Note that, while you – and perhaps your supervisor - may have chosen to honor your commitments to practice ethically, and notify your organization and other stakeholders, the "long project" may entail costs and resources that cannot be renewed. Again, there are some incentives for the organization either not to fix the error, or to fix it but not notify everyone. Your primary responsibility is to ensure that *you* practice ethically; you cannot cause the organization to do so. So, alternative responses were identified for *you* to take so that you can ethically respond to your discovery of the error, and you would choose a) or b). Your organization may respond to either option by not choosing to endorse or support your ethical practice. If this constitutes fraud – or the suppression of "uncomfortable data" (political interference [35]), for example, then you need to consider whistleblowing (again).

---

[35] Read about efforts to stop government agencies (in India) from suppressing data or evidence that is "uncomfortable" here:
https://economictimes.indiatimes.com/news/economy/policy/economists-allege-political-interference-in-statistical-data/articleshow/68418232.cms?from=mdr , or here:

5. Make and justify a decision.

Decision: *option b*: "correct the method (and any system that uses it), re-run all data, and confer with peers or supervisors about how best to notify all stakeholders" is supported by both ASA and ACM standards. This is the heart of Professional integrity and accountability (ASA Principle A) and general ethical principles (ACM Principle 1). Consulting a peer or supervisor is not necessarily an essential feature of your response, unless the decision to correct the program (common to both options a and b) requires approval. The difference between options a) and b) is only in the extent of collaboration – since assumptions *are* violated, then methods or the data (or both) need to be rethought. The complexity of the system, your role in the organization or team, or some combination of these will meant that you must (option b) or do not need to (option a) consult or confer with others to make the necessary corrections. The notification of stakeholders will be similarly complicated depending on the extent of the corrections, and conferring or consulting may also help map out the strategy for dissemination of the correction, but both the corrections and the notifications are still essential to the response in the case. Option b) is preferred over option a) in situations where the notification aspect of the solution may very well require approval or supervisory consultation, if the error you discovered has been in place long enough to have featured in actual decisions already having been made. Given the potential for the complexity of revising the code, peers or supervisors may also be important to help ensure that all aspects of the system or analysis do meet all assumptions, and methods are appropriate for the problem and the data. This makes the case for option b) even stronger in this case than the previous one. Just as with the previous case, once you have identified an error in code you are using, if you continue to use that code without correcting it (even though this is easier), you are violating multiple practice standards – not being transparent or honest in your reporting (unless you also include the error in your report); not using the most appropriate methods; and not ensuring that your work is robust and that you are accountable for it.

6. Reflect on the decision.

Construct your own reflection.

---

https://www.huffingtonpost.in/entry/govt-revising-or-suppressing-uncomfortable-data-108-economists-social-scientists-raise-concern_in_5c8b29f8e4b03e83bdbf0cf5

Summary of seven case analyses Documenting your Work

The seven case analyses of Chapter 3.6 were focused on documenting your work. While the ASA GLs and ACM CE provided guidance on formulating and evaluating responses to the ethical issues these vignettes raised, DSEC and DEFW were less (or not) helpful. However, failures to properly document work do make it impossible to address some of the DSEC items, and also represent failures of transparency, which lead to difficulties in accountability (DEFW Principle 6). While those with formal training in the sciences, math and engineering may have developed the necessary habits of maintaining current and correct documentation, those coming to statistics and data science from other disciplines and backgrounds may not have developed them as concretely as they are needed. One aspect of these cases is to underscore how important documentation really is. Although it is time consuming, transparency and accountability can only be achieved with careful and consistent documentation.

Each of the cases in this chapter were based on actual events; while two cases reflect considerations of your own documentation, the others capture the difficulties we create for others in our teams, organizations, and the profession when we do not document fully. Thus, while the ethical practitioner is transparent and accountable, these cases force us to face the ethical challenges that are created when documentation is not consistent, correct, and current. Harms obviously include "requires additional time and effort" on the part of the documenter, with benefits to ourselves being similarly limited to "satisfies requirements for transparency and accountability". However, failures to document fully and correctly can lead to a total inability to utilize data or results that were obtained in prior projects – wasting the time, effort, and resources that those projects required. This effectively wastes resources (including time and effort) that may also include public trust and/or public funds. Effective documentation promotes peer review (required in both ASA GLs and ACM CE), and if errors or improvements are found, these only strengthen our work.

Questions for Discussion:

1. Review at least two cases in this chapter.

   A. Do you agree with the decisions in these case analyses? Why or why not? Which parts of the ER process are most and least acceptable for each case? Do these parts differ for different cases in this chapter? Were there different ASA or ACM Principles or elements that you identified, and if so, how do these affect your case analyses? Are the elements and/or Principles you identify represented in your reflections?

B. Discuss your results for the tests of universality, justice, and publicity for any (or all) case(s) in this chapter.

C. Discuss the relevance to each case – including the analysis and your reflection – of federal, state, or local/organizational policies regarding mentor/mentee responsibilities and relationships. How can you incorporate what you know about these policies into any part of the case analysis (prerequisite knowledge; identifying the ethical issue; determining alternative actions; making or justifying your decision; or reflection)? Note that, if there is no alignment between federal, state, or your local/institutional policies and a given case, you can comment on that (e.g., do policies in your organization focus solely on *misconduct* – and ignore detrimental or other unethical practices?). Do you feel that these policies could have helped in the reasoning, decision or justification?

D. Discuss the relevance to each case - including the analysis and your reflection – of peer review. How can you incorporate what you know about these policies into any part of the case analysis (prerequisite knowledge; identifying the ethical issue; determining alternative actions; making or justifying your decision; or reflection)? If there are no relevant policies, you should comment on that, particularly when a) data are obtained using federal monies (grants); or b) data are collected (acquired) from humans, such that Federal Regulations about the treatment of human subjects and the acquisition of tissues and data from humans are relevant.

2. Review at least two (different) cases in this chapter.

Keeping in mind that doing nothing/not responding are *not plausible responses,* are there other plausible alternatives (KSA 4) that you can think of for any case? If not, discuss that; if so, list them and discuss their evaluation (are they equally consistent with GL/CE, do they lead to similar decisions, etc.).

3. Choose any case in this chapter. Redo the analysis, but feature a different GL or CE principle in your reasoning process. Make sure the *justification* is different from what is given. Is the *decision* also different? Discuss your decision with its justification and the tests of universality, justice, and publicity.

4. Consider all seven cases in this chapter.

A. Comment on the role of the stakeholder analysis in these decisions. Were stakeholder impacts featured in justifications? Were they features in your own reflections?

B. Comment on the importance of your reflections on the decisions you were asked for: for others doing a similar task at work; for those using results from this task; and for those who might join the profession (and be learning about planning and design).

C. Consider your professional identity and the profession itself: planning/designing projects and data collection/analysis projects is obviously foundational in statistics and data science. How can these seven cases and the decisions reached via the analyses presented earn you (the decider) the trust of the practicing community or the public for the profession and its future?

Chapter 3.7
Reporting

Case 34. You discover that incorrect results (yours and/or your team's) are going to be featured in a high-profile publication.

Case 35. You follow SOP and the GLs/CE, and report your methods and results fully, but the final report is cherry-picked, with incorrect methods and results that were "edited" to suit a senior member of the team without your knowledge or agreement.

Case 36. Stakeholders (donors, funders, employers) are given a misleading summary of your methods and results.

Case 37. Your sensitivity analyses that pinpoint the next logical step in your work are omitted from a final report to funders because "we could get a grant to support the team for another 5 years to figure that out!"

Case 38. If you report your method fully and transparently, then you will lose the opportunity to patent it.

Case 39. If you report your method fully and transparently, then a reviewer might notice that you are not the original developer of this method – although the same method was published over 30 years ago and in *another* field.

Case 40. You prepare a report identifying difficulties you encountered in your evaluation (Per ACM CE) of a system your organization wants to deploy. The organization does not have a mechanism for submitting or sharing this report either internally or with stakeholders.

Case 34. You discover that incorrect results (yours and/or your team's) are going to be featured in a high-profile publication.

1. Identify and 'quantify' prerequisite knowledge:

Which DEFW and DSEC Principles or Areas seem most relevant to this vignette?

Not discussed in DSEC or in DEFW.

Which ASA and ACM Principles and elements seem most relevant to this vignette?

ASA:

Principle A. Professional integrity and accountability (A2; A5; A11)
Principle B. Integrity of the data and methods (B; B2; B5)
Principle C. Responsibilities to stakeholders (C4; C6; C8)
Principle D. Responsibilities to research subjects, data subjects, or those directly affected by statistical practices (D; D8)
Principle E. Responsibilities to members of multidisciplinary teams (E3; E4)
Principle F. Responsibilities to fellow statistical practitioners and the Profession (F; F4; F5)
Principle G: Responsibilities of leaders, supervisors, and mentors in statistical practice (G1)
Principle H. Responsibilities regarding potential misconduct (H1; H2)
Appendix. Responsibilities of organizations/institutions (APP1; APP 4; APP 10)

ACM:

1. General Ethical Principles (1.2; 1.3)
2. Professional Responsibilities (2.1; 2.2; 2.4; 2.5)

Potential result:	HARM	BENEFIT	UNKNOWN	UNKNOWABLE
Stakeholder:				
YOU	Fixing the error will create an "erratum" that will permanently be associated with the publication. Failing to fix the error may save time/seem less embarrassing but will lead to unpredictable propagation of that error throughout the community.	Correction of the error means future work using or based on the publication will be facilitated –and correct, and interpretable.	Correct reporting supports reproducible use of/valid inferences based on those reports.	
Your boss/client	Fixing the error will create an "erratum" that will permanently be associated with the publication. Failing to fix the error may save time/seem less embarrassing but will lead to unpredictable propagation of that error throughout the community. This error may be in a 'product' that the client has paid for in the past, so this error may also identify *past* errors and undermine this and other projects.	Correction of the error means future work using or based on the publication will be facilitated –and correct, and interpretable.	While correct reporting supports reproducible use of/valid inferences based on those reports, identification of error may require resources to fix/address. *Appropriate* analysis –once the error is fixed - may not yield desired results.	

Potential result: Stakeholder:	HARM	BENEFIT	UNKNOWN	UNKNOWABLE
Unknown individuals	Erroneous reports of otherwise correct statistical analysis undermine reproducibility and rigor.	Identification and correction of the error allows mitigation of risks of harms (ACM); supports valid inferences (ASA), and can limit the unpredictable propagation of the error to other parts of the workflow and community.		
Employer	Fixing the error will create an "erratum" that will permanently be associated with the publication. Failing to fix the error may save time/seem less embarrassing but will lead to unpredictable propagation of that error throughout the community. This error may be in a 'product' that the client has paid for in the past, so this error may also identify *past* errors and undermine this and other projects.	Correction of the error means future work using or based on the publication will be facilitated – and correct, and interpretable.	While correct reporting supports reproducible use of/valid inferences based on those reports, identification of error may require resources to fix/address. *Appropriate* analysis – once the error is fixed - may not yield desired results.	Identification of the error may highlight past payments or purchases/sales of incorrect/error-prone work. Liability may be incurred.

Potential result: Stakeholder:	HARM	BENEFIT	UNKNOWN	UNKNOWABLE
Colleagues	The discovered error may complicate colleagues' work. Fixing the error will create an "erratum" that will permanently be associated with the publication.	Identification and correction of the error allows mitigation of risks of harms (ACM); supports valid inferences (ASA), and can limit the unpredictable propagation of the error to other parts of the workflow.		Projects that utilize the erroneous result could be delayed until risks associated with this error, and resulting harms/ potential harms, are addressed.
Profession	Failing to fix the error may save time/seem less embarrassing but will lead to unpredictable propagation of that error throughout the community. This error may be in a 'product' that the client has paid for in the past, so this error may also identify *past* errors and undermine this and other projects.	Correction of the error means future work with the results (or method) will be facilitated –and correct, and interpretable. This supports ethical and competent practice by future practitioners.	Competent analysis (ASA) supports reproducible results; effective analysis (ACM) promotes public good.	Commitment to identification and correction of the error, consistent with GL/CE, might strengthen trust in profession. The error itself can enable innovation (to fix, avoid, or detect similar errors in future), as well as conversations about ethical practice.

Potential result:	HARM	BENEFIT	UNKNOWN	UNKNOWABLE
Stakeholder:				
Public/public trust	Even if errors are noted so they can be fixed, weak or error-prone analysis undermines public trust in reproducibility and rigor.	Identification and correction of the error allows mitigation of risks of harms (ACM); supports valid inferences (ASA), and can limit the unpredictable propagation of the error to other uses of the method/results, and decisions based on them.	Transparency supports the public trust, even if risk/harms are identified; can tend to inform the public about limitations inherent in the profession.	

Table 3.7.1. *Stakeholder analysis: You discover an error in a paper about to be published*

2. Identify decision-making frameworks.

As noted, ASA GLs tend to support a virtue approach, and every element of the Guidelines offers input that is relevant to this case. ASA Principles B (Integrity of the data and methods), D (Responsibilities to research subjects, data subjects, or those directly affected by statistical practices) and F (Responsibilities to fellow statistical practitioners and the Profession), as well as elements from those Principles, pertain. This reflects how important honest and transparent reporting is to ethical statistical practice. This case does not specify whether correcting the error will change the interpretation or impact of the results, but *that does not matter*. The ethical practitioner recognizes the multiple ways in which they are responsible for ensuring that the errors in the results are corrected before the publication of the original. Part of the importance of transparent (correct) and honest reporting is that results must be correct and interpretable in a valid way because the ethical practitioner "uses methodology and data that are valid, relevant, and appropriate, without favoritism or prejudice, and in a manner *intended to produce valid, interpretable, and reproducible results*" (ASA A2, emphasis added). Making the corrections in this case avoids "condoning or appearing to condone statistical, scientific, and professional misconduct" (ASA H2) – particularly by assuring readers that you and your team did not intend to use detrimental practices or engage in scientific misconduct. Moreover, the ethical statistical practitioner "takes responsibility for evaluating potential tasks, assessing whether they have (or can attain) sufficient competence to execute each task and that the work and timeline are feasible. Does not solicit or deliver work for which they are not qualified or that they would not be willing to have peer reviewed." (ASA A1). These results were already peer reviewed – and no error was detected by reviewers; it is your current knowledge that the results contain an error that you would not want to take full responsibility for, but if you do not correct and report the error in this publication, your name as an author will endorse the work. Note that these responsibilities to correct this error do not specifically pertain to a scientific publication. A "high-profile publication" could be within-organization, or with a media outlet. The point is, "high profile" means that the publication could impact many individuals, and also that many stakeholders could be affected. In the modern era, once published, even a retraction or correction will not necessarily always accompany the original publication[36].

---

[36] For an excellent example of how damaging it is when incorrect results are published – even when refuted immediately and objectively, you can read about how the "anti-vaccine" movement was started by frank fraud and cherry-picking: Rao TS, Andrade C. The MMR vaccine and autism: Sensation, refutation, retraction, and fraud. Indian J Psychiatry. 2011;53(2):95-96. doi:10.4103/0019-5545.82529

The ethical statistical practitioner "Avoids compromising scientific validity for expediency" (ASA E3), so Principle B dictates that the ethical statistical practitioner: "Strives to promptly correct substantive errors discovered after publication or implementation. As appropriate, disseminates the correction publicly and/or to others relying on the results." (ASA B5). Although there might be incentives to ignore the error (because the peers who reviewed, and your team, didn't recognize it so …), this option is not consistent with the GLs, nor with the virtue perspective.

In the ACM CE utilitarian perspective, the computing professional has a responsibility to analyze the work they do/are asked to do so that decisions made on the basis of the system or its subparts are *justifiable*. If an error is identified in the analysis part of a system being evaluated, then results from that system may be biased or incorrect, potentially causing predictable (once the error is identified) but fixable harms. More specifically, "A computing professional should be transparent and provide full disclosure of all pertinent system capabilities, limitations, and potential problems to the appropriate parties." (ACM 1.3)

3. Identify or recognize the ethical issue.

Both the ASA and ACM standards support the identification and fixing of the error, with the ASA GLs stipulating that corrections should be disseminated, and the ACM CE specifying similarly that limitations of the system should be disclosed (unless the error is fixed, in which it is no longer a system limitation). As articulated, this vignette states that you found an error: *an ethical issue only arises if you do not fix the error and notify any stakeholders who rely on the code or its results*. Note that making an honest error is not unethical! Only failure to correct the error, and failure to notify stakeholders, are unethical. Since the case specifies that the paper is about to be published, this set of stakeholders includes all co-authors *as well as the journal or publication venue*. If the erroneous results are featured in grant applications or other documentation, this creates additional stakeholders whose perceptions of the results must also be corrected. This notification may take time and may involve others. Thus, there might be some incentives *not* to fix the error, or to fix it but not notify everyone. Both the GLs and CE describe the ethical practitioners' responsibilities to do that notification, however, so ethical issues will arise if you do fix the error but do not notify all relevant stakeholders.

4. Identify and evaluate alternative actions (on the ethical issue).

The three decisions that can be made in any circumstance are: a) do nothing. b) consult or confer with a peer (ASA) or a supervisor (ACM); and c) report violations of policy, procedure, ethical guidelines, or law (ASA) or refuse to implement the system (ACM). In this case, however, only two options are viable because conferring with your supervisor or a peer (option b) will be tantamount to "reporting" (option c) that an error in the results was found.

Note that the case simply describes you identifying an error in results. This could be a simple typo, or it could be a mis-pasted result from another study or part of the publication in question. As noted above, *it does not matter how big of an error you found.* Option a) must be modified to "correct the error notify and all stakeholders, but do not confer"; and the only other option, b), becomes, "correct the error, and confer with peers or supervisors about how best to notify all stakeholders".

There is only an ethical issue if you do not fix the error **and** notify all stakeholders that there is an error. Thus, if you fix and report the error locally, and your supervisor(s) or organization then seek to suppress this information – particularly, not notifying clients or other stakeholders who would rely on the results, then option c) ("report the violation of CE, GLs, and relevant laws") *is* differentiated from option b) and should be considered. Note that, while you may have chosen to honour your commitments to practice ethically, and notify your supervisor/organization and other stakeholders, there may be some incentives for the organization either not to fix the error, or to fix it but not notify everyone. Incentives include the fact that the publication will always carry the erratum, or possibly that erroneous reporting of results to a funder or in a grant application may create problems or cause deadlines to be missed. Your primary responsibility is to ensure that *you* practice ethically; you cannot cause the organization to do so. So, alternative responses are identified for *you* to take so that you can ethically respond to your discovery of the error (obviously making an error is not unethical!) and you will choose a) or b). Your organization may respond to either option by not choosing to endorse or support your ethical practice. If the organization's response constitutes fraud – because the organization does not want to correct a publication, contract, or grant application that contains the error, for example, then you need to consider whistleblowing. You might also consider how important it is for you to continue to work with people who prioritize their reputation or funding over your ethical practice, particularly given the potential for harms that will accrue if anyone tries to use your method/system and it doesn't work (as they might have known if the error had been corrected). If you are unable to report such behavior (i.e., use option c), you might consider looking for a different team *or job* where such behavior does not occur. Using the Appendix and ASA Principle G, and ACM Principle 3, to guide your search for a more ethical practice context is absolutely warranted.

5. Make and justify a decision.

Decision: *option b*: "correct the program, and confer with peers or supervisors about how best to notify all stakeholders" is supported by both ASA and ACM standards. Because the case describes results that are about to be published, there very well may be multiple stakeholders and others to notify (or confer with, or both). The decision to correct the results (common to both options a and b) should not require approval but if you are not the person who generated those results, others may be involved in the correction.

6. Reflect on the decision.

Note that, once you have identified an error in results, if you continue to use that code without correcting it (even though this is easier), you are violating multiple practice standards – not being transparent or honest in your reporting (unless you also include the error in your report); not using the most appropriate methods; and not ensuring that your work is robust and that you are accountable for it. If others choose to continue using code in which you have identified an error, and they purposely ignore that error, they are violating multiple practice standards. You would have to decide if you wanted to work – and be associated with – that type of workplace and context.

Recall that sometimes, detrimental research practices are engaged in by people who were trained to practice this way –but sometimes they are engaged in to purposefully mislead. It is unethical of you to mislead anyone into thinking that the results are correct and valid; it is unethical to knowingly contribute false, irreproducible, and/or invalid results and inappropriate methods to the literature. A failure to correct this error and remain an author on the publication means that **you** endorse making this irreproducible and invalid "contribution" to the literature.

The SHA shows plainly that options a) and b) pass a *test of justice*, because your notification of others strengthens the profession and the public trust while correcting an error that was inadvertently contributed to the literature. Both options pass the *test of publicity*, but possibly only marginally; while we all want the best science and honest work, the fact that there was an error and no one on the team noticed it until after peer review (presumably multiple drafts of in-house review) is embarrassing! However, you *would* want it publicized that you are an ethical practitioner, and that is what the correction and stakeholder notification will achieve. These options pass the *test of universality* as well, since you definitely want all practitioners to be as vigilant as you are in this case.

Note also that the results contain an error that you would not want to take full responsibility for. Allowing the publication to go ahead without correction is clearly expedient – and, if the error is a simple typo or other minor adjustment that does not change the interpretation or decision, the impact on the scientific community or other stakeholders may be negligible. It is the ethical practitioner's responsibility to ensure that the impact *will be* negligible if you do not correct and report the error before the publication occurs. For errors in results with objectively negligible impact, a correction issued after publication will suffice. However, whether the error has high or low impact on stakeholders, your name as an author will endorse the work. This represents a negative impact (harm) on you as a stakeholder as much as for other stakeholders. Finally, recall that the responsibilities to correct this error articulated across every ASA Guideline element and several ACM CE elements do not specifically pertain to a scientific publication. A "high-profile publication" could be within-organization, or with a media outlet. The fact that the publication (whatever it is) is considered "high profile" means that it is anticipated to have some impact on many individuals and stakeholders. The SHA clearly shows that none of the benefits outweighs the vast array of harms that failure to correct a known error will create.

Case 35. You follow SOP and the GLs/CE, and report your methods and results fully, but the final report has incorrect methods and results that were "edited" to suit a senior member of the team without your knowledge or agreement.

1. Identify and 'quantify' prerequisite knowledge:

   Which DEFW and DSEC Principles or Areas seem most relevant to this vignette?

DEFW: *You* fulfilled Principle 6, but the documentation no longer meets that standard.

DSEC: This vignette describes dishonest representation (DSEC C3) and an unintended use (DSEC E4) of your documentation. The editing will mean that DSEC E2 will be "no" and there will be no way to ensure DSEC C3, C5, D4, or E3.

Neither DEFW nor DSEC discusses how to *respond*.

   Which ASA and ACM Principles seem most relevant to this vignette?

ASA:

Principle A. Professional integrity and accountability
Principle B. Integrity of data and methods
Principle C. Responsibilities to stakeholders
Principle D. Responsibilities to research subjects, data subjects, or those directly affected by statistical practices
Principle E. Responsibilities to members of multidisciplinary teams
Principle F. Responsibilities to fellow statistical practitioners and the Profession
Principle G. Responsibilities of leaders, supervisors, and mentors in statistical practice
Principle H. Responsibilities regarding potential misconduct
Appendix. Responsibilities of organizations/institutions

ACM:

1. General Ethical Principles
2. Professional Responsibilities

Potential result: Stakeholder:	HARM	BENEFIT	UNKNOWN	UNKNOWABLE
YOU	Incorrect and incomplete documentation means the work can no longer be used for valid or interpretable results. Your effort on the documentation has been wasted, *and* now neither the documentation nor what it describes can be used with confidence. Failing to correct the documentation may save time but *will* lead to unpredictable propagation of error throughout the workflow and into the community.	Correction of the misuse/editing means future work will be correct, reproducible, and interpretable.	Perpetuation of false reports and misinterpretation undermine ethical practice as well as any work based on the misleading documentation, while also perpetuating deceptive and detrimental research practices.	Deceptive and detrimental research practices like cherry picking (what the editing in this case represents) undermine science, the public trust in the scientific enterprise, and the entire scientific community.
Your boss/client	Incorrect and incomplete documentation means the work can no longer be used for valid or interpretable results. All effort on the documentation has been wasted, *and* now neither the documentation nor what it describes can be used with confidence.	Correction of the misuse/editing means future work will be correct, reproducible, and interpretable.	Correction of this misuse of the documentation will require resources to fix/address. *Appropriate* documentation may not yield desired results.	

Potential result: Stakeholder:	HARM	BENEFIT	UNKNOWN	UNKNOWABLE
Unknown individuals	Incorrect interpretation undermines reproducibility and rigor. Misleading documentation negatively affects all stakeholders in the long run.	Correction of the error means future work will be correct, and valid.	Competent interpretation (ASA) supports reproducible results; effective context-ualization of your work (ACM) promotes public good. Perpetuation of false reports and misinterpre-tation undermine both while also perpetuating detri-mental research practices.	Deceptive and detrimental research practices like cherry-picking (what the editing in this case represents) undermine science, the public trust in the scientific enterprise, and the entire scientific community.
Employer	Incorrect and incomplete document-ation means the work can no longer be used for valid or interpretable results. All effort on the documentation has been wasted, *and* now neither the documentation nor what it describes can be used with confidence. The incorrect documentation may be discovered to have been used by a client, or by other clients who paid for honest work, identifying past errors and undermining this and other projects	Correction of the documentation errors means future work with the product it describes (if not the employee who cherry-picked the documentation) may be facilitated – and correct, and valid.	Correction of the documentation errors may require resources to fix/address. *Appropriate analysis* –once the error is fixed - may not yield desired results.	Correction of the documentation errors may highlight past payments or purchases/sales of incorrect/error-prone work. Liability may be incurred.

Potential result: Stakeholder:	HARM	BENEFIT	UNKNOWN	UNKNOWABLE
Colleagues	Having to respond to this misuse of your documentation may complicate colleagues' work. Pointing out and correcting a team member's deceptive and detrimental research practices can be discomfiting – and may also highlight other detrimental practices that will then invite more scrutiny.	Identification and correction of the cherry-picking allows mitigation of risks of harms (ACM); supports valid inferences (ASA), and can limit the unpredictable propagation of the error to other parts of the workflow. Other colleagues may be emboldened to call out – and stop – other deceptive and detrimental practices they observe.	Competent interpretation (ASA) supports reproducible results; effective contextualization of your work (ACM) promotes public good. Perpetuation of false reports and misinterpretation undermine both while also perpetuating detrimental research practices.	Deceptive and detrimental research practices like cherry-picking (what the editing in this case represents) undermine science, the public trust in the scientific enterprise, and the entire scientific community.

Potential result:	HARM	BENEFIT	UNKNOWN	UNKNOWABLE
Stakeholder:				
Profession	Incorrect and incomplete documentation means the work can no longer be used for valid or interpretable results. If the documentation is shared, this will create invalidity in others' work. Failing to correct the documentation may save time but *will* lead to unpredictable propagation of error throughout the workflow and into the community.	Correction of the misuse/editing means future work will be correct, reproducible, and interpretable, Identification of the deceptive and detrimental research practice of cherry-picking (what the editing in this case represents) may lead to more in the profession preventing it, and fewer people trying it.	Perpetuation of false reports and cherry-picking undermine ethical practice as well as any work based on the misleading documentation, while also perpetuating deceptive and detrimental research practices.	Deceptive and detrimental research practices like cherry-picking (what the editing in this case represents) undermine science, the public trust in the scientific enterprise, and the entire scientific community.

Potential result: Stakeholder:	HARM	BENEFIT	UNKNOWN	UNKNOWABLE
Public/public trust	Deceptive and detrimental practices like cherry-picking undermine public trust in reproducibility and rigor. An insistence (by the editor) on cherry-picked documentation being "correct" –resistance to its correction - undermines public trust as well as reproducibility and rigor.	Identification and correction of the misleading documentation allows mitigation of risks of harms (ACM); supports valid inferences (ASA), and can limit the unpredictable propagation of the error to other parts of the workflow and community, to and decisions based on them.	Competent interpretation (ASA) represents reproducible results; effective contextualization of your work (ACM) promotes public good. Perpetuation of false reports and misrepresentation in documentation undermine both while also perpetuating deceptive and detrimental research practices.	Deceptive and detrimental research practices like cherry-picking (what the editing in this case represents) undermine science, the public trust in the scientific enterprise, and the entire scientific community.

Table 3.7.2. *Stakeholder Analysis: Your documentation* was *correct and was edited to be incomplete and incorrect.*

2. Identify decision-making frameworks.

Like other cases, this vignette describes a highly prevalent, deceptive, and extremely detrimental research practice, **cherry-picking**[37]. Cherry-picking is *obviously not ethical*. However, in this vignette, while *you* took the time and effort to fully and transparently document your work, someone who should know better – a senior team member - is ignoring, possibly suppressing, the honest representation of your work. Because the editing that was done renders the documentation both incomplete and incorrect, it should be resisted, and corrected, by an ethical practitioner. Note that every element of the ASA Guidelines offers input that is relevant to this case. This reflects how important honest and transparent reporting is to ethical statistical practice. Following Principles A (Professional integrity and accountability) and B (Integrity of data and methods) are not sufficient because of the responsibilities outlined in all the other Guideline Principles. This underscores how important reporting is to ethical statistical practice – it is not sufficient to do your job competently (i.e., follow Principles A and B). The virtue perspective reflects what "the ideal practitioner" would do, and in this perspective, there is zero tolerability for cherry-picking and other detrimental practices, because they undermine and basically render competent practice irrelevant.

This practice violates ASA A2, which states that the ethical statistical practitioner "uses methodology and data that are valid, relevant, and appropriate, without favoritism or prejudice, and in a manner intended to produce valid, interpretable, and reproducible results." In *and outside of* contexts where science is being done, the ethical statistical practitioner "avoids compromising scientific validity for expediency" (ASA E4). Note that the activity is not unethical practice *by you*; instead, at least one other person on the team is seeking to subvert your ethical practice and misrepresent the work. Principle B dictates that the ethical statistical practitioner "strives to promptly correct substantive errors discovered after publication or implementation. As appropriate, disseminates the correction publicly and/or to others relying on the results." (ASA B5) In addition to notifying stakeholders of the fact that the documentation is now *incorrect* and incomplete – such that results arising from it will also be error-prone and probably invalid, the cherry-picking that the editing represents must be called out and recognized for the deceptive and detrimental practice that it is. The ethical practitioner must follow ASA H2, and "avoid condoning or appearing to condone statistical, scientific, or professional

---

[37] https://en.wikipedia.org/wiki/Cherry-picking

misconduct. Encourage other practitioners to avoid misconduct or the appearance of misconduct."

Since the documentation that you originally produced was correct, you or your name may be associated with the documentation at this point, even though someone else modified it to mislead. The failure by the senior team member to respect your original work represents statistical and scientific ("falsification"[38]), if not also professional, *misconduct*. Problematically, the senior team member is transmitting to others on the team, whether knowingly or not, some value for cherry-picking. Not only are they practicing unethically, but they are also encouraging others in– condoning – statistical, scientific, and professional misconduct. This vignette thus describes both a responsibility to correct the now-misleading report and one to call attention to – and try to prevent – this transmission of deceptive and detrimental research practices by the senior team member.

Although the ACM CE does not mention cherry-picking or falsification, "professionals should be cognizant of any serious negative consequences affecting any stakeholder that may result from poor quality work and should resist inducements to neglect this responsibility" (2.1). It also specifies that "a computing professional should be transparent and provide full disclosure of all pertinent system capabilities, *limitations, and potential problems* to the appropriate parties." (1.3, emphasis added). In this case, documentation that would support reproducible and interpretable, valid results, has been modified and the new (edited) documentation cannot be said to meet these standards - and *that* needs to be transparently reported. Deceptive and detrimental practice needs to be identified and discouraged at every possible opportunity. In the vignette, the documentation is incorrect and incomplete, so whatever results arise from the method or system the documentation describes will not function validly or reproducibly. This is a predictable limitation that will invariably cause harm – even if that harm is only to the scientific record. There is also harm in allowing the senior team member to influence the practice of others by demonstrating or suggesting that cherry-picking is tolerable.

3. Identify or recognize the ethical issue.

As this vignette is presented, the practitioner has applied their expertise to document their work appropriately, but someone else has edited this work so that it is now incorrect and misleading. There is no way for the ethical practitioner to ethically ignore this and, allow this misrepresentation of their work. Both the ASA and ACM standards support reporting of the error, with

---

[38] https://ori.hhs.gov/definition-misconduct

both the ASA GLs and the ACM CE specifying that the errors (ASA) and limitations of the system (ACM, i.e., that it cannot yield valid results) should be disclosed. The ethical practitioner must respond – ethically – to correct the inappropriate editing and notify stakeholders. The ethical issues arises if you *fail to notify* all stakeholders of the cherry-picking, and also if you do not make efforts to identify and discourage the use of deceptive and detrimental practices. The failure to notify is unethical because the misleading documentation will permit inappropriate use of the work (based on the now-inappropriate documentation), false and invalid reporting based on the incorrect use of the now-misleadingly documented system or method, and decisions that may be based on the incorrect documentation or the method/system it describes. All of these harms are predictable, and can be averted by notification. The ethical practitioner has a responsibility to call attention to cherry-picking as the unethical and deceptive, detrimental practice that it is, or else they violate ASA H2, failing in the virtue perspective.

Both the ACM (Principle 4) and ASA (Purpose, Principles A & G, and Appendix) outline specific responsibilities of the ethical practitioner to follow the practice standards *and encourage others to do so as well*. The knowledge that cherry-picking is unethical does not always work to change behavior, but the ethical practitioner will *do what they can to discourage cherry-picking and any other detrimental practices*. Notification that cherry-picking is unsupportable may not prevent them from continuing this practice. However, to the extent possible given your role (and job security), an ethical issue for the practitioner also arises when they do not seek to discourage this *deceptive and detrimental* behavior. The harms that accrue when incorrect and incomplete documentation are permitted – especially when this is the result of misuse of appropriate documentation – violates ACM 1.3 and 2.1, and fails in the utilitarian perspective.

4. Identify and evaluate alternative actions (on the ethical issue).

The three decisions that we discuss for any circumstance are: a) do nothing. b) consult or confer with a peer (ASA) or a supervisor (ACM); and c) report violations of policy, procedure, ethical guidelines, or law (ASA) or refuse to implement the system (ACM).

As usual, option a) must be modified: the practitioner *was* ethical in their original documentation but to ignore the editing, and "do nothing", is clearly in violation of ethical practice standards. Option a) should change to "correct the erroneous documentation, notify all stakeholders that the edited documentation is incorrect and incomplete, and act to prevent the use or further publication of the false documentation." Option b) becomes, "confer with peers or supervisors about how best to correct the erroneous documentation, notify all stakeholders that the edited documentation is incorrect and incomplete, and act to prevent the use or further publication of the false documentation, *and* ensure that cherry-picking and other detrimental practices (including suppression of evidence of cherry-picking or spurious results) are recognized as unethical and not tolerated".

As discussed in Cases 25 and 26, the 2017 National Academies of Sciences and Engineering and the Institute of Medicine (NASEM) report, *"Fostering Integrity"*, called on the scientific community in the US to identify and act to prevent such deceptive and detrimental practices. In this vignette, your workplace or organization has SOPs in place that instruct workers to fully and transparently document their work. Thus, there is a clear policy that should have prevented the falsification of your documentation. So option c) ("report the violation of CE, GLs, and relevant laws") *is* a viable option. While cherry-picking is unethical but not illegal, deception and withholding evidence *is* illegal – particularly when there are financial conflicts of interest[39]in play. If the senior team member seeks to suppress the correction of misleading and incorrect documentation, they may be perpetrating fraud [40] (while also modeling detrimental and deceptive practices) – and both your commitment to ethical practice and your civic duty will require that you report this. Then option c) effectively becomes, "correct the erroneous documentation, *and* notify all stakeholders that the edited documentation is incorrect and incomplete, *and* act to prevent the use or further publication of the false documentation, *and* ensure that cherry-picking and other detrimental practices (including suppression of evidence of cherry-picking or spurious results) are recognized as unethical and not tolerated, *and* report the cherry-picking and suppression of its existence to the appropriate authorities (and to the ACM)." Note that the ACM does, but the ASA does not (as of July 2022), have a reporting mechanism as well as a specifically articulated responsibility to report violations of the practice standards.

5. Make and justify a decision.

Decision: All options include "correct the erroneous documentation, notify all stakeholders that the edited documentation is incorrect and incomplete, and act to prevent the use or further publication of the false documentation". This is the only ethical response to the unethical, deceptive, and detrimental behavior in this vignette, although how it is done – or what else is also done– may depend on other features of your specific context. *Options b and c* include

---

[39] https://ori.hhs.gov/education/products/ucla/chapter4/default.htm
[40] https://en.wikipedia.org/wiki/Fraud

sharing (i.e., reporting, option c) or conferring (option b) with others. Option c) "reporting" should be considered, particularly if this is not the first experience you are having (or that you know of) where this senior team member has misused and falsified documentation. Consulting a peer or supervisor as to how best to proceed (option b) may be more relevant for a more junior practitioner. As you grow in experience and seniority, you may be able to use the conferring to give some careful thought given to why no one else on the team objected to the reporting of cherry-picking in the editing of your documentation, and to how others are trained and mentored in their reporting at your organization.

There may be a culture in your team or organization that either tolerates or encourages detrimental practices, and the ethical practitioner is obliged to do what they can change this culture; "what you can do to change the culture from one that supports or even encourages detrimental practices to an ethical one" will naturally change with seniority and experience – but it will only become more important over time and will never be less important! When an otherwise ethical practitioner fails to stop others' unethical practice, they would then violate ASA H2 and "condone, or appear to condone", these unethical practices – meaning that a team or organizational culture or context where you are encouraged or directed to violate your practice guidelines is an *unethical* culture. Options b) and c) are preferred over option a) because the notification aspect of these options may prompt organization-level review of policies or incentives that exist for reporting unethically. Option b) may be more appropriate to initiate culture change, and to lead those who are ignorant that their behavior is unethical (e.g., if their practice guidelines do not decry deceptive and detrimental practices) to the understanding that *cherry-picking is unethical*. However, option c) may be the decision that has the greatest impact, preventing frank deception from propagating from your work, and also promoting ethical practice by alerting the organization to the senior member's pattern of deceptive and detrimental practice.

6. Reflect on the decision.

Construct your own reflection on this decision.

Case 36. Stakeholders (donors, funders, employers) are given a misleading summary of your methods and results.

1. Identify and 'quantify' prerequisite knowledge:

Which DEFW and DSEC Principles or Areas seem most relevant to this vignette?

DEFW: False reporting does not honor the user need (Principle 1). This is the opposite of Principle 6.

DSEC: Opposite of DSEC C3, makes DSEC D and E impossible to answer.

Which ASA and ACM Principles seem most relevant to this vignette?

ASA:

Principle A. Professional integrity and accountability
Principle B. Integrity of data and methods
Principle C. Responsibilities to stakeholders
Principle D. Responsibilities to research subjects, data subjects, or those directly affected by statistical practices
Principle E. Responsibilities to members of multidisciplinary teams
Principle F. Responsibilities to fellow statistical practitioners and the Profession
Principle G. Responsibilities of leaders, supervisors, and mentors in statistical practice
Principle H. Responsibilities regarding potential misconduct
Appendix. Responsibilities of organizations/institutions

ACM:

1. General Ethical Principles
2. Professional Responsibilities

Potential result: Stakeholder:	HARM	BENEFIT	UNKNOWN	UNKNOWABLE
YOU	Incorrect and incomplete representation cannot be used for valid or interpretable results or conclusions. Your effort on the documentation has been wasted, *and* now neither the documentation nor what it describes can be used with confidence. Failing to correct the documentation may save time but *will* lead to unpredictable propagation of error throughout the workflow and into the community.	Correction of the misuse/editing means future work will be correct, reproducible, and interpretable.	Perpetuation of false reports and misinterpretation undermine ethical practice as well as any work based on the misleading documentation, while also perpetuating deceptive and detrimental research practices.	Deceptive and detrimental research practices like misrepresentation undermine science, the public trust in the scientific enterprise, and the entire scientific community.
Your boss/client	Misrepresentation means the work can no longer be used for valid or interpretable results. All effort on the documentation has been wasted, *and* now neither the documentation nor what it describes can be used with confidence.	Correction of the misuse/editing means future work will be correct, reproducible, and interpretable.	Correction of this misrepresentation may require resources to address and transmit. *Appropriate* documentation may not yield desired results.	

Potential result: Stakeholder:	HARM	BENEFIT	UNKNOWN	UNKNOWABLE
Unknown individuals	Incorrect representation undermines reproducibility and rigor. Misleading misrepresentation negatively affects all stakeholders in the long run.	Correction of the error means future work will be correct, and valid.	Competent representation of work (ASA) supports reproducible results; effective contextualization of your work (ACM) promotes public good. Perpetuation of false reports and misrepresentation undermine both while also perpetuating detrimental research practices.	Deceptive and detrimental research practices like misrepresentation undermine science, the public trust in the scientific enterprise, and the entire scientific community.

Potential result: Stakeholder:	HARM	BENEFIT	UNKNOWN	UNKNOWABLE
Employer	Incorrect and incomplete misrepresentation means the work can no longer be used for valid or interpretable results. All effort on the documentation has been wasted, *and* now neither the documentation nor what it describes can be used with confidence. The misrepresentation may be discovered to have been used by a client, or by other clients who paid for honest work, identifying past errors and undermining this and other projects	Correction of the misrepresentation errors means future work with the product it describes (if not the employee who is misrepresenting your work) may be facilitated —and correct, and valid.	Correction of the misrepresentation errors may require resources to fix/address. *Appropriate analysis* —once the error is fixed - may not yield desired results.	Correction of the misrepresentation may highlight past payments or purchases/sales of incorrect/error-prone work. Liability may be incurred.

Potential result: Stakeholder:	HARM	BENEFIT	UNKNOWN	UNKNOWABLE
Colleagues	Having to respond to this misuse of your documentation may complicate colleagues' work. Pointing out and correcting a team member's deceptive and detrimental research practices can be discomfiting – and may also highlight other detrimental practices that will then invite more scrutiny.	Identification and correction of the misrepresentation allows mitigation of risks of harms (ACM); supports valid inferences (ASA), and can limit the unpredictable propagation of the error to other parts of the workflow. Other colleagues may be emboldened to call out – and stop – other deceptive and detrimental practices they observe.	Competent representation (ASA) supports reproducible results; effective contextualization of your work (ACM) promotes public good. Perpetuation of false reports and misrepresentation undermine both while also perpetuating detrimental research practices.	Deceptive and detrimental research practices like misrepresentation undermine science, the public trust in the scientific enterprise, and the entire scientific community.

Potential result: Stakeholder:	HARM	BENEFIT	UNKNOWN	UNKNOWABLE
Profession	Incorrect and incomplete documentation means the work can no longer be used for valid or interpretable results. If the documentation is shared, this will create invalidity in others' work. Failing to correct the documentation may save time but *will* lead to unpredictable propagation of error throughout the workflow and into the community.	Correction of the misuse/editing means future work will be correct, reproducible, and interpretable. Identification of the deceptive and detrimental research practice of misrepresentation may lead to more in the profession preventing it, and fewer people trying it.	Perpetuation of false reports and misrepresentation undermine ethical practice as well as any work based on the misleading documentation, while also perpetuating deceptive and detrimental research practices.	Deceptive and detrimental research practices like misrepresentation undermine science, the public trust in the scientific enterprise, and the entire scientific community.

Potential result:	HARM	BENEFIT	UNKNOWN	UNKNOWABLE
Stakeholder:				
Public/ public trust	Deceptive and detrimental practices like misrepresentation undermine public trust in reproducibility and rigor. An insistence (by the editor) on cherry-picked documentation being "correct" –resistance to its correction - undermines public trust as well as reproducibility and rigor.	Identification and correction of the misleading documentation allows mitigation of risks of harms (ACM); supports valid inferences (ASA), and can limit the unpredict-able propagation of the error to other parts of the workflow and community, to and decisions based on them.	Competent representation (ASA) represents reproducible results; effective contextualization of your work (ACM) promotes public good. Perpetuation of false reports and misrepresentation in documentation undermine both while also perpetuating deceptive and detrimental research practices.	Deceptive and detrimental research practices like misrepresentation undermine science, the public trust in the scientific enterprise, and the entire scientific community.

Table 3.7.3. *Stakeholder Analysis: Your correct work is misrepresented.*

2. Identify decision-making frameworks.

This vignette describes a highly prevalent and extremely detrimental research practice whereby an ethical practitioner does their work competently and accountably, but that good work is then misappropriated. Note that every element of the ASA Guidelines offers input that is relevant to this case. This reflects how important honest and transparent reporting is to ethical statistical practice. Following Principles A (Professional integrity and accountability) and B (Integrity of data and methods) are not sufficient because of the responsibilities outlined in all the other Guideline Principles. This underscores how important reporting is to ethical statistical practice – it is not sufficient to do your job competently (i.e., follow Principles A and B). The virtue perspective reflects what "the ideal practitioner" would do, and in this perspective, there is zero tolerability for cherry-picking and other detrimental practices, because they undermine and basically render competent practice irrelevant.

In this vignette it is important to note for whom the originally ethical work is being fraudulently described: donors, funders, and employers. The motivation for the misrepresentation is clear: *to deceive these stakeholders* - contrary to ASA C2. Because deception is both unethical and possibly fraudulent (illegal) in this vignette, it should be resisted by every ethical practitioner. The ethical statistical practitioner "avoids compromising scientific validity for expediency" (ASA E4) – and while it is not stated explicitly in the ASA GLs, it is plainly unethical to compromise validity *for funding* (rather than for expediency). Principle B dictates that the ethical statistical practitioner "strives to promptly correct substantive errors discovered after publication or implementation. As appropriate, disseminates the correction publicly and/or to others relying on the results." (ASA B5) In addition to notifying stakeholders of the fraud that the misrepresentation described in this case, the ethical practitioner must follow ASA H2, and "Avoid condoning or appearing to condone statistical, scientific, or professional misconduct. Encourage other practitioners to avoid misconduct or the appearance of misconduct." The misuse of your correct/ethical work constitutes statistical and scientific ("falsification"[41]) *and* professional, misconduct. The virtue perspective creates a long list of actions the ethical practitioner would take in this case.

Although the ACM CE does not mention falsification or professional misconduct, "professionals should be cognizant of any serious negative consequences affecting any stakeholder that may result from poor quality work and should resist inducements to neglect this responsibility" (ACM 2.1). The

---

[41] https://ori.hhs.gov/definition-misconduct

CE also specifies that "a computing professional should be transparent and provide full disclosure of all pertinent system capabilities, *limitations, and potential problems* to the appropriate parties." (ACM 1.3, emphasis added). In this case, correct documentation has been misrepresented; because of the specific stakeholders for whom the falsification is occurring, the deception is intended - and *that* needs to be transparently reported. Deceptive and detrimental practice needs to be identified and discouraged at every possible opportunity. Not only is the misrepresentation deceptive and unprofessional (and possibly illegal), the representation of the methods and results now incorrect and incomplete, so that using this method or system will not function validly or reproducibly. The failures of misrepresented methods and results to lead to valid outcomes is a predictable "limitation" that will invariably cause harm. There is also harm in allowing others in your organization to influence the practice of others by demonstrating or suggesting that falsification and fraud are tolerable. The overwhelming harms and negligible benefits from the SHA table clearly reflect the utilitarian perspective on the case.

3. Identify or recognize the ethical issue.

As this vignette is presented, the practitioner has applied their expertise to carry out an analysis appropriately, but others are fraudulently misrepresenting what was otherwise appropriate work. There is no way for the ethical practitioner to ignore this fraud. Both the ASA and ACM standards support reporting of the misrepresentation, and both stipulate that corrections should be disseminated and/or disclosed. Note that in this case, there is an ethical issue if you *fail to notify* all stakeholders of the misrepresentation and fraud. While some people simply do not understand computing and statistical principles sufficiently to understand importance of honest representation of technical documentation of methods and results, in this case the misrepresentation appears intended to deceive and is therefore falsification and potentially fraud. The ethical practitioner must act to call attention to unethical and deceptive, detrimental practices, or else they violate ASA H2. Both the ACM (Principle 4) and ASA (Purpose, Principles A & G) outline specific responsibilities of the ethical practitioner to follow the practice standards *and encourage others to do so as well*. The ethical practitioner will *do what they can to discourage both unethical and illegal practices.*

4. Identify and evaluate alternative actions (on the ethical issue).

The three decisions that we discuss for any circumstance are: a) do nothing. b) consult or confer with a peer (ASA) or a supervisor (ACM); and c) report violations of policy, procedure, ethical guidelines, or law (ASA) or refuse to implement the system (ACM).

As usual, option a) must be modified – in this particular vignette, if you fail to notify stakeholders that your methods and results are being misrepresented, especially to elicit or maintain funding, you are complicit in that falsification and fraud. Option a) should change to "correct the misrepresentation of your work, notify all stakeholders of the misrepresentation, and act to prevent further dissemination of the misrepresentation." Option b) becomes, "confer with peers or supervisors about how best to correct the misrepresentation of your work, notify all stakeholders of the misrepresentation, and act to prevent further dissemination of the misrepresentation, *and* ensure that cherry-picking and other detrimental practices (including suppression of evidence of cherry-picking or spurious results) are recognized as unethical and not tolerated".

As discussed in Case 25, the 2017 National Academies of Sciences and Engineering and the Institute of Medicine (NASEM) report, *"Fostering Integrity"*, called on the scientific community in the US to identify and act to prevent such detrimental practices. In this vignette, you have reported the error internally, and your supervisor then seeks to suppress this information because it conflicts with their preferred result. To the extent that the supervisor ignoring your evidence that their result is spurious includes them preventing clients or other stakeholders, who rely on the results, of the fact, then option c) ("report the violation of CE, GLs, and relevant laws") is a viable option. There is a clear policy that should have prevented the falsification of your methods and results – particularly in an effort to solicit or maintain funding. So option c) ("report the violation of CE, GLs, and relevant laws") *is* an important option. Deception and withholding evidence *is* illegal – particularly when there are financial conflicts of interest[42]in play. Anyone who seeks to mislead and misrepresent may be perpetrating fraud[43] (while also modeling detrimental and deceptive – and possibly illegal – practices). Your commitment to ethical practice *and* your civic duty will require that you report this. Thus, option c) becomes, "correct the cherry-picking, notify all stakeholders with the documentation of cherry-picked results, and act to prevent their use or further publication, and report the misrepresentation to the appropriate authorities (and to the ACM)." Note that the ACM does, but the ASA does not (as of July 2022), have a reporting mechanism as well as a specifically articulated responsibility to report violations of the practice standards.

5. Make and justify a decision.

---

[42] https://ori.hhs.gov/education/products/ucla/chapter4/default.htm
[43] https://en.wikipedia.org/wiki/Fraud

Decision: All options include "correct the misrepresentation of your work, notify all stakeholders of the misrepresentation, and act to prevent further dissemination of the misrepresentation". This is the only ethical response to this vignette, although how it is done – or what else is also done - may depend on your specific context.

Options b and c include sharing (i.e., reporting, option c) or conferring (option b) with others. Option c) "reporting" should be considered, particularly if this is not the first experience you are having (or that you know of) where this senior team member has misused and falsified documentation. Consulting a peer or supervisor as to how best to proceed (option b) may be more relevant for a more junior practitioner. As you grow in experience and seniority, you may be able to use the conferring to give some careful thought given to why no one else on the team objected to the misrepresentation of your work, and to how others are trained and mentored in their reporting –and respect for honesty - at your organization.

There may be a culture in your team or organization that either tolerates or encourages detrimental practices, and the ethical practitioner is obliged to do what they can change this culture.[44]

However, "what you can do to change the culture from one that supports or even encourages detrimental practices to an ethical one" will naturally change with seniority and experience – but it will only become more important over time and will never be less important! When an otherwise ethical practitioner fails to stop others' unethical practice, they would then violate ASA H2 and "condone, or appear to condone", these unethical practices – meaning that a team or organizational culture or context where you are encouraged or directed to violate your practice guidelines is an *unethical* culture. Moreover, if you do violate ASA H2 and condone the misrepresentation of your work – even implicitly, even by "doing nothing" to prevent it, then you are complicit in the misconduct and fraud that the misrepresenters engage in. This can have important negative impacts on your ability to solicit or maintain funding, and on your professional reputation – as well as negatively impacting the profession itself.

Options b) and c) are preferred over option a) because the notification aspect of these options may prompt organization-level review of policies or incentives

---

[44] If you doubt the negative impact of misleading reporting on stakeholders, you can read this summary of global attitudes towards key scientific reports: Dobson G. P. (2022). Wired to Doubt: Why People Fear Vaccines and Climate Change and Mistrust Science. Frontiers in Medicine, 8, 809395. https://doi.org/10.3389/fmed.2021.809395

that exist for reporting unethically. Option b) could be more appropriate to initiate culture change, and to lead those in ignorance that their behavior is unethical (e.g., if *their* practice guidelines do not decry deceptive and detrimental practices) to the understanding that *misrepresentation of otherwise ethical work is unethical*. However, because the misrepresentation seems to be done so that funders or employers will continue or start funding, option c) may be the decision that has the greatest impact, highlighting fraud as well as promoting ethical practice while preventing frank deception from propagating from your work.

6. Reflect on the decision.

Construct your own reflection on this decision.

Case 37. Your sensitivity analyses that pinpoint the next logical step in your work are omitted from a final report to funders because "we could get a grant to support the team for another 5 years to figure that out!"

1. Identify and 'quantify' prerequisite knowledge:

Which DEFW and DSEC Principles or Areas seem most relevant to this vignette?

Not addressed in DSEC or DEFW.

Which ASA and ACM Principles seem most relevant to this vignette?

ASA:

Principle A. Professional integrity and accountability
Principle B. Integrity of data and methods
Principle C. Responsibilities to stakeholders
Principle E. Responsibilities to members of multidisciplinary teams
Principle H. Responsibilities regarding potential misconduct

ACM:

2. Professional Responsibilities (2.1; 2.4)

Potential result: Stakeholder:	HARM	BENEFIT	UNKNOWN	UNKNOWABLE
YOU	Incorrect and incomplete representation of your work is cherry-picking. Your effort on the documentation has been wasted, *and* the final report is now incomplete.	You have pinpointed the next project; instead of funding to do the pinpointing, you may have an opportunity to lead the actual project (because you've determined what it should be!).	Misrepresentation of your work perpetuates deceptive and detrimental research practices.	Deceptive and detrimental research practices like cherry-picking undermine science, the public trust in the scientific enterprise, and the entire scientific community.
Your boss/client	Effort on the documentation has been wasted.	If the community believes a lot of additional work is needed to identify the next steps, then the sensitivity analyses would put the group ahead of that curve.		

Potential result: Stakeholder:	HARM	BENEFIT	UNKNOWN	UNKNOWABLE
Unknown individuals	Incorrect representation undermines credibility.		False reporting and misrepresentation perpetuate detrimental research practices.	Deceptive and detrimental research practices like misrepresentation undermine science, the public trust in the scientific enterprise, and the entire scientific community.
Employer	Effort on the documentation has been wasted.	If the community believes a lot of additional work is needed to identify the next steps, then the sensitivity analyses would put the group ahead of that curve.		
Colleagues	Misrepresentation of what is needed "next" may waste colleagues' effort.		False reporting and misrepresentation perpetuate detrimental research practices.	Deceptive and detrimental research practices like misrepresentation undermine science, the public trust in the scientific enterprise, and the entire scientific community.

Potential result: Stakeholder:	HARM	BENEFIT	UNKNOWN	UNKNOWABLE
Profession	Misrepresentation of what is needed "next" may waste colleagues' effort and misdirect others in the field. Resources may be wasted when the best direction has already been identified.		False reporting and misrepresentation perpetuate detrimental research practices.	Deceptive and detrimental research practices like misrepresentation undermine science, the public trust in the scientific enterprise, and the entire scientific community.
Public/public trust	Misrepresentation of what is needed "next" may waste colleagues' effort and misdirect others in the field. Resources may be wasted when the best direction has already been identified.		False reporting and misrepresentation perpetuate detrimental research practices.	Deceptive and detrimental research practices like misrepresentation undermine science, the public trust in the scientific enterprise, and the entire scientific community.

Table 3.7.4. *Stakeholder Analysis: Your correct and complete work is omitted from a report.*

2. Identify decision-making frameworks.

This vignette describes a detrimental research practice whereby an ethical practitioner does their work competently and accountably, but that good work is then misrepresented. In this vignette it is important to note for whom the ethical work is being misrepresented: **funders**. The motivation for the misrepresentation is clear: *to hide advances from these stakeholders so funding can be applied for to make those same advances.* Since the results of sensitivity analyses should be borne out with additional empirical work, the results that were omitted from the final report – unless they were specifically part of the previous grant – are not final results (true "advances").

Note that the sensitivity analyses function to complete the report; they are described as identifying the logical next step (beyond what the original project has achieved). Thus, the sensitivity analyses – by showing a next logical step can be derived based on the current project – actually strengthen the credibility of the current work. Including those analyses, and this argument, in the current project report will strengthen the report and possibly the team's reputation in the funders' view. The choice to omit that weakens the report, but also suggests a tendency towards unethical behavior in the team members who chose to omit the results.

While the ethical statistical practitioner "avoids compromising scientific validity for expediency" (ASA E4), the most unethical behavior of this vignette (by those who opted to omit material that strengthens the report) would be the grant application for five years of support to "discover" what has already been discovered. Principle B dictates that the ethical statistical practitioner "strives to promptly correct substantive errors discovered after publication or implementation. As appropriate, disseminates the correction publicly and/or to others relying on the results." (ASA B5) The vignette does not specify that the omitted sensitivity analyses were part of the promised work products of the grant, and if not, then their omission would not necessarily be an error or misrepresentation within the report. However, claiming a need for additional funding to support the execution of these already-completed sensitivity analyses *is* unethical. The ethical practitioner must follow ASA H2, and "Avoid condoning or appearing to condone statistical, scientific, or professional misconduct. Encourage other practitioners to avoid misconduct or the appearance of misconduct." The misrepresentation of your completed analyses as "needed" –particularly to obtain funding – *would* constitute statistical and scientific ("falsification" [45]) *and* professional, misconduct. Thus, the virtue

---

[45] https://ori.hhs.gov/definition-misconduct

perspective on this case is clearly that the ethical practitioner must correct the misrepresentation, and if it can't be done for the final report, then it must be done to prevent false claims that a new grant is needed to discover what has already been discovered.

Although the ACM CE does not mention professional misconduct, it does specify that "making deliberately false or misleading claims, fabricating or falsifying data, offering or accepting bribes, and other dishonest conduct are violations of the Code." (ACM 1.3) In this case, "professionals should be cognizant of any serious negative consequences affecting any stakeholder that may result from poor quality work and should resist inducements to neglect this responsibility" (ACM 2.1). Here, correct documentation has been misrepresented because your full report was truncated; but if the sensitivity analyses were not part of promised work, then this omission is not misleading. It would, of course, be misleading if a new grant were submitted claiming a need for additional support to re do the analyses already done (that were omitted from the current final report, as in the vignette). Thus, the utilitarian perspective clearly focuses on preventing falsely applying for funding to derive results that have already been derived. The utilitarian perspective also highlights general harms that arise from misrepresentation of work, and knowingly misleading stakeholders.

3. Identify or recognize the ethical issue.

According to the vignette, we do not know if the sensitivity analyses were actually part of promised work. If they were, then their omission from the report is unethical and that omission *must* be corrected, with the promised results reported to the relevant stakeholders (i.e., the funder). However, if the sensitivity results were not part of the promised work, then there is no ethical issue unless a new proposal requests additional funding to support the completion of this already-completed work. The ethical practitioner must act to call attention to unethical and deceptive, detrimental practices, or else they violate ASA H2. Both the ASA and ACM practice standards state explicitly and repeatedly that professionals should not yield to pressures to behave contrary to their respective practice standards (**ASA**: A; A1, A2, A4, A9, A12; C1, C2; E4; G1; H2; **ACM**: 1.2; 1.3; 2.1; 2.2; 2.3; 3.4; 4). Moreover, both the ASA (ASA Principle G) and ACM CE (Principle 3) also suggest that the practitioner and those in leadership positions have an obligation to ignore the person(s) who want you to violate the ethical practice standards. The ethical practitioner will do what they can to discourage unethical and detrimental practices, but the vignette does not necessarily describe frankly unethical practice (yet) and the practitioner is not ethically obligated to notify the stakeholder of the sensitivity analyses *unless* that was a part of promised work.

4. Identify and evaluate alternative actions (on the ethical issue).

The three decisions that we discuss for any circumstance are: a) do nothing. b) consult or confer with a peer (ASA) or a supervisor (ACM); and c) report violations of policy, procedure, ethical guidelines, or law (ASA) or refuse to implement the system (ACM).

This may be the only vignette in the book where "do nothing" is actually a viable response: however, option a) "do nothing" is only viable if the sensitivity results were not part of the promised work. Option a) is not viable if they were; then option a') becomes "correct the final report, notifying stakeholders of the correction".

Only in the case that the sensitivity results *were* part of the promised work does option b) become, "confer with peers or supervisors about how best to correct the final report, notifying stakeholders of the correction". In that case, option c) ("report the violation of CE, GLs, and relevant laws") *is not* necessary in this vignette because the recipient of the final report is the funder – the authorities to whom omission of important promised work would need to be reported.

5. Make and justify a decision.

Decision: Option a) "do nothing" is hopefully the best- only- option in this case, assuming that the sensitivity results were not part of the promised work. However, as suggested by the fact that all of the vignettes in Section 3 are based on actual events, our considerations should go further. As noted, option a) is not viable if the sensitivity analysis results were purposefully omitted; but whether option a') or option b) is preferred will depend on your role on the team and your ability to file a revised final report with the funder.

There may be a culture in your team or organization that either tolerates or encourages "whatever it takes to secure grant funding". The ethical practitioner is obliged to do what they can change this culture – no matter what their career stage or role (junior or senior). The ASA Guidelines Appendix strongly supports workplace cultures where this is not the case. ASA Principle G and ACM Principle 3 also offer guidance to leaders, supervisors, and mentors to specifically promote an ethical practice culture and to encourage work, hires, and policies that are consistent with the ethical practice standards.

A team or organizational culture or context where you are encouraged or directed to violate your practice guidelines is an *unethical* culture. It is true that "what you can do to change the culture from one that supports or even encourages detrimental practices to an ethical one" will naturally change with seniority and experience. However, the ACM CE and ASA GLs offer every practitioner at every career stage and role both a concrete set of ethical obligations, and a way to share the perspective with other team members (ASA Principle E) and statistical practitioners (ASA Principle F). That can happen at every career stage. When an otherwise ethical practitioner fails to stop others' unethical practice, they would then violate ASA H2 and "condone, or appear to condone", these unethical practices. Thus, if you do violate ASA H2 and condone the misrepresentation of your work, particularly to solicit or maintain funding, then you are complicit in the misconduct that would otherwise be limited to the behavior of others.

6. Reflect on the decision.

Construct your own reflection on this decision.

Case 38. If you report your method fully and transparently, then you will lose the opportunity to patent it.

1. Identify and 'quantify' prerequisite knowledge:

Which DEFW and DSEC Principles or Areas seem most relevant to this vignette?

Not discussed in DSEC or in DEFW.

Which ASA and ACM Principles seem most relevant to this vignette?

ASA:

Principle A. Professional integrity and accountability
Principle B. Integrity of data and methods
Principle C. Responsibilities to stakeholders
Principle F. Responsibilities to fellow statistical practitioners and the Profession

ACM:

1. General Ethical Principles

Potential result: Stakeholder:	HARM	BENEFIT	UNKNOWN	UNKNOWABLE
YOU	Documentation takes time, adds effort (to fully document plans/ evaluations/systems); reporting this will result in inability to patent	Full documentation creates transparency and accountability		
Your boss/client	Reporting takes extra time and may alert competitors to innovation or results before client is ready	Full documentation creates transparency and accountability		
Unknown individuals		Creates transparency and accountability; open publishing of the method/ work increases access worldwide.		
Employer	Reporting takes time and may alert competitors to innovation/ results before client is ready; patents cost money	Demonstrates commitment to transparency & accountability; could add/add to a new patent/IP to improve organizational reputation	ACM: New harms/risks may be detectable more quickly (and addressed) with full documentation. ASA: replications are facilitated.	Harms may arise from loss of IP, if others implement this IP, it could limit business/- profit

Potential result: Stakeholder:	HARM	BENEFIT	UNKNOWN	UNKNOWABLE
Colleagues	Documentation and reporting may add time (in reviewing) to colleagues' work. A patent would prevent them from using the IP themselves while documentation makes it fully accessible.	Full documentation creates transparency and accountability, and supports transparency, accountability, and replicability. Promotes peer evaluation.		
Profession	Documentation takes time, adds effort (to fully document plans/ evaluations/systems); publication of unprotected work may facilitate use of work that is unauthorized or does not acknowledge original creator; full documentation could possibly increase piracy	Reporting notifies the profession of improvements and innovations and can move the field forward. Publication/ reporting increases visibility of the profession. Reports can simplify evaluations of the work (ASA) or system, and updates (ACM); add transparency and accountability. Promote and facilitates peer evaluation.		

Potential result: Stakeholder:	HARM	BENEFIT	UNKNOWN	UNKNOWABLE
Public/public trust	Without full documentation of work/thought processes, public trust in systems or their results, and the profession (ACM/ASA) are compromised. Perpetuating black box mentality does not help, and explicitly does not inform, society.	Transparency and accountability – and the potential to understand how decisions are made based on automation and systems - supports public trust in the profession and its work.		

Table 3.7.5. *Stakeholder Analysis: Full reporting may preclude ability to patent a new method*

2. Identify decision-making frameworks.

If the work is fully documented and *reported*, then intellectual property (IP) rights may be lost. However, *documentation is not the issue in this case*. It is the *reporting and communication* of that documentation that would result in the loss of a patent opportunity. Thus, we focus on reporting and communication, assuming that the work has been fully and transparently documented (as is clearly mandated in the ethical practice standards).

ASA GLs describe the ethical statistical practitioner as promoting reproducibility in all of their work (ASA A2; ASA B2 and possibly B6; ASA C6). However, this can actually be accomplished by the individual or team that produces the IP, and does not necessarily require peer review or other types of reporting or communication beyond the team – as long as they are fulfilling their ethical obligations to provide constructive feedback (ASA F2). Further, patent applications require honest and transparent communication. If the IP does not yield reproducible results, then it may not yet be suitable for submission for patenting. The transparent documentation of the work for the patent application is crucial for ethical practice: attempting to mislead patent evaluators with incomplete, incorrect or otherwise dishonest representation is frankly unethical. Thus, the virtue perspective offers support for how to resolve this apparent quandary without creating a lost IP opportunity.

The SHA supports transparent and complete documentation of your work as consistent with the utilitarian perspective of the ACM CE, but does not suggest that striving towards IP is *inconsistent* with meeting CE Principles. Since patents include full and public reporting of all relevant documentation, once the patent is granted, that documentation (which is full and transparent, as per practice standards) will then become publicly available. Again, attempting to mislead patent evaluators with incomplete, incorrect or otherwise dishonest representation is frankly unethical. The utilitarian perspective supports resolution of this apparent quandary without creating a lost IP opportunity – and without endorsing incomplete or dishonest representation in the patent application.

3. Identify or recognize the ethical issue.

Both the virtue and utilitarian perspectives, as well as practice standards of the ASA and ACM, support safeguarding IP with patents and other means as necessary, providing that – as evaluable by relevant stakeholders - the work is completely and transparently documented. To avoid the real harm that losing the opportunity to patent your work will cause, the utilitarian perspective supports patenting/safeguarding your work. Note that the ACM CE is silent on whether there is an obligation to share (rather than patent) work, while the ASA GLs (F5) *might seem* to suggest that patenting might not be considered ethical practice ("Promotes reproducibility and replication, whether results are "significant" or not, by sharing data, methods, and documentation to the extent possible.") However, there is no actual implication that appropriate patenting is not consistent with ethical statistical practice, since F5 specifies that the ethical obligation is to share "to the extent possible" rather than identifying an absolute responsibility to share everything no matter what. The communication of the work will be accomplished through the patent application safeguarding process, and once a patent is issued, then the work will have been fully documented, and will be shareable (to the extent permitted by law). Thus, patenting the work is *not unethical*, because public sharing of its complete documentation will (eventually) occur. It is possible to veer into unethical reporting, however, when reports of or about what the patent is for are purposefully misleading. In this case, the ethical practitioner must honestly and transparently *document* their work, and ensure that the IP application is also transparent. The ultimate report can be in the form of the issued patent, rather than the documentation (prior to application for the patent).

4. Identify and evaluate alternative actions (on the ethical issue).

Clearly, the same three decisions that can be made in any circumstance (a) do nothing b) consult or confer with a peer (ASA) or a supervisor (ACM); or c) report violations of policy, procedure, ethical guidelines, or law (ASA) or refuse to implement the system (ACM)), do not fit with this case. Instead, option a) becomes "report, and do not patent, your work"; and option b) becomes "patent, and then report, your work". There is no option c) because, as the work *is* fully documented, and *patenting work is not unethical*.

5. Make and justify a decision.

Decision: Applying for a patent is a legitimate way to safeguard your work; it ensures that the appropriate author/creator is recognized and also requires full and transparent documentation. In organizations where IP is valued, applying for patents prior to reporting on your work fulfills ASA GL Principle C, "Responsibilities to stakeholders" as supporting the funder/client (your employer) by ensuring their investment is protected. The objective of the ethical obligations around *documentation*, to support replication and reproducibility (ASA) as well as honest peer review (ACM) as appropriate, can be achieved even when a patent is applied for. Once the patent is awarded, the documentation will become public. Applying for the patent ahead of publication or reporting can add benefits to the employer (and the practitioner) while also potentially minimizing the real harm represented by loss of IP. It is essential to ensure that documentation provided to the patent issuers is complete and transparent, because the patent issuer becomes a stakeholder in that case.

6. Reflect on the decision.

Construct your own reflection on this decision.

Case 39. If you report your method fully and transparently, then a reviewer might notice that you are not the original developer of this method – although the same method was published over 30 years ago and in *another* field.

1. Identify and 'quantify' prerequisite knowledge:

   Which DEFW and DSEC Principles or Areas seem most relevant to this vignette?

Not discussed in DSEC or in DEFW.

   Which ASA and ACM Principles seem most relevant to this vignette?

ASA:

Principle A. Professional integrity and accountability
Principle B. Integrity of data and methods
Principle C. Responsibilities to stakeholders
Principle E. Responsibilities to members of multidisciplinary teams
Principle F. Responsibilities to fellow statistical practitioners and the Profession
Principle H. Responsibilities regarding potential misconduct

ACM:

1. General Ethical Principles

Potential result: / Stakeholder:	HARM	BENEFIT	UNKNOWN	UNKNOWABLE
YOU	You falsify your cv/resume, making a *deliberately misleading* claim, if you ignore prior work and claim to be the original innovator.	You gain a publication, but only by falsifying your cv/resume and taking responsibility as well as credit for work others did.	Perpetuation of false attribution/ authorship undermine your credibility as well as that of everyone on the author list, while also perpetuating deceptive and detrimental research practices.	Deceptive and detrimental research practices like claiming credit for an already published idea undermine science, the public trust in the scientific enterprise, and the entire scientific community.
Your boss/client	Perpetuation of false attribution/ authorship undermines your credibility as well as that of *everyone on the author list.*			
Unknown individuals	Making a *deliberately misleading* claim, by taking responsibility as well as credit for work others did, is proof that you are dishonest.		Perpetuation of false attribution/ authorship undermines your credibility as well as that of everyone on the author list, while also perpetuating deceptive and detrimental research practices.	Deceptive and detrimental research practices like claiming credit for an already published idea undermine science, the public trust in the scientific enterprise, and the entire scientific community.

Potential result: Stakeholder:	HARM	BENEFIT	UNKNOWN	UNKNOWABLE
Employer	Perpetuation of false attribution/ authorship undermines your credibility as well as that of *everyone on the author list.* By extension, the employer is implicated as an endorser of detrimental practices.		Perpetuation of false attribution/ authorship undermines your credibility as well as that of everyone on the author list, while also perpetuating deceptive and detrimental research practices.	Deceptive and detrimental research practices like claiming credit for an already published idea undermine science, the public trust in the scientific enterprise, and the entire scientific community.
Colleagues	Perpetuation of false attribution/ authorship undermines your credibility as well as that of *everyone on the author list.*		Perpetuation of false attribution/ authorship undermines your credibility as well as that of everyone on the author list, while also perpetuating deceptive and detrimental research practices.	Deceptive and detrimental research practices like claiming credit for an already published idea undermine science, the public trust in the scientific enterprise, and the entire scientific community.

Potential result: Stakeholder:	HARM	BENEFIT	UNKNOWN	UNKNOWABLE
Profession	Perpetuation of false attribution/ authorship undermines your credibility as well as that of *everyone on the author list.* Your refusal to acknowledge prior work damages the originator and decrements the respect all professionals should have for prior work, undermining the credibility of the profession in making contributions to documents/ publications.		Perpetuation of false attribution/ authorship undermines your credibility as well as that of everyone on the author list, while also perpetuating deceptive and detrimental research practices.	Deceptive and detrimental research practices like claiming credit for an already published idea undermine science, the public trust in the scientific enterprise, and the entire scientific community.
Public/public trust	Deceptive and detrimental practices like falsely claiming to be the originator undermines public trust in intellectual property and the value of innovation.			Deceptive and detrimental research practices like claiming credit for an already published idea undermine science, the public trust in the scientific enterprise, and the entire scientific community.

Table 3.7.6. *Stakeholder analysis: Your new method is actually not new.*

2. Identify decision-making frameworks.

Following the virtue perspective, the ethical statistical practitioner "accepts full responsibility for their own work, does not take credit for the work of others, and gives credit to those who contribute. Respects and acknowledges the intellectual property of others." (ASA A5) and "Is transparent about assumptions made in the execution and interpretation of statistical practices including methods used, limitations, possible sources of error, and algorithmic biases. Conveys results or applications of statistical practices in ways that are honest and meaningful." (ASA B2). Clearly, failing to identify and credit the prior accomplishments does not respect or acknowledge the contributions of others, but also does not convey the work in a way that is honest. The virtue perspective on this case is clear: the ethical practitioner does not try to hide or obfuscate their work, but is transparent and honest in all communication. The ethical practitioner avoids "condoning or appearing to condone statistical, scientific, and professional misconduct" (ASA H2), and trying to mislead a reviewer – and ultimately, readers of the paper if it is eventually published – is scientific and professional misconduct. Moreover, the ethical practitioner "takes responsibility for evaluating potential tasks, assessing whether they have (or can attain) sufficient competence to execute each task and that the work and timeline are feasible. Does not solicit or deliver work for which they are not qualified or that they would not be willing to have peer reviewed." (ASA A1). Going ahead with your publication or patent preparation when you have discovered that you (and your team) are not the originator of the work violates ASA H2 and makes it impossible for the practitioner to "follow, and encourages all collaborators to follow, an established protocol for authorship. Advocate for recognition commensurate with each person's contribution to the work. Recognize that inclusion as an author does imply, while acknowledgement may imply, endorsement of the work." (ASA A6) The virtue perspective is clear that this case reflects unethical behavior.

The SHA supports acknowledgement of the true originator of an idea because making false and misleading claims about innovation creates only two real harms for every single stakeholder: making false and misleading claims about being the originator of previously published work undermines your credibility *as well as that of everyone on the author list,* and making such claims is a deceptive and detrimental practice that this act will perpetuate. Although there are only two real harms in the SHA table, they do accrue to every stakeholder and may have unknown/unpredictable further harms.

3. Identify or recognize the ethical issue.

The ethical issue is that continuing to prepare to submit a publication when you know your work is not as innovative as publication requires fails to honor prior work and intellectual contributions. It also falsifies your resume or cv by suggesting yours is the first publication of the method, and confuses the scientific knowledge base by failing to acknowledge the intellectual history of the ideas (if your paper is published). The discovery that the work is not original is an important facet of the contextualization of the work you have done; *failing to disclose this is unethical behavior*. Importantly, in this vignette, the individual appears to be contemplating falsifying or making misleading statements so that a reviewer does not notice that this is not original work. In addition to representing deceptive practice, this constitutes *plagiarism*, one of only three ways to earn a conviction for scientific misconduct in the US. Whenever you act in a manner that does not fulfill your obligation to practice in good faith, it contravenes the GLs and CE.

4. Identify and evaluate alternative actions (on the ethical issue).

The same three decisions that we consider in each vignette are: a) do nothing b) consult or confer with a peer (ASA) or a supervisor (ACM); or c) report violations of policy, procedure, ethical guidelines, or law (ASA) or refuse to implement the system (ACM). Option a) becomes "recognize the prior work as context of the current work". Option b), which is implied in option a) but involves other people, becomes "b. confer with peers/supervisor about how best to differentiate your new method from previous ones". Additional work may be required to ensure that prior work is considered and any innovations in the "new" method contextualized against that historical context. Moreover, if you confer with others who are not authors already on the work, then you will need to ensure that they are given the opportunity to meet the ICMJE criteria for authorship[46] - to ensure that all intellectual contributions to this work whether in the past or present, are appropriately recognized.

---

[46] The four criteria for authorship published by the International Committee of Medical Journal Editors (ICMJE) are:
- Substantial contributions to the conception or design of the work; or the acquisition, analysis, or interpretation of data for the work; AND
- Drafting the work or revising it critically for important intellectual content; AND
- Final approval of the version to be published; AND
- Agreement to be accountable for all aspects of the work in ensuring that questions related to the accuracy or integrity of any part of the work are appropriately investigated and resolved.

Since you are the person responsible for preparing the publication, having an option c) would only be relevant if you are part of a team, and you notify the team that the prior work exists, and they seek to suppress that information. In that case, it would be important – in order to avoid the harms as listed in the SHA – *not to submit the work at all*. That might seem absurd, particularly if you are a junior member of a research organization where publications are essential for career success. However, the failure to acknowledge prior work is particularly and especially detrimental to new practitioners because it signals a total disregard for your reputation and the work of others. To risk establishing a reputation for unethical publication practices will virtually ensure that all the harms in the SHA accrue to you -early in your career. No benefits – not even a publication – can offset that. This recommendation assumes that, because of how the vignette is structured, you are the author responsible for submission. If you were to withdraw from the effort of preparing the publication (e.g., because other co-authors seek to suppress the fact that the method was already published – and/or if it is *they*, and not you, who hope the reviewers will not notice that the method was already published), it would seem like you had wasted the effort on getting to that point but to leave your name on – even if you do meet all criteria for authorship – is your endorsement of the failure to recognize prior work, violating ASA H2 and ACM 1.3.

5. Make and justify a decision.

Decision: As noted, the option you choose will depend on your situation. If you are more senior on your team, you will be able to choose option a) and simply recognize the prior work as context of the current work yourself. However, more junior team members may require consultation or collaboration, in which case option b) will be a better choice. Option c is naturally the worst-case scenario because no one likes to dedicate time and effort to a project and then abandon it at the end! However, if you are a junior person, and do suggest recognizing prior work, and your team seeks to suppress that, then you might simply wait to submit until you are part of another team. As long as you are able to ensure it won't be published in its current unethical form, you can hold onto it until you get onto a team or develop sufficient experience and authority to make the necessary revisions to contextualize the work yourself, and *then* publish the paper.

These criteria represent an international consensus, and even though you might not be publishing in medical journal, they are still useful and important; and are discussed here: http://www.icmje.org/recommendations/browse/roles-and-responsibilities/defining-the-role-of-authors-and-contributors.html

A team may conclude that they would rather publish whatever they have without acknowledging prior work, trusting reviewers to not search very far back in time or across other disciplines. If the context of option c) applies for you in a case like this, your best response is to ensure that your name is *not* included on a publication that fails to acknowledge prior work. Although this seems unfair to you, given all your work on whatever led to the paper, to leave your name on what is submitted is *your endorsement of the failure to recognize prior work*, violating ASA H2 and ACM 1.5. By leaving your name on you also make a misleading and false statement on your CV or resume, namely, that you are the (or at least one) author of the method, if you list the paper there.

6. Reflect on the decision.

Construct your own reflection on this decision.

Case 40. You prepare a report identifying difficulties you encountered in your evaluation of a system your organization wants to deploy or an analysis that was done. The organization does not have a mechanism for submitting or sharing this report (or peer review of any type) either internally or with stakeholders.

1. Identify and 'quantify' prerequisite knowledge:

Which DEFW and DSEC Principles or Areas seem most relevant to this vignette?

Not discussed in DSEC or in DEFW.

Which ASA and ACM Principles seem most relevant to this vignette?

ASA:

Principle A. Professional integrity and accountability
Principle B. Integrity of data and methods
Principle C. Responsibilities to stakeholders
Principle E. Responsibilities to members of multidisciplinary teams
Principle F. Responsibilities to fellow statistical practitioners and the Profession
Principle G. Responsibilities of leaders, supervisors, and mentors in statistical practice
Appendix. Responsibilities of organizations/institutions

ACM:

Principle 2. Professional Responsibilities
Principle 3. Professional Leadership Principles

Potential result: Stakeholder:	HARM	BENEFIT	UNKNOWN	UNKNOWABLE
YOU	The lack of a system prevents people (like you) from benefitting from peer evaluation. You may have satisfied the GL/CE to carry out peer review, but you are unable to disseminate the results.	You or others in the organization have an opportunity to create a new system to promote dissemination of these reports (and also, peer review itself) – strengthening practice.		
Your boss/client	The lack of a system prevents independent input/evaluation from reaching the boss or client.			
Unknown individuals	The lack of a system prevents independent input/evaluation from identifying limitations and risks – undermining the rationale for peer review inclusion in practice standards.			Cultures that "require" workers to do work that does not benefit from review limit growth and skills building opportunities that would otherwise be built in.
Employer	The lack of a system prevents independent input/evaluation from sharing identified limitations and risks that were identified via a peer review process.			Workers that do not benefit from review are limited in opportunities for growth and skills building that would otherwise be built in.

Potential result: Stakeholder:	HARM	BENEFIT	UNKNOWN	UNKNOWABLE
Colleagues	Colleagues either end up doing all the reviewing and circulating in informal ways or no one on the team gets any independent or expert input.			Cultures that "require" workers to do work that does not benefit from review limit growth and skills building opportunities that would otherwise be built in.
Profession	The lack of a system prevents independent input/evaluation from identifying limitations and risks – undermining the rationale for peer review inclusion in practice standards.			Cultures that "require" workers to do work that does not benefit from review limit growth and skills building opportunities that would otherwise be built in.
Public/public trust	Lack of a reporting mechanism results in workers who do work that does not benefit from review; the purposes for peer review inclusion in practice standards, and benefits of honest review, are all lost.			

Table 3.7.7. *Stakeholder Analysis: No way to share peer review/evaluation*

2. Identify decision-making frameworks.

The virtue perspective on this case suggests that the ethical statistical practitioner engages in peer review as a necessary part of strengthening other practitioners and the profession itself. When there is no opportunity for peer review to be shared, its benefits are lost. Many of the ASA Principles outline a responsibility to communicate – especially if errors or problems are discovered. The virtue perspective on this case is that failing to promote the dissemination of evaluation is not consistent with the GLs.

The utilitarian perspective on this case is featured in the SHA table: there are no benefits except one, namely, that the lack of a system is an opportunity for the ethical practitioner to create one *de novo* – in order to promote ethical practice more widely and to enable the organization to benefit from all of the features of peer review that led to its inclusion in both the ASA GLs and the ACM CE. This is a true benefit for the community and profession – but, does require an expenditure of time and effort on the practitioner's part.

3. Identify or recognize the ethical issue.

In this case, it is difficult to recognize the ethical issue because it is *diffuse*: the lack of any system for disseminating or sharing the peer evaluation that the practice standards describe as being part of the responsibilities of ethical practitioners makes it difficult, but not impossible, for those evaluations to be shared. As has been noted in other cases, notifying stakeholders about limitations (ACM 2.5) or considerations (ASA Principles A, B, C, F) is an essential mechanism. If there is no system in place, then some reports or evaluations may be shared within the institution while others can easily be suppressed. Thus, the lack of a system for reporting might be due to purposeful intention to mislead or suppress constructive evaluation. This creates additional ethical challenges for practitioners who seek to follow the practice standards. Note that preventing designated stakeholders from obtaining the proper evaluation is *unethical and may also be illegal*, but not having a formal system in place is not unethical. It is important to recognize when a lack of reporting is due to constraints (e.g., business and government inability to share all information freely) or an intention to deceive stakeholders. Even in the former situation, there should be a mechanism for internal reporting.

4. Identify and evaluate alternative actions (on the ethical issue).

The three decisions that we have consistently discussed in each case are: a) do nothing. b) consult or confer with a peer (ASA) or a supervisor (ACM); and c) report violations of policy, procedure, ethical guidelines, or law (ASA) or refuse to implement the system (ACM).

Option a) must be modified to simply be "disseminate your report informally (via secured transmission channels as appropriate)" (i.e., do nothing to create a formal system for sharing reports). This enables the ethical practitioner to always share their reports with stakeholders, following any relevant rules or limitations that would invariably also have been integrated into a formal dissemination system. Option b) would be, "consult with peers or supervisors to create a formal system" – achieving the only benefit in the SHA table. Option c), reporting violations of guidelines or policy, would not be feasible since "sharing" the result of peer evaluation may be implied to be informal – and/or already assumed to be included in that assignment. That is, *if* you perform an evaluation, *then* the review is only complete once shared. The only ethical issue arises when reports from evaluations or reviews are suppressed, or evaluators are prevented from sharing their reports with the appropriate stakeholders. If limitations or risks (or actual harms) are found in an evaluation and an organization (or supervisor) seek to suppress that report, it may constitute fraud. Also, just because a formal system for disseminating reports exists does not preclude unethical fraud-committers from preventing your notification of other stakeholders about problems or risks.

5. Make and justify a decision.

Note that all alternatives include the reviewer making whatever effort is necessary to ensure that the report is, in fact, disseminated. This can be more challenging when there is already some long-standing dissemination method (and this particular report is being suppressed), but would be straightforward to implement in a new system. Thus, option b) would be preferred over option a), to the extent that it is possible to create and implement a new system for dissemination. Notifying all stakeholders that an evaluation is under way would help to preclude unethical fraud-committers from preventing the notification of stakeholders about problems or risks that are identified in any evaluation or report. Note that, if there are confidentiality (or national security) considerations that prevent wide sharing, or the dissemination of a report across all stakeholders and those who might be affected by statistical practice, the ethical practitioner "follows applicable policies, regulations, and laws relating to their professional work, unless there is a compelling ethical justification to do otherwise." (ASA A11).

6. Reflect on the decision.

Construct your own reflection on this decision.

Summary of seven case analyses for Reporting

The seven case analyses of Chapter 3.7 were focused on reporting. This is as much a function of analysis and interpretation (ASA) or analysis/evaluation (ACM) as it is about documentation and communication about your work, and working on teams (next Chapter). You may also have noticed that, while most of the 'bad behavior' leading to the need for your response and its justification can be seen to be frankly unethical once you understand the practice standards, people who are not familiar with the GLs and CE may capitulate to incentives to cut corners or cherry-pick. The deceptive and detrimental research practices that the NASEM 2017 report *(Fostering Integrity in Research)* call out *exist* because of these incentives, and not simply because some people are dishonest (although unfortunately, dishonesty is sometimes a factor!).

When discussing how report ethically in Section 2, general principles from the GLs or CE, and specific questions from the DSEC and DEFW, could be used for guidance. However, in response to the ethical violations such as those in this Chapter, DSEC and DEFW were *not* very helpful in identifying whether or not an ethical problem existed, determining what exactly the ethical problem is or would be (if you failed to act), nor did they offer any suggestions for what to do. The practice standards, on the other hand, clearly did. You will have noticed a consistent message in the analyses of these vignettes: deceptive and detrimental research practices exist, and may be most obvious in reporting – that is also an important source for the harms that these practices do cause, for all stakeholders. More to the point, the ethical statistical practitioner and data scientist have clear obligations to recognize and do what they can to stop, and prevent, these deceptive and detrimental research practices. Failure to notify stakeholders of cherry-picking in reports is an important consideration in Cases 34-37, and ensuring the recognition of, and appropriate credit for, prior work (Case 39) was observed to not just violate the GLs and CE and demonstrate disrespect for others in your profession, but it also constitutes fabrication on your cv or resume. Thus, what some refer to as "detrimental" practices are actually incredibly harmful; and all of the SHA tables are filled with significant harms that are never offset with any meaningful benefits. These are important considerations whenever anyone suggests cutting corners – or engaging in "minimally detrimental practices" – justified, perhaps, by saying that "no one will find out" or "we would be the only ones who could be harmed, and those harms are minor". The SHA tables show that even if the first of these is true, the second *is not*.

Each of the cases in this chapter were based on actual events and all comments in the analyses that relate to "cultures" and/or "contexts" describe work environments and not specific anthropological cultures. As NASEM (2017) noted, detrimental practices are just as damaging as frank misconduct, if not more so, with ethical lapses specific to reporting embodying great potential for serious harms to others, including unknown individuals (anyone who wants to utilize a false or misleading report), the profession, and the public trust. There may be some incentives –or at least, no clear disincentives - to engage in these practices ("everyone does this!" or "no one will care!"), complicating the ethical practitioners' efforts to draw attention to them, correct them, and to prevent them recurring and especially from continuing to be incentivized.

Questions for Discussion:

1. Review at least two cases in this chapter.

A. Do you agree with the decisions in these case analyses? Why or why not? Which parts of the ER process are most and least acceptable for each case? Do these parts differ for different cases in this chapter? Were there different ASA or ACM Principles or elements that you identified, and if so, how do these affect your case analyses? Are the elements and/or Principles you identify represented in your reflections?

B. Discuss your results for the tests of universality, justice, and publicity for any (or all) case(s) in this chapter.

C. Discuss the relevance to each case – including the analysis and your reflection – of federal, state, or local/organizational policies regarding conflict of interest. How can you incorporate what you know about these policies into any part of the case analysis (prerequisite knowledge; identifying the ethical issue; determining alternative actions; making or justifying your decision; or reflection)? Note that, if there is no alignment between federal, state, or your local/institutional policies and a given case, you can comment on that (e.g., do policies in your organization focus solely on *misconduct* – and ignore detrimental or other unethical practices?). Do you feel that these policies could have helped in the reasoning, decision or justification?

D. Discuss the relevance to each case - including the analysis and your reflection – of responsible authorship and publication. How can you incorporate what you know about these policies into any part of the case analysis (prerequisite knowledge; identifying the ethical issue; determining alternative actions; making or justifying your decision; or reflection)? If there are no relevant policies, you should comment on that, particularly when a) data are obtained using federal monies (grants); or b) data are collected (acquired) from humans, such that Federal Regulations about the treatment of human subjects and the acquisition of tissues and data from humans are relevant.

2. Review at least two (different) cases in this chapter.

Keeping in mind that doing nothing/not responding are *not plausible responses*, are there other plausible alternatives (KSA 4) that you can think of for any case? If not, discuss that; if so, list them and discuss their evaluation (are they equally consistent with GL/CE, do they lead to similar decisions, etc.).

3. Choose any case in this chapter. Redo the analysis, but feature a different GL or CE principle in your reasoning process. Make sure the *justification* is different from what is given. Is the *decision* also different? Discuss your decision with its justification and the tests of universality, justice, and publicity.

4. Consider all seven cases in this chapter.

A. Comment on the role of the stakeholder analysis in these decisions. Were stakeholder impacts featured in justifications? Were they features in your own reflections?

B. Comment on the importance of your reflections on the decisions you were asked for: for others doing a similar task at work; for those using results from this task; and for those who might join the profession (and be learning about planning and design).

C. Consider your professional identity and the profession itself: planning/designing projects and data collection/analysis projects is obviously foundational in statistics and data science. How can these seven cases and the decisions reached via the analyses presented earn you (the decider) the trust of the practicing community or the public for the profession and its future?

Chapter 3.8
Engaging in team science/ working with others

Case 41. You notice a pattern of bullying by a senior team member.

Case 42. You are asked to do some coding/analysis by someone who is prevented from acknowledging that you helped. Your contribution cannot be recognized.

Case 43. You are told that you "only need to read/review your own work" and you are not allowed to see the final/full document or work product.

Case 44. You complete the analysis plan/system design, oversee its operation, and draft the report. Team members/collaborators do not respond to your report, so you continue to revise it over time until you are satisfied with your contributions, but do not have much from others. You suddenly receive a "new draft" of the document that includes none of the work you did, nor any of the documentation of the system or work that you had distributed to your team. The "new draft" appears to have many errors in the design/analysis and other reporting elements, but you are asked to "approve" the new draft – and agree to be/remain a co-author on the report – within the next two days.

Case 45. Someone on your team suggests a technical method to overcome a lack of consent from data contributors and collect their data even if they do not consent.

Case 46. You recognize the potential for "dual use" of your code, data, and/or results.

Case 47. You inadvertently discover that a proprietary "new method" that you were told to prepare for publication/patent application was actually published decades ago and was apparently unnoticed.

Case 41. You notice a pattern of bullying by a senior team member.

1. Identify and 'quantify' prerequisite knowledge:

NB:

"Workplace Bullying is repeated, health-harming mistreatment of one or more persons (the targets) by one or more perpetrators. It is abusive conduct that is:

Threatening, humiliating, or intimidating, or
Work interference — sabotage — which prevents work from getting done, or
Verbal abuse."[47]

If you believe you or someone else in your work group/at work is the victim of bullying, please check resources like the Workplace Bullying Institute (https://www.workplacebullying.org/) or your human resources and ombudsperson to ensure that your health and safety are protected!

Which DEFW and DSEC Principles or Areas seem most relevant to this vignette?

Not discussed in DSEC or in DEFW.

Which ASA and ACM Principles and elements seem most relevant to this vignette?

ASA:

Principle A. Professional integrity and accountability (A; A8; A9; A11; A12)
Principle G. Responsibilities of leaders, supervisors, and mentors in statistical practice (G; G2)
Principle H. Responsibilities regarding potential misconduct (H; H1; H2; (potentially H3-H8))
Appendix. Responsibilities of organizations/institutions (APP1; APP 11)

ACM:

1. General Ethical Principles (1.4)
2. Professional Responsibilities (2.1)

---

[47] https://www.workplacebullying.org/individuals/problem/being-bullied/

Potential result: Stakeholder:	HARM	BENEFIT	UNKNOWN	UNKNOWABLE
YOU	Not trying to stop bullying does not promote a respectful work environment. If the bully is not stopped, then they might start bullying you next.	**No benefit** can arise from not trying to stop bullying.	Not trying to stop bullying implicitly condones this behavior and fails to secure a respectful work environment for you and others.	Bullying in the workplace has been documented to drive diversity in practitioners and opinions down. In addition to implicitly condoning this behavior, not trying to stop bullying can ultimately limit creativity and innovation.
Your boss/client	Not trying to stop bullying implicitly promotes a hostile work environment. Bullied employees do not do their best work. However, it can be difficult to document and prove charges of bullying, creating negative feelings throughout the workplace.	**No benefit** can arise from not trying to stop bullying.	Not trying to stop bullying implicitly condones this behavior and fails to secure a respectful work environment for you and others.	Bullying in the workplace has been documented to drive diversity in practitioners and opinions down. In addition to implicitly condoning this behavior, not trying to stop bullying can ultimately limit creativity and innovation.

Potential result: Stakeholder:	HARM	BENEFIT	UNKNOWN	UNKNOWABLE
Unknown individuals	Bullying violates rights and acts against social/public good. Irreproducible results and low-quality work may result from bullied statisticians and data scientists. These weaknesses at/in work can then increase risks of bias and harms to individuals, science, and the public.	**No benefit** can arise from bullying.		Failures to stop bullying may support new "norms" that are not consistent with ethical practice guidelines, resulting in actual harms to individuals, as well as in harms to human rights.
Employer	Not trying to stop bullying implicitly promotes a hostile work environment. Bullied employees do not do their best work. However, it can be difficult to document and prove charges of bullying, creating negative feelings throughout the workplace.	**No benefit** can arise from not trying to stop bullying.	Not trying to stop bullying implicitly condones this behavior and fails to secure a respectful work environment for you and others. If senior workers are the bullies, it can be difficult to justify taking action against them; this may only intensify the perception (or actuality) of a hostility and disrespectful work environment.	Bullying in the workplace has been documented to drive diversity in practitioners and opinions down. In addition to implicitly condoning this behavior, not trying to stop bullying can ultimately limit creativity and innovation. Workplace bullies create liabilities for their employers.

Potential result: Stakeholder:	HARM	BENEFIT	UNKNOWN	UNKNOWABLE
Colleagues	Bullied employees do not do their best work. Not trying to stop bullying implicitly promotes a hostile work environment. This hostility and the disrespect of the bully may pervade relationships with other colleagues.	**No benefit** can arise from not trying to stop bullying.	Bullying undermines what might otherwise be a respectful work environment.	Bullying in the workplace has been documented to drive diversity in practitioners and opinions down.
Profession	Bullying violates rights and acts against social/public good. Bullies in the profession will undermine interest from other professionals to join groups with rumors of, or acknowledged, bullies. Bullied statisticians and data scientists may not do their best work, which increases risks of bias and harms to other individuals (stakeholders in this weakened work), science, and the public.	**No benefit** can arise from bullying.		Failures to stop bullying may support new "norms" that are not consistent with ethical practice guidelines. Bullying and other forms of intimidation have been documented to negatively impact creativity as well as the diversity of practitioners in statistics and data science.

Potential result:	HARM	BENEFIT	UNKNOWN	UNKNOWABLE
Stakeholder:				
Public/public trust	Bullies in the workplace often do not directly affect the public or public trust. However, bullying supervisors can limit or impair the practitioners' abilities to do their best work. Irreproducible results and low-quality work lower, and do not augment, informedness.	**No benefit** can arise from bullying.		

Table 3.8.1. *Stakeholder Analysis template: A senior team member exhibits a pattern of bullying*

2. Identify decision-making frameworks.

Both ACM and ASA practice standards have explicit considerations for working with others, and both are concrete in their total condemnation of any sort of harassment, bias, intimidation, and other coercive or negative personal behaviors. Following the virtue perspective would lead to both "ideal" behaviors by working team members and also an intolerance for professional misconduct; ASA A8 specifies that the ethical practitioner, "Promotes the dignity and fair treatment of all people. Neither engages in nor condones discrimination based on personal characteristics. Respects personal boundaries in interactions and avoids harassment including sexual harassment, bullying, and other abuses of power or authority", while A9 states that the ethical practitioner goes further: *"takes appropriate action when aware of such unethical practices by others."* (ASA A9; emphasis added.) In the vignette, the practitioner is not the bully, the practitioner *observes the bullying*. ASA A9 articulates the ethical practitioner's responsibility to **take appropriate action**, because bullying is specifically condemned in ASA A8 (among other reasons).

From a utilitarian perspective, the SHA highlights that **only harms** accrue to all stakeholders from bullying in the workplace. The ACM CE states that "the dignity of employers, employees, colleagues, clients, users, and anyone else affected either directly or indirectly by the work should be respected throughout the process" (2.1) – identifying bullying as unethical as it violates this CE Principle. The CE also specifies that ethical computing professionals *"should take actions to resolve the ethical issues they recognize*, including, when reasonable, expressing their concern to the person or persons thought to be violating the Code." (ACM 4.1; emphasis added.)

3. Identify or recognize the ethical issue.

Bullying and intimidation are not illegal (as of May 2022), but do represent a distinct lack of respect for those who are being harassed. Both the ASA and ACM specifically point out that dignity must be respected/harassment is unethical. As noted, this vignette describes the observation of, and not engagement in, bullying. Both practice standards charge the ethical practitioner with a responsibility to take action against frankly unethical and unprofessional behavior like bullying. ASA H2 specifically describes the responsibility to "Avoid condoning or appearing to condone statistical, scientific, or professional misconduct. Encourage other practitioners to avoid misconduct or the appearance of misconduct.", like ACM 4.1, so the ethical issue arises for the practitioner if they do not take whatever action is possible to stop that bullying. Note that bullying and intimidation are unethical (and professional misconduct), but so is observing this and doing nothing. While an individual might not be able to address the bully, the individual can still take action by supporting the person being harassed and possibly by notifying appropriate personnel within the organization (e.g., ombudsperson, human resources staff).

4. Identify and evaluate alternative actions (on the ethical issue).

The three decisions that we have consistently discussed in each case are: a) do nothing. b) consult or confer with a peer (ASA) or a supervisor (ACM); and c) report violations of policy, procedure, ethical guidelines, or law (ASA) or refuse to implement the system (ACM).

Clearly, the practice standards make option a) impossible because the ethical issue for the practitioner in this vignette *only arises if you do nothing*. The fact that the person doing the bullying in this vignette is a "senior member of the team" complicates the potential alternative actions to take. We might instead change option a) from "do nothing" (obviously unethical!) to "advise the persons being bullied to seek redress or change to another team/job". Another version of option a) could be, "assist the persons being bullied in collecting evidence of the bullying pattern". Note that your observation (i.e., the vignette) constitutes independent verification of the bullying victims' experiences, which can be fortifying for them as well as strengthening any case they may seek to bring.

Option b), "consulting with a peer", actually resulted in a globally shared blog post describing in 2017[48] the experiences of multiple victims of serial sexual harassment/sexual assault, and ultimately both of the individual perpetrators did lose their jobs as a result[49]. Consulting a peer in these cases actually identified the *serial* and predatory nature of both the harasser and the assaulter; and it may also have helped the victims feel a bit better – none of them could have anticipated that these conversations would ultimately have the result they did. But if your organization has an ombudsperson and/or human resources departments, where you can make anonymous reports of bullying and other inappropriate, disrespectful, and potentially illegal behaviors, you should report what you observed and also, try to convince the victims you observed being bullied to make a report as well. Reporting to your leadership may not be possible, although the ACM CE suggests that this is what you should do, so

[48] Read the original report here https://medium.com/@kristianlum/statistics-we-have-a-problem-304638dc5de5
[49] https://www.dailydot.com/debug/google-ai-researcher/http://www.citypages.com/news/minnesota-professor-brad-carlin-resigns-amid-sexual-harassment-complaints/482944771

that the leadership can reinforce their policies against harassment, intimidation, coercion and bullying; or create policies explicitly prohibiting such behaviors if they do not exist (which is the obligation of the leader, per ACM Principle 3). Thus, consulting with a supervisor may also be an important aspect of option b).

Option c), report the violation, may not be a viable option. One can only report violations of policy when they exist; so, if you work in an organization that does not (yet) have a policy against harassment (of any sort), intimidation, or bullying, then reporting this behavior to the organization will not be effective; in the event that the harassment or bullying are not illegal, then reporting will be similarly ineffective. The ACM –and not the ASA – specify that acting in a way that violates the CE should be reported **to that individual** and also to the ACM. In 2017-2018[50], this type of reporting (of sexual harassment and assault- which is, of course, illegal in the US) actually resulted in the professional organizations to which the behavior was reported stripping the perpetrators of their Fellowship and expelling them from the society; they also updated their code of conduct and procedures to ensure the code is upheld[51]. However, such reporting may not be available to you in your organization or your professional society (e.g., if you belong to the ASA, which has no such policy as of May 2022). Finally, what did happen when the behavior was ultimately reported to these organizations was that those predators were removed from their positions where predation was facilitated. This of course benefits the entire community directly –safeguarding others – and indirectly, by removing a real barrier to participation that fear of predation causes.

One alternative that has arisen in other case analyses is option c'), "consider a different team, boss, or even job". Typically, people who quit will have "exit interviews" and while this might be the most indirect way to report the unprofessional behavior, leaving the team (or organization) might lower at least some impediments to taking action that arise because the bully is a "senior member of the team". If they're not part of your team anymore, then you might be freer to report them, or "when reasonable, expressing (your) concern to the person or persons thought to be violating the Code". If you shift teams, the new team leader may ask you why you left the last team. While not a "formal" report, notifying this new team leader that you recognize that bullying is unacceptable – and having demonstrated that you're willing to move because

---

[50] Read about the 2017 case here https://www.bloomberg.com/news/articles/2017-12-16/google-researcher-accused-of-sexual-harassment-roiling-ai-field
[51] https://bayesian.org/wp-content/uploads/2019/03/ISBA-Code-of-Conduct-20-02-19.pdf

you will not accept such behavior – may yield benefits. Options c) and c') might only be plausible for some people – and in specific contexts where reporting is feasible, policies exist, there are enough teams to make a move, or some combination of these features.

5. Make and justify a decision.

Decision: All of the options include you *doing something*. Option a) might be a precursor to each of the other options, because without evidence, reporting would be difficult. Depending on your role in the organization, and your relationship with the senior bully (e.g., if you report to them), helping or advising the victims to collect evidence - and adding yours to theirs - might be the limit of what you are comfortable doing. This option may be effective, and it *is* consistent with the practice standards. Options b) and c) are only feasible if there are individuals with whom you can effectively consult or confer (b) or to whom you can meaningfully report (c). Option c') could be viewed as the most extreme, and may not be feasible in many or most cases – there has to be another job and an exit interview process for c') to be a real option.

Doing what you can, whatever is appropriate, to notify others of the unacceptable behavior going on so as to actively discourage it, is an important feature of both the ACM and ASA standards. A combination of conferring, consulting, and discussing unprofessional and unacceptable behavior may – as it did in 2017 – lead to the discovery of a wider pattern which will then be easier to report and for authorities/leaders to act on. Seeking any means that you can find within your organization to report the behavior is also important – so if your circumstances permit it, the decision that is most consistent with the ethical practice standards is to do both options b and c, to the extent each is possible. The most important aspect of the decision is that if you do nothing, you violate ethical practice standards and also become complicit in the professional misconduct of the bully.

6. Reflect on the decision.

Seniority of staff can complicate your efforts to follow ethical practice standards; if they are your boss or supervisor, or friends with your boss/supervisor, it can seem daunting to report to them that their friend or close colleague is behaving in an unprofessional (and unethical) way. While not specified, the ACM preamble does state that "(t)he entire computing profession benefits when the ethical decision-making process is accountable to and transparent to all stakeholders. Open discussions about ethical issues promote this accountability and transparency." It will be supremely uncomfortable to initiate a conversation about unacceptable behavior in the workplace –

especially when it is exhibited by senior team members; however, it is essential to keep in mind that *doing nothing is also unacceptable*. In Chapter 3.9 I discuss how important it is to consider exactly how you will go about initiating, and then responding within, a conversation around an ethical challenge and/or the response you propose that will address that ethical issue. Like all the CE and GLs, there is no algorithm or rule about how to go about initiating and then completing conversations that may be uncomfortable.

In this vignette, however, the violation of the practice standards and applicable laws, and probably also policies in your workplace, are clear. It should also be clear that *to do nothing is unethical* and *unacceptable*. Each of the alternative responses to the vignette passes the *tests of publicity, universality*, and *justice*. Although it might seem counter to your career trajectory to let it be known that you blew the whistle on a bully in the workplace, the *test of publicity* would instead let it be known that you do not condone or tolerate unethical behavior and professional misconduct (and that is an *excellent* thing to publicize!). If everyone in the workplace either supported or helped in the documentation of misconduct or just bullying (i.e., option a), it would eventually become impossible for bullies to disrespect colleagues in this way; and reporting such misconduct might result more directly in removing those bullies from the workplace entirely. Therefore, all options as discussed do pass the *test of universality*. In addition to taking direct action to stop bullying in the workplace, all of the options tend to create a more uniformly respectful workplace for all – so all options as discussed also pass the *test of justice*.

Case 42. You are asked to do some coding/analysis by someone who is prevented from acknowledging that you helped. Your contribution cannot be recognized.

1. Identify and 'quantify' prerequisite knowledge:

Which DEFW and DSEC Principles or Areas seem most relevant to this vignette?

Not discussed in DSEC or in DEFW.

Which ASA and ACM Principles seem most relevant to this vignette?

ASA:

Principle A. Professional integrity and accountability
Principle E. Responsibilities to members of multidisciplinary teams
Principle F. Responsibilities to fellow statistical practitioners and the Profession
Principle G: Responsibilities of leaders, supervisors, and mentors in statistical practice
Principle H. Responsibilities regarding potential misconduct

ACM:

1. General Ethical Principles
2. Professional Responsibilities

Potential result: Stakeholder:	HARM	BENEFIT	UNKNOWN	UNKNOWABLE
YOU	If you fulfil this request, you will be violating ASA H2, and you may end up limiting the completeness of the documentation of whatever you do, preventing others from fully evaluating the work that includes your contribution. Your agreeing to the request will both violate the GLs and CE, and also lead to the requestor violating them as well.	Although the requestor may benefit from not needing to do whatever you were asked to do, no real benefits can possibly accrue when you condone, or appear to condone, violations of the GLs and CE.		
Your boss/client	If the requestor has insufficient skill or time (or both) to accomplish their assignment, your completing any part of it for them will mislead the client/boss. Cultures that "require" workers to exceed their capabilities negatively impact workers as well as the work they do.	Although the work would 'get done', no real benefits can possibly accrue when you condone, or appear to condone, violations of the GLs and CE.		

Potential result:	HARM	BENEFIT	UNKNOWN	UNKNOWABLE
Stakeholder:				
Unknown individuals	Requestor will violate ASA A5 (gives credit to those who contribute), ASA E3 (promotes transparency in all statistical practices) and ASA H2 (avoids condoning or appearing to condone statistical, scientific, or professional misconduct).			Cultures that "require" workers to exceed their capabilities negatively impact workers as well as the work they do.
Employer	If the requestor has insufficient skill or time (or both) to accomplish their assignment, your completing any part of it for them will mislead the employer.	Although the work would 'get done', no real benefits can possibly accrue when you condone, or appear to condone, violations of the GLs and CE.		Identifying that someone in your company tried to "pass off" someone else's work as their own could undermine trust in your company and limit business/profit.

Potential result: Stakeholder:	HARM	BENEFIT	UNKNOWN	UNKNOWABLE
Colleagues	Colleagues must respond to all relevant aspects of what they get at their point in a workflow. If the requestor simply forwards your work on, there will be no way for colleagues to fully understand it or ask questions for clarification.	Although the work would 'get done', no real benefits can possibly accrue when someone else's work is passed off as your own to colleagues.		A "solution" – or results - that come from analyses, and your colleagues' work based on that analysis, could be misused, and go undetected. Your colleagues may end up building in bias, unfairness or other – unauditable- problems because they assumed your analysis was appropriate for the data.

Potential result: Stakeholder:	HARM	BENEFIT	UNKNOWN	UNKNOWABLE
Profession	The requestor will violate multiple ASA GL Principles. Recycling someone else's work suggests incompetence; failing to acknowledge them is unethical. *stakeholders may lose faith in the profession* when practitioners prioritize expedience over competence.		If you respond and do so correctly, there is no way for you to know if your work will be used or represented correctly. Blame for misuse or errors/bias will fall on the profession generally, when it was the requestor's unethical request that caused the GL/CE violations.	Cultures that "require" workers to exceed their capabilities negatively impact workers as well as the work they do.
Public/public trust	Ineffective design/poor planning – especially if they lead to misuse/abuse can undermine public trust in systems and the profession			

Table 3.8.2. *Stakeholder Analysis: Request for your help – which cannot be acknowledged*

2. Identify decision-making frameworks.

The virtue perspective on this case suggests that the requestor is, at a minimum, requesting that you condone a violation of ASA A5 ("Accepts full responsibility for their own work, does not take credit for the work of others, and gives credit to those who contribute. Respects and acknowledges the intellectual property of others"). It is not unethical for someone to ask for help (and in fact this is consistent with ASA GL A1 ("takes responsibility for evaluating potential tasks, assessing whether they have (or can attain) sufficient competence to execute each task, and that the work and timeline are feasible. Does not solicit or deliver work for which they are not qualified, or that they would not be willing to have peer reviewed."), but ASA E3 notes that the ethical practitioner "ensures that all communications regarding statistical practices are consistent with these Guidelines. Promotes transparency in all statistical practices." Concealing your contribution undermines the transparency, and violates ASA H2 ("avoids condoning or appearing to condone statistical, scientific, or professional misconduct"). Note that it is the prevented acknowledgement of your contribution that leads to all of these violations: if you accept the request and violate ASA H2, you will also be acting contrary to the very purpose of the ASA GLs, "All statistical practitioners are expected to follow these Guidelines and to encourage others to do the same".

The utilitarian perspective on this case, usually driven by considerations of the ACM CE, are actually driven by the violations of the ASA GLs that will accrue if you do accept the request for help. Those violations all harm the profession, and by extension, the public trust in the profession. The ACM CE (1.3) charges computing professionals with the responsibility to be honest and trustworthy, which the requestor of your unacknowledged assistance is asking you to violate. Although technically the work you're requested to help with will get done if you agree (and presumably would not be completed if you refuse), no real benefits can possibly accrue to any stakeholder by concealing the contributors to the work. ACM CE 2.1 specifies that "computing professionals should insist on and support high-quality work from themselves and from colleagues." While there is obviously a technical dimension to "high quality work", there is also an ethical dimension, specifically, ACM 1.5 which charges the ethical practitioner to "Computing professionals should therefore credit the creators of ideas, inventions, work, and artifacts, and respect copyrights, patents, trade secrets, license agreements, and other methods of protecting authors' works." The utilitarian perspective in this case supports refusing this request.

3. Identify or recognize the ethical issue.

In this case, it is difficult to recognize the ethical issue. A key word to focus on is "prevented"; it suggests that this individual (the requestor) recognizes that they should acknowledge your work (i.e., if they would follow the ASA and ACM standards), but that someone is actively impeding their ability to follow the guidelines. It is clearly not unethical for you to contribute to other people's work or projects by doing coding or analysis, but the prevention of that contribution to be recognized is worrying. If the individual asks for your help because the work is outside of their skill set, then *they – not you* (whether you agree or refuse to contribute) – are behaving contrary to the practice standards (ASA: A1, A5; ACM: 2.6). In that case, then if you do agree to contribute, you would be violating ASA H2 by condoning that professional misconduct: the individual would in fact be passing your abilities off as their own (unethical!), but by facilitating it, you are in fact, engaging in unethical behavior. If the individual notified the original requestor that they could not do the work themselves and they were told to do it anyway –but prevented from getting help, then *they* are violating ASA H2 by condoning that professional misconduct. If you do help, then you are violating H2 as well.

4. Identify and evaluate alternative actions (on the ethical issue).

The three decisions that we have consistently discussed in each case are: a) do nothing. b) consult or confer with a peer (ASA) or a supervisor (ACM); and c) report violations of policy, procedure, ethical guidelines, or law (ASA) or refuse to implement the system (ACM).

This may be the one case where the practice standards do not make option a) impossible, although it can be modified to simply be "refuse this request". The fact that the person doing the requesting in this vignette is "prevented" from following the practice standards themselves complicates the potential alternative actions to take. Since the ethical issue only arises if you agree to contribute, the alternatives need to focus on that: simply refusing to condone this misconduct –whether the true misconduct is on the part of the requestor or whomever is preventing them from acknowledging your contribution – does not require you to confer with anyone (option b). However, if the requestor works on your team and this is not the first time that they have accepted a work assignment that they are incapable of completing without secret help from you or others, then option b) would then entail conferring with a peer or supervisor about how to better assign tasks to this person –or, how to ensure their skills are brought up to the needed levels – will ensure that no one is asked to condone this sort of misconduct from the requestor in the future. Thus, the second time this happens, simply refusing the request (option a) will actually be the same as "do nothing" and will *not* be a viable alternative. Option c), reporting violations of guidelines or policy, would be the *only possible ethical option* in the case that the requestor would have shared confidential or secure information with you in order to ask for your help –and that is true whether or not this is the first time they request help. Otherwise, if the request for help does not violate security/confidentiality protocols, then "reporting" (option c) and "conferring" (option b) are actually the same thing, because you'd be notifying supervisors of the individual's inappropriate acceptance of tasks outside of their skill set (while also seeking ways to prevent such assignments, or to improve their skills so this situation is avoided in the future).

So the options would be: a) refuse this request; b) refuse the request and confer with a peer or supervisor about how to better assign tasks to this person/promote their professional growth. In the event that the requestor has breached security by making the request, option c) becomes, "refuse the request and report violations of guidelines and policy." Note that option c) does not actually preclude conferring about how better to assign tasks – but it may be that the breach of policy leads to the requestor getting reassigned or fired.

5. Make and justify a decision.

Note that all alternatives include the decision to *refuse the request* for help, which might seem to violate ASA Principle F (specifically ASA F2). However, Principle F itself states, "Responsibilities to other practitioners and the profession include honest communication and engagement that can strengthen the work of others and the profession." Concealment of your contributions may assist the other practitioner on this one task, but is not honest, and does not strengthen the profession. You do not want to act in a manner that encourages misconduct (i.e., in violation of ASA H2) *by the original requestor*, who specified that no one can help or that help cannot be acknowledged. You also do not want to encourage the practitioner who asks for your uncredited assistance. Both of these requests are detrimental practices and should be discouraged actively. The ethical courses of action open to you in a case like this one (options a, b, c above) are not the only options. While no plausible *ethical* alternatives will include you accepting the request – specifically because your contributions will not be acknowledged, additional considerations that may be relevant include helping to request an extension on time or resources for the practitioner-requestor to get the task done. If the practitioner making the request does have sufficient skills, and simply insufficient time and/or resources, then there may be options to help secure additional resources. If your organization tends to punish people who learn or work more slowly – driving some to request help that cannot be acknowledged as in this vignette, then that culture needs to change. In that case, option c) might change from "refuse the request and report violations (of security policy)" to "refuse the request and report the organization's policy of giving workers tasks they are not qualified or sufficiently resourced to complete". If you are unable to affect that needed change, then it may be necessary to find work elsewhere: a workplace culture of pushing people to the point where they must violate ethical practice standards[52] should never be supported.

6. Reflect on the decision.

Construct your own reflection on this decision.

---

[52] Read more about the Volkswagen Emissions case: https://en.wikipedia.org/wiki/Volkswagen_emissions_scandal

Case 43. Your supervisor tells you that you "only need to read/review your own work" and you are not allowed to see the final/full document or work product.

1. Identify and 'quantify' prerequisite knowledge:

Which DEFW and DSEC Principles or Areas seem most relevant to this vignette?

This situation is inconsistent with DEFW Principle 2 because the practice standards ("codes of practice") both document the importance of peer review. Principle 5 may be limited if you are unable to confirm your approach is the most robust one possible. However, DEFW Principle 6, ensure your work is transparent and be accountable, can and should be followed in this case because your work will, at some point, be reviewed.

By limiting your recourse to reviews, DSEC items relating to considering sources of bias may not be addressable. While *you* may have considered sources, other perspectives (DSEC C1; C2; D1; D5) may not do so/have done so. Most other DSEC items are unaffected by this supervisor's behavior.

Which ASA and ACM Principles seem most relevant to this vignette?

ASA:

Principle A. Professional integrity and accountability
Principle B. Integrity of data and methods
Principle C. Responsibilities to stakeholders
Principle D. Responsibilities to research subjects, data subjects, or those directly affected by statistical practices
Principle E. Responsibilities to members of multidisciplinary teams
Principle F. Responsibilities to fellow statistical practitioners and the Profession
Principle G. Responsibilities of leaders, supervisors, and mentors in statistical practice
Principle H. Responsibilities regarding potential misconduct
Appendix. Responsibilities of organizations/institutions

ACM:

1. General Ethical Principles
2. Professional Responsibilities
3. Professional Leadership Responsibilities

Potential result: Stakeholder:	HARM	BENEFIT	UNKNOWN	UNKNOWABLE
YOU	Without review, you may have missed an opportunity to innovate, or to apply a new method. You may have made an error. You are solely responsible for the accuracy of your work.	You are solely responsible for the accuracy of your work – meaning that you can take full responsibility for it, as the independent practitioner that you are.		
Your boss/client	Peer review takes time, and if errors are found, they will take even more time to fix. Key harms or risks of harms may be missed. Independent confirmation that the system/analysis will operate as intended supports valid interpretation and limits misuse opportunities –those safeguards are missing without peer review.	Trusting all practitioners to take full responsibility for independent work is faster and possibly cheaper than requiring peer review. By preventing anyone from reviewing the complete document, you can say anything you want.		

Potential result: Stakeholder:	HARM	BENEFIT	UNKNOWN	UNKNOWABLE
Unknown individuals	System and analysis planning can inadvertently create harms or risks of harms; without peer review, these may be missed.			
Employer	Peer review takes time, and if errors are found, they will take even more time to fix. Key harms or risks of harms may be missed. Independent confirmation that the system/analysis will operate as intended supports valid interpretation and limits misuse opportunities –those safeguards are missing without peer review.	Trusting all practitioners to take full responsibility for independent work is faster and possibly cheaper than requiring peer review.		

Potential result:	HARM	BENEFIT	UNKNOWN	UNKNOWABLE
Stakeholder:				
Colleagues	Peer review takes time, and if errors are found, they will take even more time to fix. Key harms or risks of harms may be missed. Independent confirmation that the system/analysis will operate as intended supports valid interpretation and limits misuse opportunities –those safeguards are missing without peer review. Colleagues' work is also unreviewed, possibly compounding problems.	Trusting all practitioners to take full responsibility for independent work is faster and possibly cheaper than requiring peer review.		A "solution" – or results - that come from unreviewed analyses, and your colleagues' work based on those analyses, could be misused, and go undetected if the original solution was not correct or made unsupported assumptions. Your colleagues may end up building in bias, unfairness or other – unauditable- problems because they assumed your work did not need review (i.e., was perfect).

Potential result: Stakeholder:	HARM	BENEFIT	UNKNOWN	UNKNOWABLE
Profession	Key harms or risks of harms may be missed in any work such that when discovered later, the profession is blamed. Independent confirmation that the system/analysis will operate as intended supports valid interpretation and limits misuse opportunities –those safeguards are *missing* without peer review.	Trusting all practitioners to take full responsibility for independent work – i.e., eliminating peer review – suggests that all practitioners are fully competent and possibly strengthens the perception of the profession.	If a statistician or data scientist designs or employs inappropriate methodology, but generates "interpretable results", the outcomes will be error-prone and could be misused. Blame for misuse or errors/bias will fall on the profession, particularly since peer review – per practice standards – could easily have been implemented.	
Public/public trust	If peer review identifies weaknesses or limitations – especially if they did lead, or could have led, to misuse/abuse it can undermine public trust in systems and the profession.	If peer review identifies weaknesses or limitations – especially if they could have led to misuse/abuse – then those will be fixed before those harms/risks of harms can accrue to the public. This can *bolster* public trust in systems and the profession.		

Table 3.8.3. *Stakeholder Analysis: Supervisor limits your access to the final work product*

2. Identify decision-making frameworks.

ASA GL Principle F describes the responsibilities that all statistical practitioners have towards others in the profession and those who are practicing. Specifically, the ethical statistical practitioner "helps strengthen the work of others through appropriate peer review" (ASA F2). However, the virtue perspective on this vignette suggests contributions from every ASA GL Principle and the Appendix. The potential for the ethical practitioner to assure the validity and interpretability, and reproducibility of the work to which they are contributing, is threatened in this case. Moreover, the imposed limitations are contrary to ethical practice by leaders (ASA Principle G) and the cultivation of an ethical workplace (ASA Appendix). In some circumstances, leaders or organizations must compartmentalize work due to security issues (ranging from proprietary business to national security considerations). While the ethical practitioner performs all of their tasks to the best of their abilities and with all of the ASA Ethical Guidelines in mind, active prevention of full understanding of the project is worrisome. The ethical practitioner should carefully consider whether the repression of details of the project are actually warranted (e.g., by proprietary or security considerations) or are instead efforts to circumvent ethical practice by compartmentalizing contributions.

ACM CE Principle 2.5 ("accept and provide appropriate professional review") is consistent with its overall utilitarian perspective to identify and try to mitigate potential harms (ACM 1.2). However, some of the key harms that accrue when no one but "the supervisor" is allowed to review a completed document is that they may – unethically – cherry-pick results or somehow misrepresent what was originally carefully and correctly reported. As with the virtue perspective, the inability to fully comprehend the potential for harms is worrisome from the utilitarian perspective.

3. Identify or recognize the ethical issue.

This case is complicated by the fact that, while you as an ethical practitioner can follow the GLs and CE, and accept work you are competent to complete, and do so in a transparent and accountable manner, your supervisor (and/or your work context) are failing to follow the specific practice standards that exist simply to promote everyone in the profession doing their best work. We don't know the origins of the "need only to review your own work" – is it a need arising from policy, security, or the supervisor's desire to subvert otherwise ethical practice? The ethical issues the limitation on peer review creates are: 1. ethical practitioners have an obligation to *provide* competent peer review to help others in their profession (and this is being blocked); and 2. Ethical practitioners have an obligation to *seek* competent peer review to help them(selves) maintain currency and expertise in the profession (and this is being blocked). Thus, the act of limiting peer review is itself contrary to the GLs and CE because it effectively limits the development or maintenance of expertise -as well as potentially preventing the identification of misleading representation of work that was done (by you) in an ethical and appropriate manner. The lack of competent peer review, at worst, may render identifiable mistakes invisible, mislead stakeholders, and lead to harms or risks of harms that could have been prevented. Thus, there are real harms that the policy (or this supervisor) may be creating – and hiding. You are not being prevented from doing all your ethical best to follow the GLs or CE, but there are aspects of ethical practice that are impeded in this case. The most obvious ones are ASA E3 "Ensures that all communications regarding statistical practices are consistent with these Guidelines. Promotes transparency in all statistical practices.", ASA F2 "Helps strengthen, and does not undermine, the work of others through appropriate peer review or consultation. Provides feedback or advice that is impartial, constructive, and objective.", ASA H1 "Knows the definitions of and procedures relating to misconduct in their institutional setting. Seeks to clarify facts and intent before alleging misconduct by others. Recognizes that differences of opinion and honest error do not constitute unethical behavior." The supervisor is preventing ethical practitioners from understanding their contributions in context, which may result in otherwise preventable harms (ACM 1.2).

4. Identify and evaluate alternative actions (on the ethical issue).

The same three decisions we usually use are applicable in this case: a) do nothing; b) consult or confer with a peer (ASA) or a supervisor (ACM); and c) report violations of policy, procedure, ethical guidelines, or law (ASA) or refuse to implement the system (ACM). In this case –unlike any others! – "do nothing" is a viable option; since no aspect of ethical practice besides stewardship of the profession is threatened by the refusal to give or let you obtain peer review, option a) is: "do the "peer review" yourself" (i.e., do not continue to request independent review). While it seems oxymoronic, you can simply adapt an existing rubric for peer review – e.g., for reviewers of grants (e.g., https://grants.nih.gov/grants/peer/critiques/rpg.htm), or of manuscripts (https://publons.com/blog/how-to-write-a-peer-review-12-things-you-need-to-know/), or of any other type of document that is relevant for your particular work. Then, because you are following the GLs and CE, your work is complete and transparent so you should be able to document whether and how your work meets "peer review" criteria that you can also be transparent about. The other options need to be modified, but not because you need to consult on how to formulate a response (b) or to report this behavior (c). Instead, you might modify option b) to "consult with a peer or supervisor to create a mechanism by which peer review can be effectively (and lawfully) implemented in your workplace or team; and meanwhile, do the "peer review" yourself". It may be that no peer review is possible due to security concerns; this makes consultation essential. Finally, option c) can be modified from "report" to "inquire", because impeding professional development may be contrary to the GLs and CE, but it is possible that your organization has important (possibly legal) reasons to do so with respect to peer review. Thus, option c) could become, "inquire about policies that could support –or be modified to support – the creation of a mechanism by which peer review can be effectively (and lawfully) implemented in your workplace or team; and meanwhile, do the "peer review" yourself".

5. Make and justify a decision.

Decision: Note that option a) is the essential response, whether or not you are able to accomplish options b or c. All three alternatives include doing the peer review yourself. As an independent practitioner, you are most likely already reviewing your work before you turn it in/hand it on to the next person on your team or in the workflow. The practitioner in this case may have been asking specifically for help with professional development or to generate some kind of independent documentation of their professional status/expertise. By selecting a peer reviewing tool and utilizing it on your own work, you are documenting your commitment to your own professional growth and development. Once this practitioner has moved out of the situation or team where peer review is not available, they would be able to document the growth as well as initiative in furthering their own skill set, to any future employer or team leader. Ethical practitioners must consider the potential for negligent (at best) and nefarious (at worst) policies that prevent their understanding the use and ultimate representation of their work. If these policies are in place for legitimate reasons (e.g., security, proprietary confidentiality), the ethical practitioner must hope that all of the contributors are equally ethical practitioners. If the policies are in place for less legitimate (or frankly illegitimate) reasons – like, a desire to assure that ethical practices by some practitioners do not interfere with unethical or illegal activities – then the ethical practitioner must contemplate whether that workplace is the best environment for them to be working.

6. Reflect on the decision.

Construct your own reflection on this decision.

Case 44. You complete the analysis plan/system design, oversee its operation, and draft the report. You suddenly receive a "new draft" of the report that excludes all of the work you did, nor does any of the documentation of the system or work from your original report appear. You have other urgent project deadlines, but briefly scan the document. You can tell without carefully reading it that the "new draft" has obvious errors in the design/analysis, results, and other reported elements, but you are asked to "approve" the new draft – and agree to be/remain a co-author on the report – within the next two days.

1. Identify and 'quantify' prerequisite knowledge:

Which DEFW and DSEC Principles or Areas seem most relevant to this vignette?

Not discussed in DSEC or in DEFW.

Which ASA and ACM Principles seem most relevant to this vignette?

ASA:

Principle A. Professional integrity and accountability
Principle B. Integrity of data and methods
Principle C. Responsibilities to stakeholders
Principle E. Responsibilities to members of multidisciplinary teams
Principle F. Responsibilities to fellow statistical practitioners and the Profession
Principle G. Responsibilities of leaders, supervisors, and mentors in statistical practice
Principle H. Responsibilities regarding potential misconduct
Appendix. Responsibilities of organizations/institutions

ACM:

1. General Ethical Principles
2. Professional Responsibilities
3. Professional Leadership Responsibilities

NB: The International Committee of Medical Journal Editors (ICMJE) recommends that authorship be based on the following four criteria:

1. Substantial contributions to the conception or design of the work; or the acquisition, analysis, or interpretation of data for the work; AND
2. Drafting the work or revising it critically for important intellectual content; AND
3. Final approval of the version to be published; AND

4. Agreement to be accountable for all aspects of the work in ensuring that questions related to the accuracy or integrity of any part of the work are appropriately investigated and resolved.[53]

---

[53] Read more about these criteria at http://www.icmje.org/recommendations/browse /roles-and-responsibilities/defining-the-role-of-authors-and-contributors.html

Potential result: / Stakeholder:	HARM	BENEFIT	UNKNOWN	UNKNOWABLE
YOU	You falsify your cv/resume, making a *deliberately misleading* claim, if you agree to be listed as an author without meeting criteria for authorship. If you endorse work you know to be errorful, you demonstrate a lack of competence or care (or both).	You gain a publication, but only by falsifying your cv/resume and taking responsibility as well as credit for work you know to have errors in it.	Perpetuation of false attribution/ authorship undermine your credibility as well as that of everyone on the author list, while also perpetuating detrimental research practices.	Detrimental research practices like accepting authorship credit without meeting any criteria undermine science, the public trust in the scientific enterprise, and the entire scientific community.
Your boss/client	Perpetuation of false attribution/ authorship undermines your credibility as well as that of *everyone on the author list.*			

Potential result: Stakeholder:	HARM	BENEFIT	UNKNOWN	UNKNOWABLE
Unknown individuals	Making a *deliberately misleading* claim, by agreeing to be listed as an author without meeting criteria for authorship, suggests you are unreliable and dishonest.		Perpetuation of false attribution/ authorship undermines your credibility as well as that of everyone on the author list, while also perpetuating detrimental research practices.	Detrimental research practices like accepting authorship credit without meeting any criteria undermine science, the public trust in the scientific enterprise, and the entire scientific community.
Employer	Perpetuation of false attribution/ authorship undermines your credibility as well as that of *everyone on the author list*. By extension, the employer is implicated as an endorser of detrimental practices.		Perpetuation of false attribution/ authorship undermines your credibility as well as that of everyone on the author list, while also perpetuating detrimental research practices.	Detrimental research practices like accepting authorship credit without meeting any criteria undermine science, the public trust in the scientific enterprise, and the entire scientific community.
Colleagues	Perpetuation of false attribution/ authorship undermines your credibility as well as that of *everyone on the author list*.		Perpetuation of false attribution/ authorship undermines your credibility as well as that of everyone on the author list, while also perpetuating detrimental research practices.	Detrimental research practices like accepting authorship credit without meeting any criteria undermine science, the public trust in the scientific enterprise, and the entire scientific community.

Potential result: Stakeholder:	HARM	BENEFIT	UNKNOWN	UNKNOWABLE
Profession	Perpetuation of false attribution/ authorship undermines your credibility as well as that of *everyone on the author list*. You appear either not to care about the errors, not to know about the errors, or both; undermining the credibility of the profession in making contributions to documents/ publications.		Perpetuation of false attribution/ authorship undermines your credibility as well as that of everyone on the author list, while also perpetuating detrimental research practices.	Detrimental research practices like accepting authorship credit without meeting any criteria undermine science, the public trust in the scientific enterprise, and the entire scientific community.
Public/public trust	Detrimental practices like falsifying authorship contributions undermines public trust in reproducibility and rigor. Endorsing errorful work undermines public trust in the discipline/ profession.			Detrimental research practices like accepting authorship credit without meeting any criteria undermine science, the public trust in the scientific enterprise, and the entire scientific community.

Table 3.8.4. *Stakeholder analysis: "authorship" credit but no actual contribution*

2. Identify decision-making frameworks.

Following the virtue perspective, the ethical statistical practitioner avoids "condoning or appearing to condone statistical, scientific, and professional misconduct" (ASA H2). Endorsing errorful work is clearly condoning statistical misconduct, but would also be scientific misconduct because it lends your authority to a false/not-reproducible report. Moreover, the ethical practitioner "takes responsibility for evaluating potential tasks, assessing whether they have (or can attain) sufficient competence to execute each task and that the work and timeline are feasible. Does not solicit or deliver work for which they are not qualified or that they would not be willing to have peer reviewed." (ASA A1). Agreeing to be added as an author on a paper or report on which none of the ICMJE criteria for authorship can pertain violates ASA H2 and makes it impossible for the practitioner to "accept full responsibility for their own work, not take credit for the work of others, and give credit to those who contribute. Respect and acknowledge the intellectual property of others." (ASA A5). Since none of the practitioner's work was included in this draft, responsibility cannot honestly or ethically be accepted for any of its content, and the virtue perspective clearly prohibits these behaviors.

ACM 1.3 outlines the responsibility to be honest and trustworthy. To agree to be listed as an author without meeting the criteria is not just unethical, it also not in your interests in a utilitarian perspective: you *and your reputation* will be harmed by your permanent association with a poorly done study – and once published, your name will always be associated with the document. The only "benefit", if it can be characterized as such, is that your role in the overall project will appear to have lasted from the start to the "finish". This is not a benefit *to you* – because absolutely no benefit accrues to you from inclusion as a co-author whatsoever: as the statistician/data scientist on the project, any errors in the report will be attributed to you, even if all you did was "approve" an error-prone document. The vignette suggests that it is not possible to prevent this error-prone document from being published (or, if not being submitted for publication, then prevent it from being used to support any decisions). Those future harms cannot be prevented but the harms accruing to you, your reputation, and the profession can all be prevented if you do not agree to be included as a co-author. The SHA table not only reflects only harms that accrue in this case, but also these are highlighted harms (i.e., significant ones) across multiple stakeholders. The utilitarian perspective clearly prohibits the ethical practitioner from endorsing this report.

3. Identify or recognize the ethical issue.

The ethical issue here is: you are being asked to "endorse" and approve a document that is both incorrect and also has had no actual input from you. Not only do you not qualify for authorship on this document – because everything you contributed was removed – but also, you can see from even a superficial perusal that robust and reproducible practices were *not* used, and the results are unlikely to be reproducible or valid. Thus, allowing the team to include you as a co-author fully violates authorship criteria and also, by agreeing to be labelled a co-author, you are violating ASA H2 "condoning (*not* "appearing to condone"!) statistical, scientific, and professional misconduct". The inclusion of non-contributors as co-authors is scientific and professional misconduct – NASEM (2017) characterizes this behavior as "detrimental research practices" that are just as damaging, if not more so, to the scientific enterprise and the public trust as are fabrication, falsification, and plagiarism (also referred to as "FFP", the only three aspects of scientific misconduct that are prosecutable[54]. Note that, whether or not the document is intended as a contribution to the scientific knowledge base or a within-organization document, condoning inappropriate authorship credit is still unethical and professional (if not scientific) misconduct. Further, endorsing incorrect work is statistical misconduct. According to the case, you are asked to **approve** – not edit or contribute to – the document. To do so would be both detrimental *and unethical,* because it represents falsification of your role on the paper as well as condoning misconduct (statistical, scientific, and professional!). The US Office of Research Integrity notes that falsification includes behaviors such that "the research is not accurately represented in the research record.[55]" This policy was intended for the scientific research record (i.e., as peer reviewed publications), but it is also applicable for within-organization research records.

Keep in mind that sometimes, detrimental research practices are engaged in by people who were *trained to practice this wa*y –but sometimes they are engaged in to purposefully mislead. You may not know which of these motivations is in effect. It is unethical of you to mislead anyone into thinking that you fulfilled the role of author when you did not; it is unethical to knowingly contribute false, irreproducible, and/or invalid results and inappropriate methods to the literature -whether peer-reviewed or within-organization. Even though you are

---

[54] In December 2000 the Office of Science and Technology Policy defined "research misconduct" as "fabrication, falsification, or plagiarism in proposing, performing, or reviewing research, or in reporting research results". For more about FFP, see Steneck 2009 – Chapter 2, https://ori.hhs.gov/content/chapter-2-research-misconduct-federal-policies or https://oir.nih.gov/sourcebook/ethical-conduct/research-ethics/nih-policies/investigation-allegations-research-misconduct

[55] https://ori.hhs.gov/definition-misconduct

not technically an "author" (not having contributed to this document), your endorsement and claim that you *are* an author – if you do accept that characterization from this team – mean that *you* are effectively making this irreproducible and invalid "contribution" to the literature. Critically, there are two important aspects of ethical practice in play here: unethically claiming authorship that you cannot justify, and unethically endorsing work you know is incorrect.

4. Identify and evaluate alternative actions (on the ethical issue).

Our standard three options start with "Do nothing, approve the draft and agree to remain a co-author", which is absolutely not appropriate. Thus, this option must be changed to

> a- Do not approve the draft and notify the team that your name must be removed as a co-author. This means the team can move forward and seek to publish (or disseminate) what you know to be errorful, but since this is not your work and your name is not on it, you may only be able to ensure you are not associated with the work and hope a reviewer catches the error you spotted in your superficial review.
>
> b- notify the team that you cannot turn the paper around in 2 days because of your other deadline and the errors you perceive, and also that the draft as it is should not include you as a co-author since you have not met any criteria qualifying you as an author (on this draft). Leave the errors for the other (actual) co-authors, or reviewers, to identify.
>
> c -Work overtime to make sure you correct all errors and add back in your original material into the draft – and then specify to team that the corrections and original material must remain in the draft in order for you to qualify as a co-author.

5. Make and justify a decision.

Note that only two of the three alternative actions are within your control: options a) and b) are similar in that you are refusing to be included as a co-author. They differ in the level of activity you are willing to engage in to alert the other authors that the manuscript has errors in it. These options reflect your honest communication as well as your commitment to ethical practice. They also signal the impropriety of submitting the work at all, because you have identified errors that render the report inconsistent with ethical practice standards.

Option c requires *both* that you take time that you might need in order to finish your other work to fix the errors *and* qualify yourself for authorship (according to ICMJE criteria), but it also requires that the other authors accept your edits and your conditions of authorship. This is a huge amount of work for you without any guarantee that your efforts will be taken seriously. The ethical practitioner does what they can to ensure that others also practice ethically, but there are limits to how far you can push others towards ethical practice of statistics and data science. Note that all options require that you are either not a co-author because you do not qualify, or that you do in fact qualify and so are included as a co-author. Depending on how well you know, and how well you have worked with, the co-authors, you will need to determine how likely it is that your edits will be included in the "final" version of the document. This may evolve as you grow in your career and reputation for ethical and robust, reproducible practice.

This case highlights the need for all ethical practitioners to encourage ethical statistical and data science practice, irrespective of their career stage (ASA Principles E, F, G). Detrimental practices are highly prevalent and without ethical practitioners making noise and raising awareness, these are likely to continue to plague the scientific community. It is less obvious how detrimental practices (identified by NASEM 2017) affect businesses, but the ethical practitioner has the same responsibilities – to the profession and to stakeholders, which ultimately include the employer and organization – irrespective of their workplace context.

6. Reflect on the decision.

Construct your own reflection on this decision.

Case 45. Someone on your team suggests a technical method to overcome a lack of consent from data contributors and collect their data even if they do not consent.

1. Identify and 'quantify' prerequisite knowledge:

Which DEFW and DSEC Principles or Areas seem most relevant to this vignette?

This situation is inconsistent with DEFW Principle 2, since "relevant legislation" as well as "codes of practice" require that human subjects give consent and are notified about the uses to which their data will be put.

In this case, the answer to the first question in the DSEC set, A1 (have human subjects given informed consent) would be "no" – because if this method were used, you would not be able to answer in the affirmative for anyone who did not consent.

Which ASA and ACM Principles seem most relevant to this vignette?

ASA:

Principle A. Professional integrity and accountability
Principle B. Integrity of data and methods
Principle C. Responsibilities to stakeholders
Principle D. Responsibilities to research subjects, data subjects, or those directly affected by statistical practices
Principle E. Responsibilities to members of multidisciplinary teams
Principle F. Responsibilities to fellow statistical practitioners and the Profession
Principle G. Responsibilities of leaders, supervisors, and mentors in statistical practice
Principle H. Responsibilities regarding potential misconduct
Appendix. Responsibilities of organizations/institutions

ACM:

1. General Ethical Principles
2. Professional Responsibilities
3. Professional Leadership Responsibilities

Potential result: Stakeholder:	HARM	BENEFIT	UNKNOWN	UNKNOWABLE
YOU	Using data with unknown provenance, particularly if some of it is known not to have been contributed with informed consent, violates the data contributor rights as well as ethical practice standards.	While a violation of multiple specific elements of the professional practice standards and some laws, the sampling might be less biased/more representative if everyone is 'forced' to contribute data.	Failing to obtain consent to use data and using stolen data could lead to unpredicted harms, bias, or unfair results, as well as risks to data contributors with the statistician or data scientist bearing responsibility for misuse, unauthorized access, or losses of that data.	Using stolen data and data that was not contributed with informed consent may suggest to others/other system developers that collecting stolen/breached data is OK, even though this directly violates practice standards.
Your boss/client	Data with unknown provenance can lead to unpredicted harms, bias, or unfair results, and can create risks for the data contributors that **were** foreseeable.	Ignoring the provenance of data, while a violation of multiple specific elements of the professional practice standards, is simpler and cheaper than ensuring data are obtained with proper consent.		

Potential result: Stakeholder:	HARM	BENEFIT	UNKNOWN	UNKNOWABLE
Unknown individuals	Data that is stolen or accessed without authorization and consent (i.e., from breaches) may expose data contributors to risks, as well as to (further) misuse by others.	There are no benefits that accrue when stolen data are utilized, and no one stops the use of that kind of data.		Using stolen data and data that was not contributed with informed consent may suggest to others/other system developers that collecting stolen/breached data is OK, even though this directly violates practice standards.
Employer	Using data with unknown provenance could lead to unpredicted harms, bias, or unfair results, and can create risks for the data contributors that were foreseeable, thus incurring liability to the employer.	Ignoring whether individuals do or do not consent to contribute their data, while a violation of multiple specific elements of the professional practice standards, may lead to a more representative sample than if only data obtained with proper consent are used.		

Potential result:	HARM	BENEFIT	UNKNOWN	UNKNOWABLE
Stakeholder:				
Colleagues	If others on the team are not aware that the data provenance is mixed (and some is obtained even though the contributor did not consent to give it), colleagues may mistakenly share –i.e., further the misuse of- the data.	While a violation of multiple specific elements of the professional practice standards and some laws, the sampling might be less biased/more representative if everyone is 'forced' to contribute data.		
Profession	The profession may appear untrustworthy when it is discovered that practitioners created and then used a method to ensure they could take your data even if no consent was given. Using data with unknown provenance is a violation of multiple specific elements of the professional practice standards.	There are no benefits that accrue when methods of obtaining informed consent are circumvented, and no one stops the use of that kind of data.	Failing to obtain consent to use data and using data obtained without consent could lead to unpredicted harms, bias, or unfair results, as well as risks to data contributors - with the statistician or data scientist bearing responsibility for misuse, unauthorized access, or losses of that data.	Using data that was not contributed with informed consent may suggest to others/other system developers that collecting stolen/breached data is OK, even though this directly violates practice standards. This decrements professional integrity in a concrete way.

Potential result:	HARM	BENEFIT	UNKNOWN	UNKNOWABLE
Stakeholder:				
Public/public trust	Public sentiment about the security of their data will continue to worsen, and people will become less inclined to contribute or sharing data if public concerns about data security, or the lack of honesty in data collection systems, continue.	There are no benefits that accrue when stolen data are utilized, and no one stops the use of that kind of data.		

Table 3.8.5. *Stakeholder Analysis: Team member suggests a method that gets data even from those who do not consent*

2. Identify decision-making frameworks.

The ethical statistical practitioner *"uses data only as permitted by data subjects' consent when applicable or considering their interests and welfare when consent is not required.* This includes primary and secondary uses, use of repurposed data, sharing data, and linking data with additional data sets." (ASA D5, emphasis added). Moreover, the ethical practitioner "Does not conduct statistical practice that could reasonably be interpreted by subjects as sanctioning a violation of their rights. Seeks to use statistical practices to promote the just and impartial treatment of all individuals.)" (ASA D11). **Circumventing** a mechanism whereby humans are given a choice to either contribute *or say no* is frankly unethical and may be illegal. Obviously, failing to protect basic human rights by using data you do not have consent to use violates basic human rights (which the virtue perspective unilaterally rejects based on multiple elements of the ASA Ethical Guidelines), but also violates ASA GL Principle H2 ("Avoid condoning or appearing to condone statistical, scientific, or professional misconduct. Encourage other practitioners to avoid misconduct or the appearance of misconduct.") as well as federal laws in some cases. Note that, if you tell data contributors that they do have the option to consent or not consent to contribute data, but you ignore their choice, this constitutes misleading these stakeholders, and may also represent a falsification of the research record unless the fact that individuals chose not to contribute data, but you took it anyway *is reported*. Not all statistical and data science applications will end up in the research record – in those cases, the definitions of scientific misconduct are less relevant and taking people's data even though they choose not to give it is clearly *statistical and professional* misconduct. The case clearly states that there is a consent mechanism, and that is being actively evaded, which the ethical practitioner would never do.

Not only is there active violation of basic human rights not to participate in data collection at work here, but also, if such a mechanism is created or deployed, then it obfuscates the provenance of the data. That is, you will not be able to tell what parts of the data were legitimately obtained, and what is unethically sourced. Thus, the provenance of the data will be unknown. This contributes to much of the harms in the SHA table.

The ACM CE, with its utilitarian perspective, focuses on limiting harms and in this case, there are many harms, and no benefits, to using data that was contributed to your analysis without consent; the harms are all compounded – and with no benefits –when the choices and exercise of basic human rights not to contribute data are circumvented. Moreover, a system that collects data without obtaining consent may include methodologies or other features that are insecure or otherwise create risks (in addition to risks of confidentiality and privacy breaches) that are not addressed – because the manner in which data are collected may not be legitimate or sufficiently specified. In this case, you simply cannot know what the provenance is of the data in the sense that some contributors consented, and others did not. The method for *circumventing* non-consenters to take their data anyway might be characterized as a "strength" of a computing system, but it is clearly a violation of basic human rights and principles of autonomy, offsetting the technical strength with profound cultural, social, legal, and ethical *limitations* (i.e., inconsistent with ACM CE Principles).

Such a mechanism prioritizes the system or data user over human rights to autonomy, creating harms to all stakeholders that must include tricking users of the data into believing they were collected with consent, since an individual who inserts a method to circumvent refusals to give consent is unlikely to want that fully documented. No real benefits –certainly none that can offset the harms to all stakeholders – can accrue, so the utilitarian perspective as well as the ACM CE point out that this method is clearly not ethical. This potential method will create additional violations of human rights, as well as of ACM Principles – causing all those who use the data from this dishonest system to violate human rights and other CE Principles as well.

There are distinct responsibilities in both the virtue and utilitarian perspectives for leaders and supervisors to avoid designing or even encouraging such systems, and for organizations to avoid using or encouraging them.

3. Identify or recognize the ethical issue.

Both the ASA and ACM practice standards state explicitly that professionals must respect laws, basic human rights, *and* ethical practice standards to ensure data are collected with the knowledge *and consent* of contributors to the extent possible. The primary ethical issue in this case is that both GLs and CE require that data be obtained *with consent and the contributors' knowledge of what the data will be used for*, but this method violates these principles for those who choose not to contribute. It is frankly unethical to use data that was not obtained with informed consent, but it is **worse** to pretend that you or your system respects their choice not to consent, because in addition to using unethically sourced data, this case describes a system that actively lies to a key stakeholder, the data donor. The case describes actively circumventing the consent process, so there is no way this is an unwitting error or honest mistake.

4. Identify and evaluate alternative actions (on the ethical issue).

As usual, the three decisions that can be made in any circumstance are: a) do nothing; b) consult or confer with a peer (ASA) or a supervisor (ACM); and c) report violations of policy, procedure, ethical guidelines, or law (ASA) or refuse to implement the system (ACM). Clearly, "do nothing" is both totally inappropriate in this case, and also a violation of the practice standards relating to respecting data contributors. We can change option a) from "do nothing" to "stop the use of the method that circumvents non-consent and do not proceed with any use of the data except what was actually contributed with informed consent". Then option b) would become, "stop any use of the data except what was actually contributed with informed consent, and consult or confer with a peer (ASA) or a supervisor (ACM) as to how best to stop the use of the method that circumvents non-consent". Option c) must also be modified, because simply reporting the fact that the data collection violates the GLs (ASA) and basic human rights, or refusing to implement the system (that collects data when consent is not given) (ACM) might not ensure that the data with unknown provenance is both not used and also, that its collection ceases. So, option c) needs to change to "report violations of policy, procedure, ethical guidelines, or law to a peer (ASA) or a supervisor (ACM), *and* stop the collection of non-consented data, *and* stop any use of the data except what was actually contributed with informed consent."

5. Make and justify a decision.

Decision: *Note that all options include* "stop the use of the method that circumvents non-consent and do not proceed with any use of the data except what was actually contributed with informed consent". What else is done (nothing else in option a); *conferring* on how best to stop the method in option b); and not conferring but reporting to relevant authorities the existence of that method in option c).

Clearly the non-consented data collection – as well as its use – must stop because both are unethical. If data with unknown provenance is commonly collected in your work context, it might be impossible for you choose even option a) and stop the mechanism as well as all use of data with unknown provenance. In that case, it is impossible for you to do *anything* ethical except notify relevant stakeholders and authorities of the situation, and then seek another team, other colleagues, or a different job.

Option b) would be the *most desirable option* in that consulting with a colleague/supervisor will ensure at least some kind of notification that, or publicity for the fact that, data collection without consent is frankly unethical, as is the use of such inappropriately obtained data. There may not be policies in your workplace against such interference/violation of the ASA GLs or ACM

CE, so reporting (option c) may not be a viable option, making it a slightly less desirable option than b. However, in contexts where such policies do exist, it may still be difficult to report such behaviors; but you (the practitioner) can – and should - still demonstrate your professional competence and follow the relevant practice standards. Option c may not be as much about "reporting" a person or practice as it is about "educating" people or your company/ organization about the practice standards, and pointing out that this system both violates basic human rights and also the ethical practice standards for statistics and data science. It is important *to the profession* to stop detrimental and unethical practice, and also to support the next practitioner who ends up in a similar situation (or better, to prevent such unethical data collection in the future). So, it is definitely worth considering politely notifying relevant individuals that the data collection system violates ethical practice standards. Recall that both the ASA and ACM practice standards state explicitly and repeatedly that professionals do not yield to pressures to behave contrary to their respective practice standards (**ASA**: A; A1, A2, A4, A9, A12; C1, C2; E4; G1; H2; **ACM**: 1.2; 1.3; 2.1; 2.2; 2.3; 3.4; 4). In circumstances where you share, report on, or describe the system, the consent mechanisms that may limit the amount of data but also limit the likelihood of stealing data from those who do not consent to give it should be highlighted. The harms and benefits tradeoffs can be outlined in order to underscore that obtaining consent, not circumventing non-consent, is the only way to meet ASA GLs/ACM CE, limit harms/risks of harms, and also enable the team to answer DEFW and DSEC questions.

6. Reflect on the decision.

Construct your own reflection on this decision.

Case 46. You recognize the potential for "dual use" of your team's code, data, and/or results.

1. Identify and 'quantify' prerequisite knowledge:

NB: the European Union (EU) and United States define *"dual use"* as "goods, software and technology that can be used for both civilian and military applications", which are controlled specifically to "contribute to international peace and security and prevent the proliferation of Weapons of Mass Destruction (WMD)". The US also includes "terrorism" among the non-civilian purposes or applications of such technology. This case may require background information like that published in the Bulletin of the World Health Organization (WHO): Selgelid MJ. (2009). Governance of dual-use research: an ethical dilemma. *Bulletin of the World Health Organization* 2009; 87:720-723. doi: 10.2471/BLT.08.051383. See also, Uncensored exchange of scientific results. Journal Editors and Authors Group Proceedings of the National Academy of Sciences Feb 2003 100 (4) 1464; DOI:10.1073/pnas.0630491100

NB: A *conflict of interest* is defined as any consideration –or interest - that competes with other considerations and affects a decision or behavior surreptitiously or unfairly. For example, if you are a judge of gymnastics at the Olympics, you would be expected to be impartial and unbiased. If one of your relatives is a gymnast and you are supposed to judge them, your interest in family success could, in fact - *or just in the minds of observers* - bias your judging in favour of your family and/or against other gymnasts. Professional conflicts of interest may arise when you are in a position to provide peer review of grants or papers from individuals at the same organization as you. **When your work qualifies as "dual use", it can create a conflict of interest.** Even if you do not intend to use your work in a military application, its potential to be useful that way might limit your ability to capitalize on the work. The identification of dual use means you will need oversight that your work might not ordinarily require, and this can result in limitations (e.g., restrictions on continuing the work at all, or on publishing or patenting it, or on other core aspects of its use).

Just as we have seen with ASA H2, acting *or the appearance of acting* are equally to be avoided; in ASA H2 it is "condoning or appearing to condone" unethical behavior. In the case of a conflict of interest, there may be bias, or observers (i.e., anyone who isn't you) may *perceive* bias -whether or not you actually act with bias. This may seem highly arbitrary and abstract, but "interest" and "bias" can be abstract. In our roles as statistical and data science practitioners, our biases can have profound effects on multiple stakeholders. When the public perceives unfairness or bias, it harms their perception of the profession and undermines the public trust in our work. Thus, avoiding both actual conflicts of interest *and the perceptions of conflicts* of interest by outside observers are equally important. Moreover, your work may clearly qualify as dual use, or it might just be perceived as potentially dual use. Either way, this designation can mean limitations on your work – and those may lead some to ignore or otherwise try to prevent the duality of use from being detected. Ultimately, the dual use regulations are intended to protect life and property; thus, it is unethical to act in a way that circumvents these regulations.

Sometimes conflicts of interest are financial; for example, when research into the effects of sugar on health is sponsored by sugar manufacturers. In the event that the research sponsored by sugar manufacturers yields results suggesting that sugar is *beneficial* to health, it suggests that the research –or interpretation of results - were unduly influenced by the financial considerations and were not done appropriately. Similarly, there may be financial inducements to avoid a dual-use designation to your work. Financial conflicts are typically managed by ensuring that decisions about purchasing or funding are not made by those in conflict; contractually ensuring that funders have no say in, and make no contribution to, any research report; or refusing the funding/financial incentive. Also, all financial considerations – including employer's name, or funding/financial relationships, are typically disclosed in all work. However, with a dual use designation, these conflict managements may be moot (in cases where the regulation makes it impossible to capitalize on financial gains based on the work).

Which DEFW and DSEC Principles or Areas seem most relevant to this vignette?

Not discussed in DSEC or in DEFW.

Which ASA and ACM Principles seem most relevant to this vignette?

ASA:

Principle A. Professional integrity and accountability
Principle B. Integrity of data and methods
Principle C. Responsibilities to stakeholders
Principle D. Responsibilities to research subjects, data subjects, or those directly affected by statistical practices

ACM:

1. General Ethical Principles
2. Professional Responsibilities
3. Professional Leadership Responsibilities

Potential result: Stakeholder:	HARM	BENEFIT	UNKNOWN	UNKNOWABLE
YOU	Takes time, adds effort to fully document the dual use. The lack of documentation of dual use may suggest to your supervisor that you are not very good (thorough) at your job.	Documentation simplifies evaluations of the work (ASA) or system, and updates (ACM); adds transparency and accountability.	ACM: harms/risks of dual use may not be detectable or addressed with incomplete documentation. ASA: Replications may not be facilitated because work has dual use characterization.	
Your boss/client	Identification of work as dual use may limit how it can be used/shared or protected as intellectual property.	Documentation adds transparency and accountability of statistician/ computing professional work.		
Unknown individuals	Minimal documentation will not support review or replication.	Creates transparency and accountability		
Employer	Identification of work as dual use may limit how it can be used/shared or protected as intellectual property.	Documentation adds transparency and accountability of statistician/ computing professional work.		

Potential result: Stakeholder:	HARM	BENEFIT	UNKNOWN	UNKNOWABLE
Colleagues	Incomplete documentation may fail to notify colleagues who want to extend or use the work that it is dual use, and potential dangerous.	ACM: harms/risks of dual use may not be detectable or addressed with incomplete documentation. ASA: Replications may not be facilitated because work has dual use characterization.		
Profession	Identification of work as dual use may limit how it can be used or shared. Incomplete documentation may fail to notify other professionals who want to extend or use the work that it is dual use, and potential dangerous.	ACM: harms/risks of dual use may not be detectable or addressed with incomplete documentation. ASA: Replications may not be facilitated because work has dual use characterization.		
Public/public trust	Incomplete documentation may fail to notify the public that the work is dual use, and potential dangerous. Without full documentation of work/thought processes, public trust in systems or their results, and the profession (ACM/ASA) are compromised.	ACM: harms/risks of dual use may not be detectable or addressed with incomplete documentation. ASA: Replications may not be facilitated because work has dual use characterization.		

Table 3.8.6. *Stakeholder Analysis template: Documenting dual use work*

2. Identify decision-making frameworks.

ASA GLs describe the ethical statistical practitioner as fully documenting their work so as to be transparent and not to mislead any stakeholder (ASA C2). As a virtue approach, this suggests that *full* documentation is the ideal, and essential to ethical practice - even if it ends up highlighting dual use of the work. The fact that work can be used in both civilian and military or terrorist applications is an inherent and important aspect of the work, so that must be included in any documentation to the extent it is legitimate. The ethical practitioner "knows when work requires ethical review and oversight." (ASA D1) and also "does not knowingly conduct statistical practices that exploit vulnerable populations or create or perpetuate unfair outcomes." (ASA A3). Transparent and complete reporting of the dual use nature of your work accomplishes both of these. However, in consideration of the potential harms that the work may represent or facilitate– i.e., the purpose for regulations about dual use – two other ASA GL principles must be balanced: ASA C6: "strives to make new methodological knowledge widely available to provide benefits to society at large. Presents relevant findings, when possible, to advance public knowledge" vs ASA A11: "Follows applicable policies, regulations, and laws relating to their professional work, unless there is a compelling ethical justification to do otherwise." Neither of these affects the obligation to fully and transparently document your dual-use work as such; however, ASA Principles C and D both offer caution about how widely to disseminate or share the documentation of work that may qualify (or is designated) as dual use. The conflict of interest (benefits to you for creating/sharing your work vs harms accruing from a potential military application) must be adjudicated independently – i.e., this conflict of interest must be managed appropriately (ASA A7; B1; C8).

The SHA supports transparent and *complete* documentation of your work as consistent with the utilitarian perspective of the ACM CE and specifically, ACM: 1.1: Contribute to society and to human well-being; and 1.2: avoid harm. While "major" (dark grey highlighted) harms could potentially accrue to you and to the profession if you identify your work as dual-use and it ends up not being disseminable, other harms are minor. Importantly, benefits of the dual-use designation accrue to all stakeholders because potential harms due to military applications are avoided or managed appropriately

It is clear, though, that ensuring complete and transparent documentation of the work – which is what the dual-use designation will be based on – is essential for any benefits to be realized.   The benefits identified in the SHA are minimized by incomplete documentation -particularly when documentation is purposefully incomplete so as to avoid others' recognizing the duality of use. It is, therefore, essential to completely and transparently document your work. The sharing (reporting and communication) of the documentation may be limited by laws or rules, but the transparency and completeness of documentation is not limited. A failure to completely document your work, even if it ensures designation as dual use (and thus limits your ability to benefit from the work) may have the result of only being noticed as having militaristic applications by those who would misuse it.

3. Identify or recognize the ethical issue.

It is important to recognize that *this case does not describe anything unethical*: innovation that has dual uses is not unethical itself. As with many cases, unethical behaviors can often follow from ethical work. Given the financial, public safety, and in some cases, national security risk considerations that prompted the laws governing dual use in the first place, *it is unethical and possibly unlawful if you do not report the dual use nature of your work to all stakeholders* – including relevant authorities. Acting to hide the potential for dual use is unethical; specifically failing to document the work fully to prevent a dual use designation represents a clear conflict of interest and knowing misleading of stakeholders.

As we have seen in many other cases, our efforts to practice ethically may be thwarted or impeded by others in our working environments. In the case of dual use work, this interference – even if it is very common in your workplace – is absolutely intolerable. The real harms and risks of harms may be totally unknown or unknowable *to you*, but the rules and laws established internationally must be prioritized. This is one reason why most ethical practice standards, including the ASA GL and ACM CE (and DEFW) specify an ethical obligation to be familiar with rules, laws, and policies that may be relevant to the practitioner's work. Like collecting data without consent, there is simply no incentive or benefit that can possibly justify the harms or risks of harm that attend dual use work. Thus, it is unethical for *anyone* to try to suppress your notification of the relevant stakeholders and authorities of the dual use nature of your work – and it may be unlawful. As a concerned citizen as well as someone trying to practice ethically, you may be obligated – by the practice standards as well as applicable laws - to blow the whistle on anyone who seeks to prevent you from notifying the relevant stakeholders and authorities of the dual use nature of your work. Again, managing the conflict of interest requires the determination of dual use by parties *other than* those who may be biased, or may be perceived to be biased, in the determination of duality of use.

4. Identify and evaluate alternative actions (on the ethical issue).

Although we examine the same three decisions that can be made in any circumstance: a) do nothing b) consult or confer with a peer (ASA) or a supervisor (ACM); or c) report violations of policy, procedure, ethical guidelines, or law (ASA) or refuse to implement the system (ACM), none of these fit our case. Instead, option a) becomes "stop working on the project and notify the team of the dual use potential." Option b) becomes, "stop working on the project and confer with a peer or supervisor and/or lawyer about the applicable laws relevant to the "other" uses of the work; notify the team of the dual use potential and the relevant regulations." Option c) becomes relevant only if you notify the team of the dual use potential and they seek to suppress that information.

5. Make and justify a decision.

Decision: As indicated, all options include "stop working on the project and notify the team of the dual use potential." As we saw in other cases about documentation (Chapter 3.6), "fully document your work" requires inclusion of all relevant aspects of it, and the potential for dual use is simply part of the complete documentation. While option a) is clearly consistent with practice standards, option b) is preferred. As indicated, option c) is only a relevant response in case others seek to suppress your full and complete documentation of your work – which must include consideration of its potential for both civilian and military or terrorist use.

As was noted in the previous case, how your documentation can be "safeguarded" may require consultation with a colleague or supervisor. The objective of the documentation, to be complete, transparent and accountable, as well as supporting replication (ASA) and honest peer review (ACM) as appropriate, can be achieved with either options a) or b). However, there have been debates about whether dual use research can be shared. The sharing does not affect the requirement for documentation, but it does complicate your choices for sharing. In addition to conferring with relevant stakeholders about proprietary aspects of the work, and how best to document – and if possible, share – the work, you should also discuss the wider ethical considerations of the channels for publication or sharing[56].

---

[56]For starting points on the discussions about publishing/sharing documentation about dual use work, see Selgelid MJ. (2009). Governance of dual-use research: an ethical dilemma. *Bulletin of the World Health Organization* 2009; 87:720-723. doi: 10.2471/BLT.08.051383. See also, Uncensored exchange of scientific results. Journal Editors and Authors Group. *Proceedings of the National Academy of Sciences* Feb 2003 100 (4) 1464; DOI:10.1073/pnas.0630491100

6. Reflect on the decision.

Construct your own reflection on this decision.

Case 47. You inadvertently discover that a proprietary "new method" that you were told to prepare for publication/patent application was actually published decades ago and was apparently unnoticed until you found it.

1. Identify and 'quantify' prerequisite knowledge:

Which DEFW and DSEC Principles or Areas seem most relevant to this vignette?

Not discussed in DSEC or in DEFW.

Which ASA and ACM Principles seem most relevant to this vignette?

ASA:

Principle A. Professional integrity and accountability

Principle B. Integrity of data and methods

Principle E. Responsibilities to members of multidisciplinary teams

Principle G. Responsibilities of leaders, supervisors, and mentors in statistical practice

Principle H. Responsibilities regarding potential misconduct

Appendix. Responsibilities of organizations/institutions

ACM:

1. General Ethical Principles

Potential result: Stakeholder:	HARM	BENEFIT	UNKNOWN	UNKNOWABLE
YOU	You falsify your cv/resume, making a *deliberately misleading* claim, if you ignore prior work and claim to be the original innovator.	You gain a publication, but only by falsifying your cv/resume and taking responsibility as well as credit for work others did.	Perpetuation of false attribution/ authorship undermine your credibility as well as that of everyone on the author list, while also perpetuating deceptive and detrimental research practices.	Deceptive and detrimental research practices like claiming credit for an already published idea undermine science, the public trust in the scientific enterprise, and the entire scientific community.
Your boss/client	Perpetuation of false attribution/ authorship undermines your credibility as well as that of *everyone on the author list*.			
Unknown individuals	Making a *deliberately misleading* claim, by taking responsibility as well as credit for work others did, is proof that you are dishonest.		Perpetuation of false attribution/ authorship undermines your credibility as well as that of everyone on the author list, while also perpetuating deceptive and detrimental research practices.	Deceptive and detrimental research practices like claiming credit for an already published idea undermine science, the public trust in the scientific enterprise, and the entire scientific community.

Potential result: Stakeholder:	HARM	BENEFIT	UNKNOWN	UNKNOWABLE
Employer	Perpetuation of false attribution/ authorship undermines your credibility as well as that of *everyone on the author list*. By extension, the employer is implicated as an endorser of detrimental practices.		Perpetuation of false attribution/ authorship undermines your credibility as well as that of everyone on the author list, while also perpetuating deceptive and detrimental research practices.	Deceptive and detrimental research practices like claiming credit for an already published idea undermine science, the public trust in the scientific enterprise, and the entire scientific community.
Colleagues	Perpetuation of false attribution/ authorship undermines your credibility as well as that of *everyone on the author list*.		Perpetuation of false attribution/ authorship undermines your credibility as well as that of everyone on the author list, while also perpetuating deceptive and detrimental research practices.	Deceptive and detrimental research practices like claiming credit for an already published idea undermine science, the public trust in the scientific enterprise, and the entire scientific community.

Potential result: Stakeholder:	HARM	BENEFIT	UNKNOWN	UNKNOWABLE
Profession	Perpetuation of false attribution/ authorship undermines your credibility as well as that of *everyone on the author list*. Your refusal to acknowledge prior work damages the originator and decrements the respect all professionals should have for prior work, undermining the credibility of the profession in making contributions to documents/ publications.		Perpetuation of false attribution/ authorship undermines your credibility as well as that of everyone on the author list, while also perpetuating deceptive and detrimental research practices.	Deceptive and detrimental research practices like claiming credit for an already published idea undermine science, the public trust in the scientific enterprise, and the entire scientific community.
Public/public trust	Deceptive and detrimental practices like falsely claiming to be the originator undermines public trust in intellectual property and the value of innovation.			Deceptive and detrimental research practices like claiming credit for an already published idea undermine science, the public trust in the scientific enterprise, and the entire scientific community.

Table 3.8.7. *Stakeholder analysis: A new method that you were directed to prepare for publication is actually not new.*

2. Identify decision-making frameworks.

Following the virtue perspective, the ethical statistical practitioner "Accepts full responsibility for their own work, does not take credit for the work of others, and gives credit to those who contribute. Respects and acknowledges the intellectual property of others." (ASA A5) and avoids "condoning or appearing to condone statistical, scientific, and professional misconduct" (ASA H2). Moreover, they "take responsibility for evaluating potential tasks, assessing whether they have (or can attain) sufficient competence to execute each task and that the work and timeline are feasible. (they) do not solicit or deliver work for which they are not qualified or that they would not be willing to have peer reviewed." (ASA A1). Going ahead with your publication or patent preparation when you have discovered that you (and your team) are not the originator of the work violates ASA A5 and H2 and makes it impossible for the practitioner to "follow, and encourage all collaborators to follow, an established protocol for authorship. Advocate for recognition commensurate with each person's contribution to the work. Recognize that inclusion as an author does imply, while acknowledgement may imply, endorsement of the work." (ASA A6) The violation of ASA A1 comes from the fact that peer review of work may identify that your claim of original work will be uncovered to be false because it is in fact not original. In this case, however, the work is characterized as "proprietary". This suggests that there are financial interests in being, or claiming to be, original and innovative. Failing to notify stakeholders that this work is not actually original to the proprietor might effectively mislead them, violating ASA C2 ("regardless of personal or institutional interests or external pressures, does not use statistical practices to mislead any stakeholder").

The SHA supports acknowledgement of the true originator of an idea because making false and misleading claims about innovation creates only two real harms for every single stakeholder: making false and misleading claims about being the originator of previously published work undermines your credibility *as well as that of everyone on the author list*, and making such claims is a deceptive and detrimental practice that this act will perpetuate. Although there are only two real harms in the SHA table, they do accrue to every stakeholder and may have unknown/unpredictable further harms. In this case, the 'benefits' may be financial (because the case characterizes the work as "proprietary"). No matter how much financial benefit may accrue to a few stakeholders, false claims mislead, and may harm, other stakeholders. When weighing harms and benefits, it can be important to prioritize the public good (ACM Preamble) and, "When the interests of multiple groups conflict, the needs of those less advantaged should be given increased attention and priority." (ACM 1.1).

3. Identify or recognize the ethical issue.

The ethical issue is that continuing to prepare to apply for publication or patent when you know your work is not as innovative as publication/patent requires fails to honor prior work and intellectual contributions. It also falsifies your resume or CV by suggesting yours is the first publication/patent for the work, and confuses the scientific knowledge base by failing to acknowledge the intellectual history of the ideas (if your paper is published). The discovery that the work is not original is an important facet of the contextualization of the work you have done; *failing to disclose this is unethical behavior*. Note that this is true whether the objective is to publish in the scientific record or to describe your work within the organization, or to specific stakeholders.

Less obviously unethical is the failure to engage in a good faith[57] search for prior work in the first place. The good faith effort is "what a reasonable person would determine is a diligent and honest effort under the same set of facts or circumstances." In this case, it does not state whether the prior work was in the same domain or in a different domain from yours. Obviously, it is impossible for you to be familiar enough with the history of all potentially relevant domains to recognize when your work is a replication of work from another field. However, there is a tendency in some newer disciplines – like data science is – to focus only on the most-recent past, e.g., the previous five years of publications. Since the discipline is new, journals and other publication outlets that are dedicated to the discipline may also be new, so going back five or so years may be perceived to capture anything that could be relevant. This does not clear matters up much if "everyone" in the new discipline considers a search in just the previous five years' worth of work to be sufficiently "diligent and honest". Moreover, it is unethical and inappropriate to identify the issue (i.e., that the work is not, in fact, original) and then simply go along with characterizing it as original simply because it didn't turn up in a search of the previous <short time frame> of work. Your identification of the prior work is the core of this case: identifying prior work that renders a claim of "originality" false is an error that must be corrected (the ethical statistical practitioner "strives to promptly correct substantive errors discovered after publication or implementation. As appropriate, disseminates the correction publicly and/or to others relying on the results" (ASA B5)).

A difficulty with the cultural practice of treating only the previous <recent interval of time as relevant> for a new discipline is that no discipline is *so* new and so different from any other discipline that there would not be a plausible

---

[57] https://en.wikipedia.org/wiki/Good_faith

expectation that related –and much older – disciplines, e.g., statistics, computer science, math, and engineering in the case of data science, would contain relevant materials that should be searched for an effort to contextualize new work to truly qualify as "good faith". This case clearly states that the original work is decades old, and two or more decades ago there might have been too little computing power to effectively implement a method that today seems trivial to implement –accounting for the fact that the old method was never utilized or mentioned after its initial publication, which you have now found. One important aspect about the search beyond what the discipline currently expects (five years in this example) is that, once you have seen that this expectation is groundless – e.g., after you see this case or experience something like it – then a five-year window on your future searches will be known, to you and your team, as merely convenient, and *not* "diligent and honest". Thus, if this is not the first time you have encountered relevant prior work outside the 'standard' time window or in the 'typical' places (disciplinary journals) to search, then it was clearly not a good faith effort to use those culturally acceptable, yet not-especially-reasonable, search parameters in any subsequent searches. Thus, while it is plainly unethical to fail to report and recognize the prior work, the determination of whether your original search was truly done in good faith depends on your experience. When you execute a search for prior work and you do not do it diligently or honestly, that *is* unethical. Whenever you act in a manner that does not fulfill your obligation to practice in good faith, it contravenes the GLs and CE. By contrast, acknowledging that an independent investigation decades ago led to the same innovation can strengthen confidence in your 'new' discovery.

4. Identify and evaluate alternative actions (on the ethical issue).

The same three decisions that we consider in each vignette are: a) do nothing b) consult or confer with a peer (ASA) or a supervisor (ACM); or c) report violations of policy, procedure, ethical guidelines, or law (ASA) or refuse to implement the system (ACM). As we have seen, option a) is clearly unethical so requires modification. Option a) becomes "recognize the prior work as context of the current work". Option b), which is implied in option a) but involves others, becomes "b. confer with peers/supervisor about how best to differentiate your new method from previous ones". This is simpler to do with manuscripts on methods, and harder with patents[58]. Additional work may be

---

[58]In a patent application, existing work ("prior art") must be identified in a specific way. For US Patent information, see https://www.uspto.gov/patent/initiatives/prior-art-search and for European patent applications see https://www.epo.org/learning-events/materials/inventors-handbook/novelty/prior-art.html

required to ensure that prior work is considered and any innovations in the "new" method contextualized against that historical context.

Since you are the person responsible for preparing the publication/patent application, option c) reporting would only be relevant if you notify the team that the prior work exists, and they seek to suppress that information. In that case, it would be important – in order to avoid the harms as listed in the SHA – to ensure your name is not included on a publication or patent that fails to acknowledge prior work, because to leave your name on – even if you do meet all criteria for authorship – is your endorsement of the failure to recognize prior work, violating ASA H2 and ACM 1.5. You want to do everything you can to prevent detrimental practices like this; it may be that all you can do is refuse to endorse (condone) this scientific misconduct (of claiming to be the sole innovator of an idea that has previously been published).

5. Make and justify a decision.

Decision: As noted, the option you choose will depend on your situation. If you are more senior on your team, you will be able to choose option a) and simply recognize the prior work as context of the current work yourself. However, more junior team members may require consultation or collaboration, in which case option b) will be a better choice. Option c is naturally the worst-case scenario because it is difficult to find a body to whom you could reasonably report a violation of the practice guidelines relating to failures to identify prior work. As noted earlier, "younger" disciplines may have already established a culture by which only very recent work is reviewed; it can be difficult to claim that a failure to thoroughly search the relevant literature was done purposefully or in order to deceive. A team may conclude that they would rather publish or patent whatever they have without acknowledging prior work, trusting reviewers to follow a similar cultural approach to not searching very far back in time or across other disciplines. If the context of option c) is what occurs for you in a case like this, your only concrete response is to ensure that your name is *not* included on a publication or patent that fails to acknowledge prior work. Although this seems unfair to you, given all your work on whatever led to the paper or patent, to leave your name on what is submitted is your endorsement of the failure to recognize prior work, violating ASA H2 and ACM 1.5. By leaving your name on you also make a misleading and false statement on your cv or resume if you list the paper or patent there. Any reader or reviewer could find the prior work and document that your current publication makes a misleading and false claim to be original work, so being "caught" – with your name on this publication – will always (your entire career) have the potential to cause harm to your reputation in addition to perpetuating the unfair acceptance of credit for someone else's idea.

6. Reflect on the decision.

Construct your own reflection on this decision.

Summary of seven case analyses for Team Science/ Teamwork

The seven case analyses of Chapter 3.8 were focused on working on teams (team science/teamwork). As you might suspect, since the DEFW and DSEC are project-specific instruments, they were not relevant in many of these cases; by contrast, since the ASA GLs and ACM CE describe the ethical practitioner in their respective fields, they do offer guidance (and specific instructions on some cases). When it comes to working on teams, modern quantitative practice requires at least some level of interactions. Sometimes teams include more than one person with your same or a similar skill set, but many times you are on a team representing your profession alone. The GLs and CE both stipulate that they are relevant for anyone who utilizes the concepts, tools, and techniques of the discipline. Both also stipulate that, while different professions may follow different guidelines, these guidelines cannot be prioritized over the GLs and CE, *especially not* when they contradict or direct you to violate the GLs and CE.

In two cases there may be concerns about secrecy, security, or protecting proprietary ideas. These would be concerns if you were working alone (i.e., not on a team), but the cases are contextualized in the "working with others" task because taking the perspectives and (where it exists) practice guidance of these others on your team can complicate your response. While the GLs and CE (and in some cases, the law) may be clear, you must still both choose and justify your response to ethical issues, and you must also communicate the response and its justification to the others on your team.

You may encounter, on your team, either support for, *or resistance to,* your efforts to follow the GLs and CE. Recall that both the ASA and ACM practice standards state explicitly and repeatedly that professionals do not yield to pressures to behave contrary to their respective practice standards (**ASA:** A; A1, A2, A4, A9, A12; C1, C2; E4; G1; H2; **ACM:** 1.2; 1.3; 2.1; 2.2; 2.3; 3.4; 4). Moreover, both the GLs (G1) and the CE (Principle 3.4) direct employers and supervisors (or those in leadership roles) to respect the efforts of the ethical practitioner to follow the practice standards. As we have seen multiple times throughout these cases, not everyone is as familiar with these standards as you are. It might be simply a matter of you refusing to engage in behaviors contrary to the GLs/CE and ciing these standards to justify your behavior. You may also have executed a complete case analysis using ethical reasoning KSAs and have a well-reasoned decision with its justification at the ready! Whenever you stand up for ethical practice, refuse to participate in unethical work, and condemn detrimental practices, it is helpful to have a careful and objective support (i.e., a case analysis like those in this book!). Not only will this help you craft your argument, but you may also convince others in your context (team, organization) to commit to following the ethical practice standards.

It is worth reflecting on what it means to be a whistleblower, or to withdraw from a project, or to refuse a request or refuse to deploy a system: the ASA GLs specify that the ethical statistical practitioner "takes appropriate action when aware of deviations from these Guidelines by others (ASA A9), and that they "Avoid condoning or appearing to condone statistical, scientific, or professional misconduct. Encourage other practitioners to avoid misconduct or the appearance of misconduct." (ASA H2). While these principles obviously pertain in every circumstance and for every task (from planning right through to reporting), it may be in your work with others that these are even more pronounced on teams because in the team setting, you might become more *"aware of such unethical practices by others".*

When discussing "working on teams" ethically in Section 2, we saw that general principles from the GLs or CE, and specific questions from the DSEC and DEFW, could be used to ensure that our general behavior is ethical. However, in response to specific challenges and ethical violations such as those in this chapter, DSEC and DEFW were *not* helpful in either identifying whether or not an ethical problem existed (or determining what exactly it is), nor were they helpful in figuring out what to do. When going about your daily work (whether alone or on teams), DSEC and DEFW do include some considerations of others. You may recall from Section 2 that DSEC considers stakeholders for their perspectives (to limit bias) or for communication – and the DSEC items are focused on what "we" have done ("have *we* considered" or "do *we* have a plan"). DEFW also recognizes the perspectives of others –possibly on your team - and includes peer review/access as part of its considerations for transparency and accountability. The cases we analyzed in Chapter 3.8 highlight the point made in Chapter 2.9: the ASA and ACM standards recognize more explicitly that, while there may be only one statistician or data scientist involved on any project, statistics and data science are rarely done by individuals outside of *teams*. Moreover, these practice standards state that ethical practice by the statistician/data scientist is essential to both cooperative and individual efforts. The cases in this chapter document that, in addition to the *expertise* of others on the team and in the wider scientific community - particularly when it is not in statistics and data science, their biases or inclinations to follow or resist following ethical guidelines will need to be taken into consideration as you seek to follow the ASA and ACM practice standards.

As with all the vignettes, the cases in this chapter were based on actual events. Given the pervasive nature of teams in the workplace, the cultural norms that exist for non-statisticians and data scientists may encourage violation of the ASA and ACM ethical practice standards or may prompt others on your team to exert their influence to cause others to violate those standards. Unlike most other SHAs, for team science/teamwork the SHAs tended to highlight benefits to others (usually the person to whose behavior you must respond) and harms to you and the profession (rather than harms to unknown individuals or to the public trust). So, while the ACM CE states that the public good is to be prioritized in all decisions, when you are making and justifying decisions relating to teamwork, the public good would be an indirect (if any) consideration. Nonetheless, the SHA is still a useful way to organize and visualize harms and benefits.

Questions for Discussion:

1. Review at least two cases in this chapter.

A. Do you agree with the decisions in these case analyses? Why or why not? Which parts of the ER process are most and least acceptable for each case? Do these parts differ for different cases in this chapter? Were there different ASA or ACM Principles or elements that you identified, and if so, how do these affect your case analyses? Are the elements and/or Principles you identify represented in your reflections?

B. Discuss your results for the tests of universality, justice, and publicity for any (or all) case(s) in this chapter.

C. Discuss the relevance to each case – including the analysis and your reflection – of federal, state, or local/organizational policies regarding research misconduct and policies for handling misconduct. How can you incorporate what you know about these policies into any part of the case analysis (prerequisite knowledge; identifying the ethical issue; determining alternative actions; making or justifying your decision; or reflection)? Note that, if there is no alignment between federal, state, or your local/ institutional policies and a given case, you can comment on that (e.g., do policies in your organization focus solely on *misconduct* – and ignore detrimental or other unethical practices?). Do you feel that these policies could have helped in the reasoning, decision or justification?

D. Discuss the relevance to each case - including the analysis and your reflection – of the statistician and data scientist as a responsible member of society. How can you incorporate what you know about these policies into any part of the case analysis (prerequisite knowledge; identifying the ethical issue; determining alternative actions; making or justifying your decision; or reflection)? If there are no relevant policies, you should comment on that, particularly when a) data are obtained using federal monies (grants); or b) data are collected (acquired) from humans, such that Federal Regulations about the treatment of human subjects and the acquisition of tissues and data from humans are relevant.

2. Review at least two (different) cases in this chapter.

Keeping in mind that doing nothing/not responding are *not plausible responses,* are there other plausible alternatives (KSA 4) that you can think of for any case? If not, discuss that; if so, list them and discuss their evaluation (are they equally consistent with GL/CE, do they lead to similar decisions, etc.).

3. Choose any case in this chapter. Redo the analysis, but feature a different GL or CE principle in your reasoning process. Make sure the *justification* is different from what is given. Is the *decision* also different? Discuss your decision with its justification and the tests of universality, justice, and publicity.

4. Consider all seven cases in this chapter.

A. Comment on the role of the stakeholder analysis in these decisions. Were stakeholder impacts featured in justifications? Were they features in your own reflections?

B. Comment on the importance of your reflections on the decisions you were asked for: for others doing a similar task at work; for those using results from this task; and for those who might join the profession (and be learning about planning and design).

C. Consider your professional identity and the profession itself: planning/designing projects and data collection/analysis projects is obviously foundational in statistics and data science. How can these seven cases and the decisions reached via the analyses presented earn you (the decider) the trust of the practicing community or the public for the profession and its future?

Chapter 3.9
Summary of Section 3

Each Chapter in Section 3 featured 5-9 cases that were analyzed following the Ethical Reasoning KSAs. All of the analyses represent a fairly superficial "matchup" of the DSEC, DEFA, ASA GLs and ACM CE. Other readers may find answers in other ASA or ACM content, and indeed the reader is invited to explore both the ASA and ACM guidance documents fully so as to confirm, or to offer alternatives, to the analyses presented here. The application of the ER KSAs follows a standard pattern and while the identification of violations of practice standard principles may have been superficial, that does not imply that they are incorrect but instead means that there may be other priorities (e.g., the public good over the dignity of the profession) that can lead to analyses that feature other principles than were presented in Section 3 analyses.

The vignettes presented in Section 3 are all based on actual experiences, unfortunately. They were chosen from a much wider range of events requiring the statistician and/or data scientist to respond to an ethical issue (again, unfortunately!). These are not meant to capture every single event, or present "the only correct method" for responding ethically, but to provide examples of how the ethical practice standards can be used to identify and then respond/justify response to such cases that you may encounter yourself.

All but the first analysis included a reflection on the analysis, and readers were invited to create their own reflection on the other cases relating to each task. Reflection, as was mentioned in Section 1, is an important aspect of the ethical reasoning procedure; not only does it allow you an opportunity to review your own analysis – to confirm you have identified the correct/all of the ethical issues, that you considered plausible alternatives, and that you satisfactorily justified the choice from among those alternatives that you made. Readers were also invited to choose any two cases in the chapter and consider whether or not they agree with the decisions I made in either analysis. This is an important consideration: As you saw, I only considered three simple alternative actions in every case. Of course, we had to modify them – particularly option a) in every case, "do nothing", because that option, while a real decision that you could actually make, is not an ethical, and therefore not a plausible, option to consider. You may disagree with me – because obviously, ignoring the unethical conduct of others, including all of those who created most of these vignettes, is nearly always going to be the easiest decision you can make.

Deciding not to respond at all when someone else behaves unethically prevents you from creating uncomfortable situations in which you have to confront a peer, colleague or supervisor. You might have to obtain evidence supporting your observation of unethical behaviors, and once you have made all the necessary effort, the individuals to whom you present your findings/observations may not agree that anything unethical has occurred. Worse, they may seek retribution of some kind – all because you made a decision not to "do nothing". My hope is that, after all of these cases and especially, all of the stakeholder analyses, you will agree that doing nothing is never an ethical option once you have identified an ethical issue. I fully understand that doing the ethical thing is not easy and may in fact have repercussions.[59] My points are that 1) doing nothing is a choice you make; 2) doing nothing is never an ethical alternative among the choices you have to respond to unethical behaviors or situations; and 3) *ignoring unethical behavior perpetuates it* - because it allows it to go unrecognized and unchallenged, and therefore implicitly *endorses and condones it*.

Another consideration I hope the reader took was the invitation to discuss the tests of universality, justice, and publicity for the cases in this chapter. If you apply these tests to the unused alternative action, "do nothing", you will see more clearly that making a choice to ignore unethical behavior fails all of these tests. However, just because "do nothing" is not an ethical choice does not make it easier to make other, ethical, choices when you are faced, as I and many (MANY) other practitioners in statistics and data science are and continue to be, with unethical behavior or pressures to violate the ethical practice standards yourself.

This book is structured so that the reader can take time to familiarize themselves with the practice standards as they apply to/pertain in their everyday work (Section 2). It is hoped that the reader will only ever need this level of familiarity with the GLs and CE, although I encourage you to remain aware of the revisions to these standards, because they are a "living" representation of "the conscience of the profession" (ACM Preamble), and so must remain reflective of current practice. While the cases and their analyses in Section 3 are all abstract representations of actual cases that I, my colleagues, or my students have encountered (and some of these cases were encountered by all of us), being a statistician/data scientist may bring you into similar or

---

[59] Discussion of repercussions include recent cases of Andreas Georgiou (https://www.economist.com/finance-and-economics/2018/06/14/the-hounding-of-greeces-former-statistics-chief-is-disturbing) and Rebekah Jones (https://apnews.com/article/florida-coronavirus-pandemic-health-04d20114301ebcd1d7494a9a061eb3e7)

identical situations. I hope the analyses will help you - either by giving you a starting point for your own analysis, or by enabling and empowering you to do the hard thing, make the right choice, and try to support the stakeholders who – if you choose option a) and do nothing – will either suffer those harms or be put at risk for those harms.

Acknowledging that "doing the right thing" is difficult, even if you are able to fully outline the same ethical reasoning steps and justify your own decision in a case you confront, I recommend that you engage in some role-playing. Specifically, this means you would both consider how you would make your point, outlining your own (or my!) analysis and the decision with its justification, and also consider how <whomever you make the point to> will respond. Recall that options b) and c), which are almost invariably the "best" options in most cases, both include you initiating a discussion with someone. Either you will be conferring or consulting with a colleague (ASA) or supervisor (ACM), or you would be reporting a violation of policy, code, or law. In any of these cases, you will need to initiate the conversation, and that can be uncomfortable. That is why I recommend role-playing – even if you do it alone and with your notebook or a whiteboard - rather than another person. You should carefully and honestly consider both sides of the conversation that you would have:

- Your perspective: Describe your analysis of the case and your decision – how detailed would you be? Will you send a report/your analysis ahead of time in an email for the other person to review before you initiate the conversation? Or will you be able to remember all the points and make them in your conversation?

- "Their" perspective: How do you think the person you have identified as the right or best person to discuss this with will "receive" your case analysis? How will you convince them (thinking about their role, their perspective) to work with you to collaboratively determine the best way to ensure such situations do not recur?

These two perspectives are obviously going to be relevant as you consider how exactly this conversation is going to go. But you should also take into account – and be prepared for the eventuality- that not everyone will agree with you that "something should be done". Then what will you do? Will your professional ethical responsibilities be satisfactorily discharged if you do the analysis, come to a justifiable decision that you are also certain is universal, publishable, and just, find the right person to discuss it with, and then ... stop there? That may in fact be what happens! (That *has* happened to me several times over my career). To prepare yourself, you might go further with your role playing and consider what would you do if the person you discuss the case with tries to dissuade you from actually making the decision in your analysis, or publicizing the fact that such a decision needs/needed to be made? Consider what you would say to someone who:

- Receives your case analysis and tries to dissuade you from making – or publicizing – that decision.
- Reads/reviews your case analysis but does not support your analysis and disagrees with your decision.
- Reads/reviews your analysis and responds with one (or more) of the following:
  - "I'm sure they didn't mean <to be unethical>"
  - "Leave it alone"
  - "That isn't your responsibility"
  - "It doesn't matter"

All options b) and many of the option c) modifications in the analyses include consideration of how what you do in response to the ethical issue you have identified should or would include publicizing or sharing the ASA GLs and/or ACM CE. The ACM CE Principle 3 includes a specific stipulation that leaders have a responsibility to "articulate, apply, and support policies and processes that reflect the principles of the Code." (ACM CE 3.4) You will have noted in Section 1 that the ASA GLs has a whole set of Principles for the employer – who may not even be a statistician! How would a non-ASA or non-ACM member (who didn't read this book) ever find out that they are actually obliged to let ASA members follow the ASA GLs? It is hoped that your reading and engaging with this book and these practice standards, will help incite a grass roots effort to inform the communities using statistics, computing, and data science to practice ethically.

Chapter 3.10
Book Summary

This book was written to introduce the reader to the concept of a "steward of the profession" and to provide the background (prerequisite) knowledge necessary for this stewardship in the form of the ethical practice standards of the ASA and ACM. A steward is defined as one to whom "we can entrust the vigor, quality, and integrity of the field" (Golde & Walker 2006: p. 5), and I would argue that anyone who is not familiar with consensus-based practice standards and how to use, or prioritize, them whenever this is necessary is unlikely to fill this role. Rather than simply being an expert, stewardship requires that you can be entrusted with the "quality and integrity of the field"; without these two key attributes, the *vigor* of a field will be diminished.

In another book on ethical reasoning for statistics and data science, I assert that "Being stewardly involves both ethical practice and professional integrity; if you are stewardly and practice data science ethically, these serve to make your professional integrity *observable*. Together, stewardship and ethical practice promote the integrity of the field." (Tractenberg, 2022, Chapter 1). The reflections you completed in the exercises in Chapter 3 will serve as observable evidence that you are committed to stewardly practice; if you also engaged in alternative analyses of the vignettes (Section 3) or the discussions of how practice standards and DSEC/DEFW support everyday practice (Section 2), then that strengthens your evidence of a commitment to stewardly practice. **Stewardship of *the profession*** (or discipline) goes far beyond *just the data* (Golde & Walker, 2005; Rios et al. 2019). While the ethical statistical practitioner/data scientist is as stewardly as possible with any data entrusted to their care – by definition of the profession – stewardship of the profession itself implies greater responsibilities. These were presented in Section 1, demonstrated in Section 2, and featured prominently in the case analyses of Section 3.

In that same book, I argued that "Awareness of the potential for unethical behavior by data scientists has led to the profession "data science" becoming a target requiring ethical guidelines of some kind (e.g., Simonite 2018; Loukides et al. 2018). Public trust in this discipline was harmed by each of these events, and legislation in the United States and European Union was created specifically to limit the potential for data science to engage in harmful behaviors." (Tractenberg, 2022, chapter 1). By agreeing to become a steward of statistics and data science, and by committing to demonstrating your stewardship, you can help to offset the negative impact of these other events – by non-stewardly practitioners – and thereby, help to promote the field. You can see by the constant news coverage of data breaches and other unethical and illegal behaviors that this discipline *desperately needs stewards*. Not only is this stewardship needed to combat the unethical and illegal behaviors proliferating worldwide, but also, the discipline(s) of statistics and data science are *dynamic*. That is why stewards concern themselves with the *vigor* of the discipline, as well as its integrity and quality. Because the discipline is dynamic, the practice standards, as well as the DSEC and DEFW, are not static – and are recognized in their respective contexts as being subject to change across time and as technology develops.

Ethical reasoning is an approach to identifying "responsible" responses to an identified ethical challenge, and choosing among plausible alternative responses in a defensible manner. Note that, as a steward, your role is to promote the quality *and* integrity *and* vigor of the discipline; each of the case analyses in Section 3 demonstrates how *indefensible* it is to "do nothing" when you have identified unethical behavior (most typically, in these examples, by others) in your practice. Ethical Reasoning (ER) KSAs, like all KSAs in every MR, are learnable and improvable. That means that, like with stewardship, there are ER-specific KSAs to learn, and the level at which you perform each KSA also changes as you become more sophisticated and capable. Although each vignette in Section 3 included my performance of KSAs 1-5, you can develop your performance of these KSAs yourself by revisiting these case analyses, or (in what I hope will be extremely rare circumstances) by using all six KSAs yourself in your practice. While the ethical practice standards for the disciplines are not static, the ER KSAs *are fixed*. It is not the KSAs of ER that change over time to accommodate the vigor of the field, but *you the user* of the KSAs – you will actually grow and change over time in your ability to use them as new cases or new applications come into being. The KSAs are intended to be both learnable and improvable – just like the stewardship KSAs. As you have seen, you can use KSAs in the planning stage, i.e., before any unethical behaviors can be executed (i.e., KSAs 1-2), and you can use all of them to evaluate what may have gone wrong in any given case, or what may be going on in a real/real-time situation.

The ethical practice of statistics and data science starts with prerequisite knowledge. Sections 1 and 2 present the relevant knowledge from the Association of Computing Machinery (ACM) in their Code of Ethics (CE), and the American Statistical Association (ASA) in their Ethical Guidelines (GLs) for Statistical Practice. Even if you are not a member of either organization or even consider yourself to be a "statistician", "computing professional", or "data scientist", these guidelines are essential prerequisite knowledge for competent, professional practice in the domain of statistics and data science (and in the domains of statistics and of data science, in case you identify with or have a job title representing only one of these disciplines). As we have discussed, both organizations recognize the importance of following these practice standards any time you utilize tools, techniques, or methods from their disciplines. Those who fail to do so will perpetuate the deceptive and detrimental practices that the NASEM 2017 report identified.

The ASA and ACM acknowledge that there cannot be a rule or algorithm for every possible situation, whereas the existence or use of the DSEC and DEFW suggest that, in any situation there is a concise list of questions that, once answered, create an 'ethical' project. By contrast, the construct of ethical reasoning (ER) suggests that there is a core of "prerequisite knowledge" that can guide all of the decisions to be made in the course of practice (ER KSA 1); and also, that a decision-making framework can further structure those decisions (ER KSA 2).

We saw in Chapter 1.4 – and throughout the rest of the book - that the ASA GLs, ACM CE, DEFW and DSEC had areas of overlap and disjuncture:

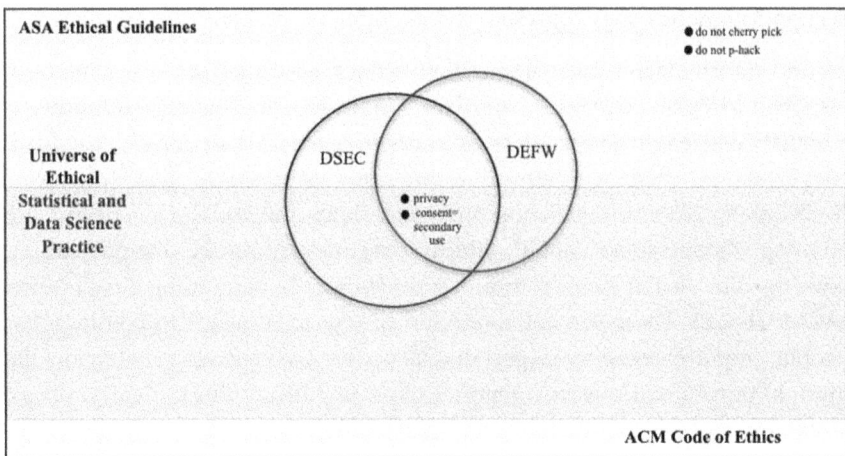

Figure 3.10.1. *Unions and disjunctions: ASA GLs, ACM CE, DSEC, DEFW*

Through exploration of applicability of each of these resources as you carry out the seven tasks of statistics and data science (Section 2), and especially through the identification of specific elements of these resources to assist in identifying and responding to ethical challenges (Section 3), the extent of disjuncture should now be clear. As I argued early in the book, the choice of just one of these resources for ethical practice will not be sufficient to support ethical statistics and data science practice fully. While DSEC and DEFW questions can help a group or an individual to pinpoint exactly how their engagement with the ACM and ASA ethical practice standards should be represented in projects, it is the practice standards themselves that do most of the work in every vignette in Section 3. That is, without a good working understanding of the CE and GLs, identifying and responding to ethical challenges that arise in the practice of statistics and data science will be impoverished, if not impossible.

As we saw in several of the vignettes, detrimental and unethical use of statistics and data science may be depressingly common in your work context. We have discussed at length that "do nothing" is never an ethical choice, and I tried to convert this into a *minimum* response for every option a) in every vignette, but it might be impossible for you choose even option a) as articulated. In that case, it is impossible for you to do *anything* ethical except notify relevant stakeholders and authorities of the situation, and then seek another team, other colleagues, or a different job. The notification could be in the form of a blog post[60] or another publication venue[61]. All of these notifications are evidence based, and represent the extent of the authors' capabilities in affecting change to correct the unethical behavior or problems that were identified. Even though these notifications are not consistent with any of the options a) discussed in Section 3 vignettes, they represent considerably more than "do nothing".

As uncomfortable as conversations about ethical case analyses – particularly in response to events at your workplace – may be, "the entire community of scientists and engineers benefits from diverse, ongoing options to engage in conversations about the ethical dimensions of research and (practice)," (Kalichman, 2013: 13). Echoing this sentiment, the ACM CE specify that ongoing conversations about ethical practice are to be supported and developed as part of every computing professional's continuing development (ACM CE 2.2). These are not suggested to you as a means to identify "bad people", nor do I mean to suggest that all people who engage in behaviors like those in our vignettes were intending to do unethical things. Clearly, *some of them were*; and your careful and thoughtful analysis of plausible alternatives, your decision and its sound justification will have absolutely no effect on such

---

[60] Blog post notifications to the wider community that unethical behaviors are going on include   https://bruegel.org/2018/06/the-european-union-must-defend-andreas-georgiou/ and https://medium.com/@kristianlum/statistics-we-have-a-problem-304638dc5de5.
[61] Table 2 in this peer reviewed and published manuscript documents the failure of alignment between institutional training and ethical reasoning objectives. https://www.tandfonline.com/doi/abs/10.1080/02602938.2011.596923

people. I would suggest that people who *choose to act unethically* do not belong to the "community of scientists and engineers" but rather are members of a "community of self" – since their behavior so adversely affects the profession, and public trust in the profession, they cannot be acknowledged to be part of the professional community. So, in your role playing –whether it is with others or with just your notebook – it might be better to consider "the other" to be a real member of the community to which Kalichman refers. If, by chance, you find that the individual with whom you confer is really not a member, you will recognize it very quickly, and you can factor that into your reflection (and possibly into your prerequisite knowledge).

I also want to take the opportunity to encourage you to embrace your membership in Kalichman's community: the ethical practitioner follows the CE and GLs. If you should find yourself on a team that does not follow these, or if violating GL and CE principles is encouraged in a workplace, that does not mean you should follow suit – or stay on such a team. The ethical quantitative practitioner is committed to *supporting the profession* – meaning that behaviors that have the result of harming the profession or public trust in the profession are not to be engaged in, condoned, or tolerated if they are directed at you or exhibited in your work environment. After reading this book, you will easily be able to identify situations where the profession, and public trust in the profession, are *valued*; and employers are beginning to appreciate that workplace pressures to violate ethical practice standards drive excellent employees to other jobs and workplaces. In your exit interview, you would be well prepared to outline the role that such pressures played in your departure, should this situation ever happen to you!

References

American Statistical Association (ASA) *ASA Ethical Guidelines for Statistical Practice-revised* (2022) downloaded from https://www.amstat.org/ASA/Your-Career/Ethical-Guidelines-for-Statistical-Practice.aspx on 20 January 2022.

Association for Computing Machinery (ACM). *Code of Ethics* (2018) downloaded from https://www.acm.org/about-acm/code-of-ethics on 12 October 2018.

Beauchamp TL, & Childress JF. (2001). *Principles of biomedical ethics*. Oxford University Press, USA.

Briggle A & Mitcham C. (2012). *Ethics and science: An introduction*. Cambridge, UK: Cambridge University Press.

Coventry LL, Finn J, Bremner AP. (2011). Sex differences in symptom presentation in acute myocardial infarction: a systematic review and meta-analysis. Heart Lung 40(6):477-91. doi: 10.1016/j.hrtlng.2011.05.001.

DeVeaux RD, Agarwal M, Averett M, Baumer BS, Bray A, Bressoud TC, Bryant L, Cheng LZ, Francis A, Gould R, Kim AY, Kretchmar M, Lu Q, Moskol A, Nolan D, Pelayo R, Raleigh S, Sethi RJ, Sondjaja M, Tiruviluamala N, Uhlig PX, Washington TM, Wesley CL, White D, Ye P. (2017). Curriculum Guidelines for Undergraduate Programs in Data Science. *Annual Review of Statistics and Its Application* 4:1, 15-30

Dow MJ, Boettcher CA, Diego JF, Karch ME, Todd-Diaz A, Woods KM. (2015). Case-based learning as pedagogy for teaching information ethics based on the Dervin sense-making methodology. *Journal of Education for Library and Information Science* 56(2-Spring): 141-157.

Golde CM, & Walker GE. (Eds.). (2006). *Envisioning the future of doctoral education: Preparing stewards of the discipline-Carnegie essays on the doctorate*. San Francisco: Jossey-Bass.

Head ML, Holman L, Lanfear R, Kahn AT, Jennions MD (2015) The Extent and Consequences of P-Hacking in Science. *PLoS Biol* 13(3): e1002106. https://doi.org/10.1371/journal.pbio.1002106

Hogan H & Steffey D. (2014). Professional Ethics for Statisticians: An Organizational History. *Proceedings of the Joint Statistical Meetings, Boston, MA*. Pp 1397-1404.

Hogan H, Tractenberg RE, Elliot AC. (2017, July). *Ethics and Big Data: Perspective of the American Statistical Association Committee on Professional Ethics*. Presented at the 61st International Statistics Institute World Statistics Congress, Marrakesh, Morocco.

Journal Editors and Authors Group (2003). Uncensored exchange of scientific results. *Proceedings of the National Academy of Sciences* Feb 2003 100 (4) 1464; DOI:10.1073/pnas.0630491100

Kalichman M. (2013). Why teach research ethics? In, National Academy of Engineering (Eds). *Practical Guidance on Science and Engineering Ethics Education for Instructors and Administrators*. Washington, DC: National Academies Press. Pp. 5-16. Accessed from https://doi.org/10.17226/18519 February 2015.

Loukides M, Mason H, Patil DJ. (10 July 2018). *Doing good data science.* Downloaded from https://www.oreilly.com/ideas/doing-good-data-science on 15 July 2018

Macrina FL. (2014). *Scientific Integrity, 4E*. Washington, DC: ASM Press.

McGoughy E. (2019). Could Brexit be void? *29(3) King's Law Journal 331-343 King's College London Law School Research Paper No. 2018-29.*

McNamara A, Smith J, Murphy-Hill E. (2018). Does ACM's Code of Ethics change ethical decision making in software development? In *Proceedings of the 26th ACM Joint European Software Engineering Conference and Symposium on the Foundations of Software Engineering (ESEC/FSE '18), November 4–9, 2018, Lake Buena Vista, FL, USA*. ACM, New York, NY, USA, 5 pages. downloaded from https://doi.org/10.1145/3236024.3264833 on 15 July 2019.

National Academy of Sciences, National Academy of Engineering, and Institute of Medicine. (2009). *On Being a Scientist: A Guide to Responsible Conduct in Research: Third Edition*. Washington, DC: The National Academies Press. https://doi.org/10.17226/12192.

National Academies of Sciences, Engineering, and Medicine. (2017). *Fostering Integrity in Research*. Washington, DC: The National Academies Press. https://doi.org/10.17226/21896.

National Academies of Sciences, Engineering, and Medicine. (2018). *Data Science for Undergraduates: Opportunities and Options*. Washington, DC: The National Academies Press. https://doi.org/10.17226/25104.

Penchava D. (2019). *Brexit and migration: our new research highlights fact-free news coverage*. Downloaded from https://theconversation.com/brexit-and-migration-our-new-research-highlights-fact-free-news-coverage-109309 on 31 March 2019.

Rios CR, Golde CM, Tractenberg RE. (2019). The preparation of stewards with the Mastery Rubric for Stewardship: Re-envisioning the formation of scholars *and practitioners*. *Education Sciences* 9(4), 292; https://doi.org/10.3390/educsci9040292

Selgelid MJ. (2009). Governance of dual-use research: an ethical dilemma. *Bulletin of the World Health Organization* 2009; 87:720-723. doi: 10.2471/BLT.08.051383.

Simonite T. (2018). *Should data scientists adhere to a Hippocratic Oath?* Accessed from https://www.wired.com/story/should-data-scientists-adhere-to-a-hippocratic-oath/

Stadler HA. (1986). Making hard choices: Clarifying controversial ethical issues. *Counseling & Human Development*, 19, 1-10. See also www.walsall-socialcareworkforce.co.uk/ckfinder/userfiles/files/refStadler.docx

Stark PB & Saltelli A. (2018). Cargo cult statistics and the scientific crisis. Downloaded from https://www.significancemagazine.com/2-uncate-gorised/593-cargo-cult-statistics-and-scientific-crisis on 18 July 2018.

Steneck NH. (2009). *ORI Introduction to the Responsible Conduct of Research, Revised*. Accessed from https://ori.hhs.gov/sites/default/files/rcrintro.pdf February 2010.

Tractenberg, RE. (2017, October 6). *Preprint*. The Mastery Rubric: A tool for curriculum development and evaluation in higher, graduate/post-graduate, and professional education. Published in the *Open Archive of the Social Sciences (SocArXiv)*, https://doi.org/10.31235/osf.io/qd2ae

Tractenberg, RE. (2019-a). *Preprint*. Strengthening the practice and profession of statistics and data science using ethical guidelines. Published in the *Open Archive of the Social Sciences (SocArXiv)*, https://doi.org/10.31235/osf.io/93wuk (also, to appear in P. Glazebrook, J. Collmann & S. Matei (Eds)., *Ethical Data Science*. Forthcoming)

Tractenberg, RE. (2019-b). *Preprint*. Becoming a steward of data science. Published in the *Open Archive of the Social Sciences (SocArXiv)*, https://doi.org/10.31235/osf.io/j7h8t (also, to appear in P. Glazebrook, J. Collmann & S. Matei (Eds)., *Ethical Data Science*. Forthcoming)

Tractenberg, RE. (2019-c). *Preprint*. Teaching and Learning about Ethical Practice: The Case Analysis. Published in the *Open Archive of the Social Sciences (SocArXiv)*, https://doi.org/10.31235/osf.io/58umw

Tractenberg RE. (2020, February 19). *Preprint*. Concordance of professional ethical practice standards for the domain of Data Science: A white paper. Published in the *Open Archive of the Social Sciences* (SocArXiv), 10.31235/osf.io/p7rj2

Tractenberg, RE. (2022). *Ethical Reasoning for a Data Centered World*. Ethics International Press.

Tractenberg RE & FitzGerald KT. (2012). A Mastery Rubric for the design and evaluation of an institutional curriculum in the responsible conduct of research. Assessment and Evaluation in Higher Education. 37(7-8): 1003-21. DOI 10.1080/02602938.2011.596923

Tractenberg RE, Russell A, Morgan G, FitzGerald KT, Collmann J, Vinsel L, Steinmann M, Dolling LM. (2015) Amplifying the reach and resonance of ethical codes of conduct through ethical reasoning: preparation of Big Data users for professional practice. Science and Engineering Ethics 21(6):1485-1507. http://link.springer.com/article/10.1007%2Fs11948-014-9613-1.